GEOLOGICAL SOCIETY SPECIAL PUBLICATION NO. 162

Uplift, Erosion and Stability: Perspectives on Long-term Landscape Development

EDITED BY

BERNARD J. SMITH, W. BRIAN WHALLEY

AND

PATRICIA A. WARKE

School of Geosciences
Queen's University of Belfast
Belfast

1999

Published by

The Geological Society

London

THE GEOLOGICAL SOCIETY

The Geological Society of London was founded in 1807 and is the oldest geological society in the world. It received its Royal Charter in 1825 for the purpose of 'investigating the mineral structure of the Earth' and is now Britain's national society for geology.

Both a learned society and a professional body, the Geological Society is recognized by the Department of Trade and Industry (DTI) as the chartering authority for geoscience, able to award Chartered Geologist status upon appropriately qualified Fellows. The Society has a membership of 8600, of whom about 1500 live outside the UK.

Fellowship of the Society is open to persons holding a recognized honours degree in geology or a cognate subject and who have at least two years' relevant postgraduate experience, or not less than six years' relevant experience in geology or a cognate subject. A Fellow with a minimum of five years' relevant postgraduate experience in the practice of geology may apply for chartered status. Successful applicants are entitled to use the designatory postnominal CGeol (Chartered Geologist). Fellows of the Society may use the letters FGS. Other grades of membership are available to members not yet qualifying for Fellowship.

The Society has its own Publishing House based in Bath, UK. It produces the Society's international journals, books and maps, and is the European distributor for publications of the American Association of Petroleum Geologists (AAPG), the Society for Sedimentary Geology (SEPM) and the Geological Society of America (GSA). Members of the Society can buy books at considerable discounts. The Publishing House has an online bookshop (http://bookshop.geolsoc.org.uk).

Further information on Society membership may be obtained from the Membership Services Manager, The Geological Society, Burlington House, Piccadilly, London W1V 0JU (Email: enquiries@geolsoc.org.uk; tel: +44 (0)171 434 9944).

The Society's Web Site can be found at http://www.geolsoc.org.uk/. The Society is a Registered Charity, number 210161.

Published by The Geological Society from:
The Geological Society Publishing House
Unit 7, Brassmill Enterprise Centre
Brassmill Lane
Bath BA1 3JN, UK
(*Orders*: Tel. +44 (0)1225 445046
 Fax +44 (0)1225 442836)
Online bookshop: http://bookshop.geolsoc.org.uk

First published 1999

The publishers make no representation, express or implied, with regard to the accuracy of the information contained in this book and cannot accept any legal responsibility for any errors or omissions that may be made.

© The Geological Society of London 1999. All rights reserved. No reproduction, copy or transmission of this publication may be made without written permission. No paragraph of this publication may be reproduced, copied or transmitted save with the provisions of the Copyright Licensing Agency, 90 Tottenham Court Road, London W1P 9HE. Users registered with the Copyright Clearance Center, 27 Congress Street, Salem, MA 01970, USA: the item-fee code for this publication is 0305-8719/99/$15.00.

British Library Cataloguing in Publication Data
A catalogue record for this book is available from the British Library.

ISBN 1-86239-047-9
ISSN 0305-8719

Typeset by Wyvern 21 Ltd, Bristol, UK

Printed by Anthony Rowe Ltd, Chippenham, UK

Distributors

USA
AAPG Bookstore
PO Box 979
Tulsa
OK 74101-0979
USA
(*Orders*: Tel. +1 918 584-2555
 Fax +1 918 560-2652)
 Email bookstore@aapg.org

Australia
Australian Mineral Foundation Bookshop
63 Conyngham Street
Glenside
South Australia 5065
Australia
(*Orders*: Tel. +61 88 379-0444
 Fax +61 88 379-4634
 Email bookshop@amf.com.au)

India
Affiliated East-West Press PVT Ltd
G-1/16 Ansari Road, Daryaganj
New Delhi 110 002
India
(*Orders*: Tel. +91 11 327-9113
 Fax +91 11 326-0538)

Japan
Kanda Book Trading Co.
Cityhouse Tama 204
Tsurumaki 1-3-10
Tama-Shi
Tokyo 206-0034
Japan
(*Orders*: Tel. +81 (0)423 57-7650
 Fax +81 (0)423 57-7651)

Contents

SMITH, B. J., WHALLEY, W. B., WARKE, P. A. & RUFFELL, A. Introduction and background: interpretations of landscape change — vii

The British Isles

JONES, D. K. C. Evolving models of the Tertiary evolutionary geomorphology of southern England, with special reference to the Chalklands — 1

JONES, D. K. C. On the uplift and denudation of the Weald — 25

WALSH, P., BOULTER, M. & MORAWIECKA, I. Chattian and Miocene elements in the modern landscape of western Britain and Ireland — 45

BATTIAU-QUENEY, Y. Crustal anisotropy and differential uplift: their role in long-term landform development — 65

Mainland Europe and Scandinavia

PUURA, V., VAHER, R. & TUULING, I. Pre-Devonian landscape of the Baltic Oil-Shale Basin, NW of the Russian platform — 75

LIDMAR-BERGSTRÖM, K. Uplift histories revealed by landforms of the Scandinavian domes — 85

MIGON, P. Inherited landscapes of the Sudetic Foreland (SW Poland) and implications for reconstructing uplift and erosional histories of upland terrains in Central Europe — 93

BASILI, R., GALADINI, F. & MESSINA, P. The application of palaeo-landsurface analysis to the study of recent tectonics in central Italy — 109

BARTOLINI, C. An overview of Pliocene to present-day uplift and denudation rates in the Northern Apennine — 119

Africa and The Middle East

BAIRD, A. W. & RUSSELL, A. J. Structural and stratigraphic perspectives on the uplift and erosional history of Djebel Cherichira and Oued Grigema, a segment of the Tunisian Atlas thrust fault — 127

BUTLER, R. W. H. & SPENCER, S. Landscape evolution and the preservation of tectonic landforms along the northern Yammouneh Fault, Lebanon — 143

ERIKSSON, M. G. Influence of crustal movements on landforms, erosion and sediment deposition in the Irangi Hills, central Tanzania — 157

Asia

PETLEY, D. N. & REID, S. Uplift and landscape stability at Taroko, eastern Taiwan — 169

FOTHERGILL, P. A. & MA, H. Preliminary observations on the geomorphic evolution of the Guide Basin, Qinghai Province, China: implications for the uplift of the northeast margin of the Tibetan Plateau — 183

OWEN, L. A., CUNNINGHAM, W. D., WINDLEY B. F., BADAMGAROV, J. & DORJNAMJAA, D. The landscape evolution of Nemegt Uul: a late Cenozoic transpressional uplift in the Gobi Altai, southern Mongolia — 201

The Americas

CONRAD, C. T. & SAUNDERSON, H. C. Temporal and spatial variation in suspended sediment yields from eastern North America — 219

COSTA, C. H., GIACCARDI, A. D. & GONZÁLEZ DÍAZ, E. F. Palaeo-landsurfaces and neotectonic analysis in the southern Sierras Pampeanas, Argentina 229

COLTORTI, M. & OLLIER, C. D. The significance of high planation surfaces in the Andes of Ecuador 239

Antarctica

SUMMERFIELD, M. A., SUGDEN, D. E., MARCHANT, D. R., COCKBURN, H. A. P., STUART, F. M. & DENTON, G. H. Cosmogenic isotope data support previous evidence of extremely low rates of denudation in the Dry Valleys region, southern Victoria Land, Antarctica 255

Index 269

Introduction and background: interpretations of landscape change

B. J. SMITH, W. B. WHALLEY, P. A. WARKE & A. RUFFELL

School of Geosciences, The Queen's University of Belfast, Belfast, BT7 1NN, UK

Research into the origins of landscape has a long and distinguished tradition. In America, landscape studies form the core of physical geology while elsewhere the study of landscape development has spread beyond the confines of geology to develop a firmer foothold in the geographical tradition, particularly in geomorphology. The development of ideas in landscape study has, however, been neither steady nor sequential. It has been characterized instead by several major conceptual shifts interspersed with periods of reinforcement or neglect. Within geomorphology, and certainly within the English-speaking world, the most important of these shifts occurred in the early 1960s with the move from denudation chronology of landscapes as a core activity to an emphasis on process-based investigation of individual landforms.

In geology, the 1960s also marked an important change in perspective. Publication of plate tectonic theories (sea-floor spreading, continental drift, subduction and mountain building) had far-reaching consequences for all earth science. In particular, plate tectonics made the study of apparently ancient processes (billions of years old) applicable to the present-day with, for example, application to earthquake studies.

Whilst the initial shift towards process studies was a response to the desire for a better understanding of the mechanisms driving landscape change, it was not long before this goal became relegated in the eyes of many of its practitioners. As a consequence, process study became an end in itself, rather than a means to an end. This movement has been reinforced in recent years by a demand for greater 'relevance' from, for example, research funding bodies resulting in an emphasis on problem solving and wealth generation that has moved away from studies at the landscape scale. Consequently, within geomorphology, process studies and applied geomorphology have come to dominate the literature – a dominance reinforced by the increasing division and separation of the discipline into process domains such as fluvial and glacial studies. This has reached the point where many researchers would now naturally identify themselves as specialists in a particular process rather than as geomorphologists.

A similar trend can be identified within geology, whereby studies of uplift, erosion and stability clearly fall into pre-plate tectonics and post-plate tectonics generations. In the 1920s, studies of denudation were popular, as evidenced by the volume of work conducted in southern England and based largely on observations of palaeovalleys, tilted strata and unconformities. Sediment provenance became important in the 1920s and 1930s and early works on the use of heavy minerals in studies of past erosion were developed. By the 1960s geologists were standing back and viewing many processes in terms of their global plate tectonic significance. However, this emphasis on processes inevitably led to the development of more labour-intensive analytical methods, as a consequence of which, the wider perspective began to be neglected. Through the 1970s and 1980s, therefore, studies of uplift developed using methods that determine heating or pressurization of the uplifted rocks. Thermal estimates of uplift, including spore or conodont alteration indexes, fluid inclusion studies and diagenetic models were developed, together with estimates of pressurization including studies of mineralogy and sonic borehole logs. However, perhaps the most significant advance in the determination of uplift rates and amounts has been in the field of apatite fission track analysis (AFTA) (Carter 1999). The widespread use of AFTA in the oil industry has provided large datasets that encompass the offshore basins formerly ignored by those concerned with land erosion (Buchanan & Buchanan 1995). Studies of erosion and palaeoerosion have also developed from the study of derived clasts to clay minerals and heavy minerals, requiring increasingly labour-intensive, and hence specialist, methods to derive local or even regional perspectives.

Specialization clearly has many advantages. If, however, it is allowed to proceed unchecked, these advantages eventually come to be outweighed by the effects of an increasingly reductionist approach whereby concentration on small-scale studies and the pursuit of ever greater detail detracts from appreciation of the large-scale picture and ultimately fragmentation and loss of the subject core. Clearly, the longer this reductionist/segregationist approach persists the more difficult it is to define and to justify the existence of the original parent discipline.

From: SMITH, B. J., WHALLEY, W. B. & WARKE, P. A. (eds) 1999. *Uplift, Erosion and Stability: Perspectives on Long-term Landscape Development.* Geological Society, London, Special Publications, **162**, vii–xi. 1-86239-030-4/99/ $15.00 © The Geological Society of London 1999.

The case for integration and collaboration

Calls for the integration of process studies and their application to the original question of landscape characterization and change are not new (Summerfield 1981; Douglas 1982). In general these calls have been hindered by problems of scale, whereby there is still an incomplete understanding of the multiplier effects of extrapolating small-scale, spatially and temporally confined process studies to larger scales of investigation. There is, however, an increasing awareness that such extrapolation is essential and that present-day landscape cannot be explained solely in terms of current processes or even those that operated in the geologically recent past.

This view has been fuelled by a growing understanding that many of the world's landscapes are much older than the Quaternary and thus require explanations involving long-term and large-scale consideration of both climatic change and tectonics. This has been emphasized by geomorphological and geological investigations over the last 20–30 years in low-latitude environments not directly affected by Quaternary glacial activity, especially in Australia and South Africa (Young 1983; Fried & Smith 1992; Gale 1992). This in turn, has been supported by an increasing number of studies in western Europe where evidence has accumulated to support the survival of pre-glacial landforms (Mitchell 1980), extensive deeply weathered Tertiary material (Fitzpatrick 1963; Hall & Sugden 1987) and pre-Quaternary weathering and associated land surfaces (Lidmar-Bergström & Åse 1988; Lidmar-Bergström 1995; Whalley et al. 1997).

One factor that has facilitated this renewed interest in long-term landscape change has been the beginning of collaboration between geologists and geomorphologists, which has greatly enhanced our ability to interpret landscape development. Central to this have been improvements in the elucidation of tectonic and erosional histories and the use of dating techniques such as apatite fission track and cosmogenics that have significantly contributed to characterization and landscape interpretation (e.g. Morisawa & Hack 1985; Thomas 1995; Widdowson 1997). In addition, a greater concern of structural geologists with questions of landscape change has fostered a much less dogmatic and simplistic view of 'uplift, erosion and subsequent deposition' within geomorphological thinking. In particular, the role of geosynclinal deposition in promoting the deformation and uplift of sediment supply areas and the possibilities for using rates and patterns of basin deposition for better understanding of continental erosion (e.g. Partridge & Maude 1987; Lidmar-Bergström 1995). Similarly, it is significant that the current motivation for linking erosion rates, structural geology and dating methodologies has not come solely from geomorphologists, but has been matched by a comparable desire within geology to integrate these components, largely driven by the demands of petroleum exploration and a consequent burgeoning of interest in sequence stratigraphy. Thus, there is clear evidence for the emergence of collaborative, interdisciplinary investigations.

The present volume

It was recognition of a need for interdisciplinary research that provided the stimulus for the meeting, jointly sponsored by The Geological Society of London, The British Geomorphological Research Group and IGCP 317 (Palaeoweathering Records and Palaeosurfaces), that led to this publication. The contributors were encouraged to examine large-scale earth surface change – as a contribution towards the setting of an agenda for the integration of process and landscape studies. However, the aim extended beyond a desire to chronicle and understand individual landscape histories. Instead, it was hoped that, by demonstrating the benefits of interdisciplinary discourse, a widening of interest in landscape studies would be encouraged. It was also hoped that the presentations would demonstrate that studies of present-day processes can be successfully placed within an evolutionary framework and geological setting, the necessity for which increases as appreciation of the antiquity of many landscapes grows.

The material in this volume represents work from both the geological and geomorphological traditions, and encompasses a wide geographical spread and many geological interests. Most importantly, however, the papers highlight the significance of recent advances in analytical technology for improving interpretation of both geologically 'ancient' and 'young' landscapes.

To set these developments in context, **Jones** revisits the 'classic' landscape studies of SE England by Wooldridge & Linton (1938, 1955) in the first two papers. In these, recent advances in understanding of the Tertiary evolutionary geomorphology of this region are supported by developments in, for example, apatite fission track dating and the establishment of inversion tectonics. However, while acknowledging that substantial progress has been made towards establishing a much clearer evolutionary geomorphology for SE England, **Jones** recognizes

that uncertainties persist especially with regard to the temporal and spatial dimensions of uplift, the significance of palaeoenvironmental conditions and the lack of widespread dating of surfaces. **Walsh et al.** and **Battiau-Queney** continue the theme of revisiting traditional models of landscape development. **Walsh et al.** use a combination of sedimentological, palaeokarstic and palaeobotanical evidence to suggest a greater antiquity for surfaces previously thought to be of late Pliocene or Pleistocene age. **Battiau-Queney** highlights the importance of such factors as continuous energy input associated with plate tectonics and inherited crustal discontinuities in the long-term landscape development of the British Isles.

The study by **Puura et al.** of the pre-Devonian landscape of the Baltic Oil-Shale Basin provides a timely reminder that investigations of complex present-day landscapes may have implications for interpretation of similar periods of continental dominance within the geological record. The value of traditional landform interpretation in the provision of a relative chronology of landscape development is also demonstrated by **Lidmar-Bergström** in her study of Scandinavia which highlights the increasingly significant role of apatite fission track analysis in interpreting denudational histories and in identifying the relative antiquity of many present-day landscapes. From Central Europe, **Migon** demonstrates the complexity and antiquity of previously glaciated areas and suggests that much of the upland topography of the region may be inherited from the Early Cenozoic.

The theme of continuing tectonic control on the development of individual landscapes and landforms is followed in a series of papers which extend from Tunisia (**Baird & Russell**) to the Lebanon (**Butler & Spencer**), Italy (**Basili et al.**), Tanzania (**Eriksson**) and to extremely active landscapes dominated by neotectonics in Taiwan (**Petley & Reid**), the Tibetan Plateau (**Fothergill & Ma**), Argentina (**Costa et al.**), Ecuadorian Andes (**Coltorti & Ollier**) and the Gobi Desert (**Owen et al.**). The close links between geomorphological processes, specific landforms and uplift in the formation of contemporary landscapes in tectonically active regions of Asia are also clearly illustrated by **Petley & Reid**, **Fothergill & Ma** and **Owen et al. Eriksson** further exemplifies these interactions with specific regard to soil erosion in Tanzania with removal of soil occurring as part of the natural long-term stripping of regolith associated with crustal uplift and tilting of tectonic blocks. These examples of relatively rapid landscape development are contrasted with the data presented by **Butler &** **Spencer** who suggest that even in tectonically active areas large-scale tectonic landforms may be preserved for many millions of years with only minor surface modification because of low denudation rates (e.g. in arid regions).

In all of these studies, the integration of structural, sedimentological and geomorphological data have contributed to large-scale landscape interpretation and identification of the spatial and temporal complexity of landscape development. The consequence of continued uplift for regional erosion and offshore sedimentation is explored by **Bartolini** with specific reference to Italy and the Adriatic whereby sedimentological data have been used to identify variability of uplift rates since the Upper Oligocene. These issues are given a North American perspective by a quantitative analysis of historic erosion and deposition rates along the northeastern seaboard of the USA by **Conrad & Saunderson** – an analysis that recommends caution regarding the extrapolation of modern sediment yields to longer-term landscape development.

The possibility of fixing absolute time-scales to landscape development through the use of cosmogenic dating is explored in detail by **Summerfield et al.** in a study of the Dry Valleys region of Antarctica. Cosmogenic dating has identified very slow rates of denudation and landscape modification in the hyper-arid conditions of Antarctica. It is sufficiently sensitive to identify differences between upland and coastal sites where slightly higher denudation rates in the latter possibly reflect the seasonal availability of liquid water and hence more active weathering. The ability to date surfaces is perhaps one of the most important technological advances in landscape interpretation. It has enabled recognition of the antiquity of many landsurfaces and has thus necessitated reassessment of former interpretations especially with regard to climate change and tectonics.

Conclusions

There is a growing awareness that the explanation of long-term landscape change remains a core objective of geomorphology and physical geology. This realization has been prompted by the coincidence of a number of factors:

- a much improved knowledge of geomorphological processes over the last 40 years, with a recognition that applied and process geomorphology alone cannot provide a conceptual framework for a discipline and that there is a need for a purpose to these studies

- the growing acknowledgement, demonstrated by some of the contributions to this volume, that many of the world's landscapes pre-date the Quaternary and Tertiary – even those that were subject to extensive glaciation
- a much improved understanding of tectonic processes, coupled with a willingness on the part of geomorphologists and structural geologists to exchange ideas and undertake collaborative work
- the advent of new techniques amenable to the dating of land surfaces and estimation of erosion rates
- the need to review concepts and practicalities of long-term landscape evolution, before first-hand experience and expertise in denudation chronology is lost

However, if long-term landscape change is to be revisited, it is clear that there must be certain prerequisites. First, earth surface studies at the landscape scale, especially within geomorphology, must avoid interpretation of the history of landscape based solely upon morphological evidence or the efficacy of proprietary models of landscape evolution. Second, given the breadth of relevant knowledge and range of analytical techniques now available, it is evident that, if landscape change is to be understood, it must be through an interdisciplinary approach.

Plate tectonics, in particular, has had a major impact on studies of land stability. The realization that these essentially geological processes were occurring in the present day brought geomorphology into perspective for many structural geologists (especially in North America), resulting in the birth of the new science of neotectonics. Improved analytical techniques such as real-time satellite imagery and seismic monitoring were the result of the obvious social and economic requirements for understanding Earth movements. Until now, the more advanced analytical methods (AFTA, heavy minerals) have been labour intensive and thus confined to local studies. Technological advances and an ever-increasing database suggest that we are now in a position to bring the detail of local studies into their regional context, using all the skills of geology and geomorphology.

By their very nature, interdisciplinary volumes place great strains on the expertise of a limited number of editors. As a consequence, publication would not have been possible if we had not been able to draw upon the experience of a large number of reviewers from a wide range of backgrounds. In addition Dr Bob Holdsworth has provided unstinting guidance and support. To all of these individuals we extend our sincerest thanks.

References

BUCHANAN, J. G. & BUCHANAN, P. G. (eds) 1995. *Basin Inversion*. Geological Society, London, Special Publications, **88**.

CARTER, A. 1999. Present status and future avenues of source region discrimination and characterisation using fission track analysis. *Sedimentary Geology*, **124**, 31–45.

DOUGLAS, I. 1982. The unfulfilled promise: earth surface processes as a key to landform evolution. *Earth Surface Processes and Landforms*, **7**, 101.

FITZPATRICK, E. A. 1963. Deeply weathered rock in Scotland, its occurrence, age and contribution to the soils. *Journal of Soil Science*, **14**, 33–43.

FRIED, A. W. & SMITH, N. 1992. Timescales and the role of inheritance in long-term landscape evolution, northern New England, Australia. *Earth Surface Processes and Landforms*, **17**, 375–385.

GALE, S. 1992. Long-term landscape evolution in Australia. *Earth Surface Processes and Landforms*, **17**, 323–343.

HALL, A. M. & SUGDEN, D. E. 1987. The modification of mid-latitude landscapes by ice sheets: the case of north-east Scotland. *Earth Surface Processes and Landforms*, **12**, 531–542.

LIDMAR-BERGSTRÖM, K. (ed.) 1995. Preglacial landforms. *Geomorphology*, **12**, 1–89.

—— & ÅSE, L. E. (eds) 1988. Preglacial weathering and landform evolution. *Geografiska Annaler*, **70A**, 273–374.

MITCHELL, G. F. 1980. The search for Tertiary Ireland. *Journal of Earth Science* (Dublin), **3**, 13–33.

MORISAWA, M. & HACK, J. T. (eds) 1985. *Tectonic Geomorphology*, Unwin Hyman, Boston.

PARTRIDGE, T. C. & MAUDE, R. R. 1987. Geomorphic evolution of southern Africa since the Mesozoic. *South African Journal of Geology*, **90**, 179–208.

SUMMERFIELD, M. A. 1981. Macroscale geomorphology. *Area*, **13**, 3–8.

THOMAS, M. F. 1995. Models for landform development on passive margins. Some implications for relief development in glaciated areas. *Geomorphology*, **12**, 3–16.

WHALLEY, W. B., REA, B. R., RAINEY, M. M. & MCALISTER, J. J. 1997. Rock weathering in blockfields: some preliminary data from mountain plateaus in North Norway. *In:* WIDDOWSON, M. (ed.) *Palaeosurfaces: Recognition, Reconstruction and Palaeoenvironmental Interpretation*. Geological Society, London, Special Publications, **120**, 133-145.

WIDDOWSON, M. (ed.) 1997. *Palaeosurfaces: Recognition, Reconstruction and Palaeoenvironmental Interpretation*. Geological Society, London, Special Publications, **120**.

WOOLDRIDGE, S. W. & LINTON, D. L. 1938. Some episodes in the structural evolution of southeast England. *Proceedings of the Geologists' Association*, **49**, 264–291.

—— & —— 1955. *Structure, Surface and Drainage in Southeast England*. George Philip & Sons, London.

YOUNG, R. W. 1983. The tempo of geomorphic change: evidence from southeastern Australia. *Journal of Geology*, **91**, 221–230.

Evolving models of the Tertiary evolutionary geomorphology of southern England, with special reference to the Chalklands

DAVID K. C. JONES

Department of Geography, London School of Economics, Houghton Street, London WC2A 2AE, UK

Abstract: This paper examines recent advances in understanding of the Tertiary evolutionary geomorphology of the southern England Chalklands, a subject of continuing controversy despite over a century of investigation. It begins by briefly outlining the classic Wooldridge and Linton model and discussing the many lines of criticism. Particular attention is paid to the explanation of discordant drainage and numerous lines of evidence are presented to explain why the 'superimposition model' has been rejected in favour of an explanation involving the development of anteconsequents. Attention is then directed to the new models of evolution advanced in the early 1980s which placed heavier emphasis on Palaeogene denudation and pulsed tectonism. These models are compared and contrasted, and attention focused on the growing recognition that the fundamental erosional surface is an etchplanated Summit Surface originating in the Palaeogene rather than a peneplanated surface developed during the Miocene and Pliocene. Significant developments of the last fifteen years are then reviewed – the establishment of inversion tectonics, lithostratigraphical division of the Chalk, apatite fission-track dating of uplift episodes and new information on the extent of the late Pliocene Red Crag incursion – and combined to produce a new evolutionary sequence which differs from previous interpretations in a number of regards. First, the concept of structural compartmentalization into morphotectonic regions removes the need for uniformity of evolution and indicates that uplift could have been variable in time and space, thereby removing the problems caused by continual adherence to the concept of mid-Tertiary tectonism. Second, the growing evidence for Pleistocene differential uplift of at least 200 m is considered. Third, an explanation of the so-called Plio-Pliocene marine bench involving modified etchplanation is advanced. The paper concludes with a discussion of remaining uncertainties.

The purpose of this paper is to review changing interpretations of the evolutionary geomorphology of extra-glacial southern England, an area that has traditionally been an important focus of palaeogeomorphological studies for two main reasons. First, the onshore presence of Upper Cretaceous Chalk and overlying Palaeogene sediments, both of which are folded, faulted and warped by Tertiary tectonic activity, has provided the firm basis for reconstructions of long-term landform evolution. Second, the area was the focus of the classic synthesis by S. W. Wooldridge and D. L. Linton (1939), *Structure, Surface and Drainage in South-East England*, which was to dominate thinking on the geomorphological evolution of southern Britain from its original publication in 1939 (it was reissued with minor modifications and additions in 1955) until the 1970s, when mounting criticisms (Hodgson *et al.* 1967, 1974; Green 1969, 1974; Worssam 1973; Jones 1974; Catt & Hodgson 1976) led to fundamental reinterpretations (Jones 1980, 1981; Small 1980) which were greatly influenced by the alternative thesis of Pinchemel (1954).

The passage of nearly two decades provides a suitable interval to evaluate the robustness of these new interpretations in the light of the very considerable recent advances in knowledge in such pertinent fields as structural geology, crustal stability, surficial sediments and weathering residues. Through necessity, however, such an assessment must begin with a brief consideration of the Wooldridge and Linton synthesis for three main reasons. First, to illustrate how thinking on tectonic movements and erosion surface development in one relatively small region has changed over time. Second, because it has to be recognized that Wooldridge and Linton's 'classic' reconstruction was, to some degree, a masterly elaboration and consolidation of previous work, as is clear from Clayton's (1980, p. 1) statement, 'all twentieth century work on South-East England has its roots in the nineteenth century, some of them very strong', so that reassessment inevitably returns to these antecedents. It is important to recognize the existence of such antecedents, especially in an age when the combination of new technology, new techniques, increasingly sophisticated computer manipulation

and the investigation of hitherto un- or under-explored areas tends to encourage neglect of, or indifference to, the contributions made by earlier workers. Third, the forty year dominance of their synthesis has left such an imprint on the literature that some authors (e.g. Embleton 1984; Goudie 1990) persist in according their model dominant status and view later reconstructions merely as criticisms that require substantiation. This is understandable, for Wooldridge and Linton's regional reconstruction was both satisfyingly simple and elegantly comprehensive, qualities yet to be matched by the findings of the subsequent more numerous but locally focused researchers.

The extra-glacial Chalklands

Much of the research into the pre-Quaternary geomorphological evolution of southern England has been focused on the irregular outcrop of the originally continuous Upper Cretaceous Chalk (Fig. 1), which forms a prominent border to the London and Hampshire–Dieppe structural basins within which are preserved Palaeogene sedimentary sequences totalling 240 m and 730 m, respectively. This soft, white limestone is now divided into a marly (grey) Lower Chalk (50–80 m) and a relatively pure White Chalk (Mortimore 1986) which varies in thickness depending on the original sedimentation conditions and the scale of post-Cretaceous erosion. A maximum of 435 m of Chalk has been recorded beneath Palaeogene sediments in the Hampshire Basin, but the thicknesses of White Chalk at outcrop is usually less than 250 m (Sellwood et al. 1986), and it has been totally removed from over the Weald–Artois Anticline, the Channel High (except for two small outliers in the southern Isle of Wight) and to the north and west. The conspicuous variations in outcrop width (Fig. 1) reflect structural control, with narrow 'hog's back' ridges developed where dips approach the vertical and broader cuestas formed where regional dips are shallow (<2°) or are effectively reduced due to the presence of secondary anticlines (see Fig. 5).

Cuesta topography is also highly variable. Summit elevations usually exceed 180 m, rising to 260 m at several locations (see Fig. 3) and reaching 297 m at Walbury Hill. These summits form characteristically undulating crestlines above prominent, steep escarpments, the boldest of which are to be found along the northern and northwestern margins of the outcrop and on the

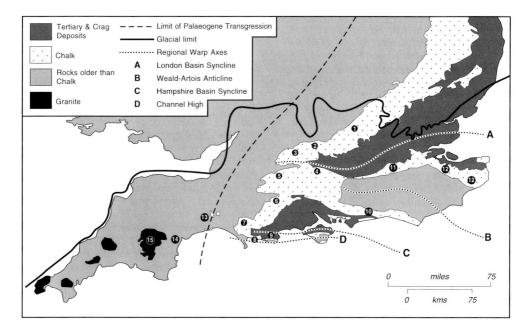

Fig. 1. Map of southern Chalklands. Numbers refer to Chalkland regions as follows: 1, Chiltern Hills; 2, Berkshire Downs; 3, Marlborough Downs; 4, Hampshire Downs; 5, Wiltshire Downs; 6, Cranborne Chase; 7, Dorset Downs; 8, South Dorset Downs; 9, Purbeck Hills/Downs; 10, South Downs; 11, North Downs; 12, Kentish North Downs. Other uplands: 13, Black Down Hills; 14, Haldon Hills; 15, Dartmoor.

northern and southern rims of the denuded Weald–Artois Anticline. In these localities escarpment relief is between 100 m and 170 m, bearing testimony to the intensity of Quaternary denudation. The conspicuous exceptions, in terms of cuesta form, are the two blocks of Kentish North Downs between the Medway and the Dover Straits, where the Chalk has been planed across to yield remarkably level crestlines at between 180 m and 192 m (Fig. 2). This 'Lenham Surface' can be traced westwards around the flanks of the London Basin (Fig. 3) and its origin has proved highly controversial.

The Chalklands are, therefore, conspicuous topographic belts between the relatively downwarped Tertiary basins and adjacent uplifted zones from which the Chalk and other strata have been removed by denudation. They display considerable morphological variation and carry a range of Palaeogene residuals and a spatially variable cover of surficial deposits and weathering residues, including blocks of silcrete known as sarsens (Summerfield 1979; Summerfield & Goudie 1980). They also form the majority of the highest ground in the region, the sole exceptions being the Lower Cretaceous Lower Greensand ramparts developed in the northwestern quadrant of the Weald (Fig. 3), which rise to 273 m at Hindhead, 280 m at Blackdown and 294 m at Leith Hill, and the summits developed on the same deposits in the Vale of Wardour which rise to 249 m near Shaftesbury. However, beyond the western margin of the Chalk outcrop occur high tablelands developed on Lower Cretaceous Upper Greensand with outliers of Chalk preserved in minor synclines. These tablelands rise in elevation northwards from c. 150 m at the East Devon coast to 315 m on the Black Down Hills, and can also be traced westwards across the River Exe to the flat-topped Upper Greensand summits of the Haldon Hills that stand at 250 m. Thus the high topography of the Chalklands is continued westwards for 50 km on the sedimentary unit immediately underlying the Chalk.

The Wooldridge and Linton interpretation

The classic model of Wooldridge and Linton (1939, 1955) was based on the recognition of three fundamental surfaces especially well developed on the Chalkland flanks of the London Basin (see Figs 2 and 3):

(i) An inclined and recently exhumed marine-trimmed surface fringing the present outcrop of Palaeogene sediments, which they called the Sub-Eocene Surface but which is now more correctly called the Sub-Palaeogene Surface.

(ii) An undulating Summit Surface above c. 210 m mantled by thick residual deposits of 'Clay-with-Flints' which they interpreted as remnants of a region-wide subaerial peneplain, thereby endorsing the view of Davis (1895) and rejecting the hypothesis of an extensive high-level marine plain at a height not far above the present Chalkland summits (Ramsey 1864; Foster & Topley 1865; Topley 1875; Bury 1910).

(iii) A conspicuous, gently inclined, erosional platform (c. 150–200 m O.D.) cut into the Summit Surface and apparently truncating the inclined Sub-Eocene Surface (Fig. 2). As this platform carries evidence for marine activity in terms of patches of sand (Headley Sands, Surrey) and solution pipe in-fills (Lenham Beds, Kent), together with fragments of fossiliferous ironstone (Netley Heath in Surrey, Rothamsted in Hertfordshire) (see Fig. 8), it was interpreted as a marine plain of Pliocene age (Wooldridge 1927; for reviews see Worssam 1963, 1973; Jones 1974, 1980, 1981; John 1980; John & Fisher 1984; Moffat & Catt 1986).

In view of the fact that the two higher surfaces – the Summit Surface and the marine platform – were deemed to be unwarped, their formation very clearly post-dated the tectonic episode that had deformed the Sub-Eocene Surface. As the major and minor folds of southeast England have traditionally been attributed to a mid-Tertiary (Oligocene and Miocene) phase of tectonic activity (e.g. Bury 1910) – a phase that Wooldridge and Linton (1955, p. 11) were to refer to as '...the main tectonic crisis, the "Alpine Storm", ...' – then the summit plain had to be the product of subaerial denudation in the later Miocene and Pliocene, hence the term Mio-Pliocene Peneplain. Their envisaged sequence of events is shown in Fig. 4.

Their reconstruction emphasized prevailing stability throughout the Tertiary, save for broad-scale flexuring in the early Palaeogene to establish the overall pattern of major structures and the subsequent late Oligocene–early Miocene brief phase of upheaval when the area was 'warped and locally corrugated' (Wooldridge & Linton 1955, p. 2) by 'the outer ripples of the Alpine Storm'. As a consequence, it was considered possible to extrapolate Neogene surfaces away from the London Basin on the basis of elevation (Fig. 3), most especially the late Pliocene marine surface which was thought to have been produced by a sea-level variously estimated at between 183 m and 210 m. Maps of the Pliocene

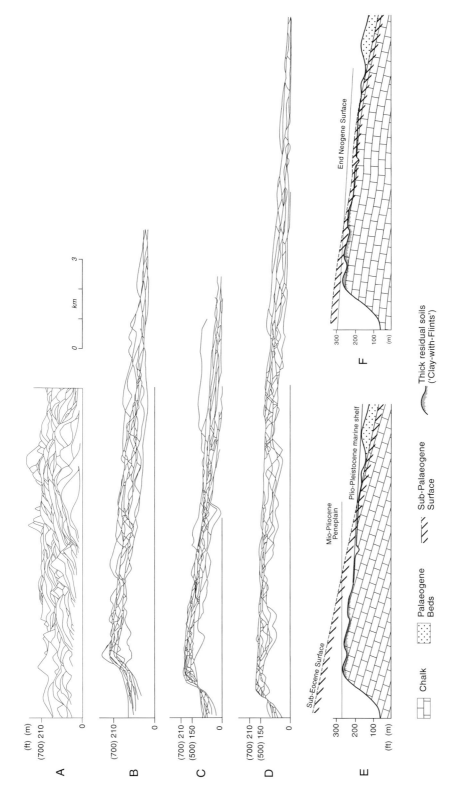

Fig. 2. Profiles of southern Chalklands. (**A**) Superimposed profiles of the Wiltshire Chalk uplands showing residual hills (monadnocks) above Mio-Pliocene Peneplain (Wooldridge & Linton 1955) reinterpreted by Green (1974) as remnants of Palaeogene Summit Surface rising above Miocene and Pliocene levels (from Green 1974). (**B**) Superimposed profiles of the central North Downs (after Wilson 1985). (**C**) Superimposed profiles of the western Kentish North Downs (after Wilson 1985). (**D**) Superimposed profiles of the eastern Kentish North Downs (after Wilson 1985). (**E**) Sketch section of Chalk cuesta showing three fundamental surfaces of Wooldridge and Linton (1955) (after Jones 1981). (**F**) Sketch section of Chalk cuesta with gross topography interpreted as developed from flexured Sub-Palaeogene Surface (after Jones 1981).

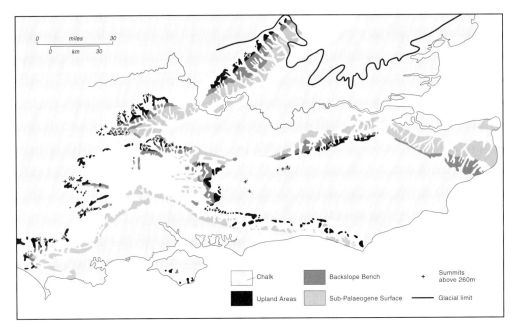

Fig. 3. Map of southern Chalklands showing distribution of main facets and highest summits. The distribution of Sub-Palaeogene Surface remnants in the Hampshire Basin is based on Small (1964) and elsewhere on Wooldridge & Linton (1955). The location of the Backslope Bench is based on Wooldridge & Linton (1955) and all remnants outside the London Basin have been disproved.

coastline (Fig. 5) were prepared on the basis of morphological evidence, the existence of land above 210 m and the widespread presence of discordant drainage, the last mentioned being of crucial importance to the region-wide reconstruction. According to Wooldridge and Linton (1938a, 1955), the growth of secondary structures in the mid-Tertiary was sufficiently rapid in the southern half of the region to disrupt the pre-existing drainage so as to create a predominantly east–west orientated network. The establishment of the present north–south discordant pattern required the eradication of much of this Neogene network, a task they credited to the late Pliocene transgression, with the present drainage superimposed from extensively developed marine shelves. The extent of the submergence was subsequently extrapolated to the Southwest Peninsula and Wales (Wooldridge 1952; Brown 1960). The existence of considerable topography above the 210 m 'strandline' in these areas required the identification of numerous surfaces between the late Pliocene datum and the perceived position of the end Cretaceous surface, with higher warped surfaces allocated to the Palaeogene and lower unwarped surfaces credited to the Neogene. As a consequence, the fundamental Summit Plain or Mio-Pliocene Peneplain of southeast England, with its occasional residual hills or 'monadnocks' rising above 250 m (Fig. 3), was envisaged to be replaced westwards by multiple surfaces created by uncompleted erosional cycles in the Miocene and Pliocene.

Criticisms of the Wooldridge and Linton model

How long the Wooldridge and Linton model of Tertiary landscape evolution continued wholly to dominate thinking is a matter of debate. Clayton (1980) argued that it was only until the early 1960s, while others suggest the early 1970s. Certainly the death of Wooldridge in 1963 heralded a progressive decline in interest amongst British geomorphologists in the long-term landform evolution of Britain, or denudation chronology as it came to be dismissively labelled, which has continued to the present day. However, the early stages of this withdrawal were accompanied by reappraisals in the late 1960s and early 1970s, resulting in publications in the early 1980s which effectively demolished the Wooldridge and Linton 'grand design' and

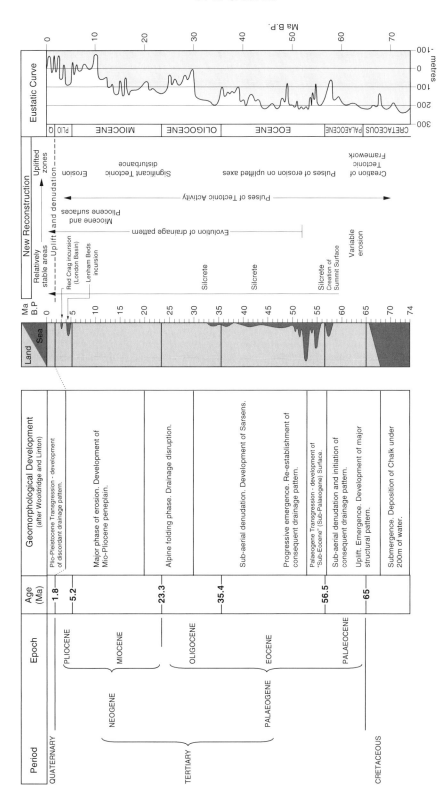

Fig. 4. The evolutionary geomorphology of southern England for the Tertiary, with dating of Palaeogene events based on Haq et al. (1987). The Wooldridge & Linton (1955) sequential interpretation on the left is to be contrasted with the contemporary 'evolutionary' interpretation on the right. The date for the start of the Quaternary (1.81 Ma) is based on arguments presented in Funnell (1995).

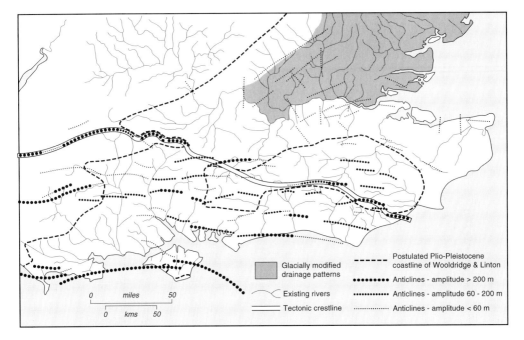

Fig. 5. The relationship between structure and drainage in SE England.

put forward the framework for a new reconstruction. It has to be stressed that this new framework was not wholly the product of new knowledge, but mainly derived from abandoning the fixation with morphological evidence in favour of the careful examination of Quaternary (surficial) deposits; much of the change reflected the reinterpretation and re-emphasis of previously known facts.

Criticisms of the Wooldridge and Linton reconstruction focused on three fundamental issues: the timing of denudation of the Chalk from above upland areas; the adoption of the Davisian approach which envisaged erosion cycles leading to peneplanation; and the belief in prevailing tectonic stability with brief interruptions of structural disturbance. The sequential or compartmentalized nature of the model (Fig. 4) was also criticized and subsequent interpretations sought to emphasize continuity of evolution. More specific criticisms included the following:

1. Stratigraphic evidence indicates structural restlessness from the mid-Jurassic to end Tertiary, including the initial development of a Weald upwarp in the Palaeocene together with the contemporaneous development of certain faults and secondary periclines (Williams-Mitchell 1956). The overlap and overstep relationships displayed by the Sub-Palaeogene unconformity are further evidence of early Tertiary movements. Wooldridge and Linton (1938b, 1955) were aware of much of this evidence but, nevertheless, chose to emphasize the importance of the brief mid-Tertiary tectonic episode, an emphasis described by T. N. George (1974, p. 117) as being 'highly selective in the Cenozoic time-span'. Reassessment resulted in the acceptance of long-term tectonic disturbance of an episodic nature, or 'pulsed tectonism' (George 1974), although there remained some disagreement as to whether or not tectonic activity reached a peak in the mid-Tertiary (see Jones 1980; Small 1980).

2. Examination of the soils and surficial deposits mantling the Chalklands failed to reveal the zonation suggested by the Wooldridge and Linton interpretation. Loveday (1962) had shown that the Clay-with-Flints was divisible into two distinct units: the Clay-with-Flints *sensu stricto* formed at the junction between the Chalk and overlying deposits (Hodgson *et al.* 1967, 1974) and a heterogeneous Plateau Drift. As both occur over the full width of Chalk backslopes and probably developed during the Pleistocene (Catt & Hodgson 1976; Catt 1983) their existence is of no relevance to Neogene landform development. Green (1974) reinvestigated the Wiltshire type area for the Mio-Pliocene Peneplain and could find no evidence

for either the peneplain or the Pliocene marine surface, but instead identified Palaeogene debris on the highest summits below which occur low relief subaerial surfaces ('partial planation surfaces') of Miocene and Pliocene age. The soils of the South Downs backslopes were found to be largely derived from the former cover of Reading Beds (Palaeocene) (Hodgson *et al.* 1974; Catt & Hodgson 1976) and the same is true of the Chiltern Hills (Jarvis *et al.* 1983). The soils of the central North Downs have been revealed to be developed on a complex melange of periglacially redistributed sediments of aeolian, fluvial and marine origin mixed with residuals from the Chalk and remnants of the former Palaeogene cover (John 1980; John & Fisher 1984; Wilson 1995). Thus all studies of soils have emphasized the important inheritance from the former Palaeogene cover while failing to provide support for the existence of a subaerial summit surface or marine platform created in the Neogene.

3. The existence of Palaeocene/Eocene residual outliers (Blackheath Beds and Woolwich Beds) near the crest of the central North Downs at elevations well above 200 m was used by Wooldridge and Linton (1955, p. 49) as evidence that the late Pliocene transgression had 'never washed this part of the Downs'. True though this may be, their presence also indicates that the Sub-Palaeogene Surface must have lain much closer to the crest of the Downs than suggested in the Wooldridge and Linton reconstruction and that it was not a simple plane but flexured or faceted to reflect the known overstep and overlap relationships (Jones 1980) (Fig. 2). The survival of these residuals can also be used to argue against intense Neogene subaerial erosion (Jones 1980; Wilson 1995).

4. The late Pliocene marine transgression, otherwise known as the Plio-Pleistocene, early Pleistocene, Calabrian, Waltonian, Red Crag or pre-Ludhamian transgression depending on the date of authorship, was of crucial importance to the Wooldridge and Linton model. The geomorphological significance of this incursion was effectively dismissed by Jones (1974) in a paper that has been fully vindicated with the passage of time. Four key points are worthy of note.

(i) A Plio-Pleistocene marine origin for the prominent platforms so well displayed on the flanks of the London Basin has never been proven. Indeed John (1974, 1980) has shown for the Netley Heath type site on the Surrey North Downs that the bench is not a wave-cut platform but a flexured facet of the Sub-Palaeogene Surface. Similar conclusions were reached following an intensive study of the extensive platforms preserved on the Chiltern backslopes (Moffat *et al.* 1986) which established that they are structurally controlled, although the authors added that 'it is possible that marine erosion before deposition of the Reading Beds (Palaeocene) played a small role in its formation' (p. 105), before concluding that the bench has 'no value as erosional evidence for a Plio-Pleistocene marine transgression' (p. 105). Neither study entirely dismissed the idea of the Plio-Pleistocene marine transgression extending over parts of the platform at a higher elevation. Moffat and Catt (1986, p. 244) remarked that the deposits at Lane End, Little Heath and Rothamsted on the Chilterns (Fig. 8) 'seem to be marine' and that they share many features with the Headley Formation of the North Downs as described by John (1980); these deposits, in turn, have been shown to have 'obvious mineralogical affinities with the Lenham Beds of Kent (John & Fisher 1984, p. 245). However, even in the case of the Lenham Beds it has to be noted that these marine deposits of uncertain Neogene age (early Pliocene?) occur in pipes set into the strikingly bevelled Chalk and that no definite link between the morphological platform and the piped sediments has ever been established. Whether there were one or two marine incursions in the Pliocene remains unclear, for fossil evidence of a late Pliocene incursion into the heart of the London Basin remains elusive, despite recent searches, and is restricted to the early finds of fossiliferous ironstones made at Netley Heath (Chatwin 1927) and Rothamsted (Dines & Chatwin 1930). The situation is aptly summed up by the words of John & Fisher (1984, p. 245) who discovered fragments of fossiliferous ironstone associated with the Headley Sand but found that 'the fossiliferous ironstone is neither *in situ* nor was it incorporated into the Headley Sand at the time of its deposition' and were forced to conclude 'whether the present association of the apparently marine Headley Sand with the Red Crag fossils is merely fortuitous is hard to say, but it is equally difficult to reject the view that the fossils derive from an unrelated deposit of which only the ironstone remains. The original geographical and altitudinal location of this former deposit ... is impossible to determine, although there is no obvious reason why it could not have been much farther to the south and at a greater elevation'. These conclusions, when combined with the reassessment of the Caesar's Camp Gravels on the nearby flat-topped Hale Plateau (186 m) as early Pleistocene fluvial deposits (Clarke & Fisher 1983), indicate the total rejection of the link between Neogene marine erosion and the

prominent platforms developed on the London Basin Chalklands. It is paradoxical that this rejection should have occurred at a time when sub-surface investigations revealed a 40 km westward extension of a finger of *in situ* Red Crag deposits to just north of Bishops Stortford (Mathers & Zalasiewicz 1988) where they lie at close to 100 m O.D., thereby indicating a local sea-level of *c.* 120 m and confirming the penetration of this incursion into the London Basin. It must be concluded, therefore, that the pre-Ludhamian Red Crag transgression resulted in a shallow marine embayment within which marine processes had insufficient energy to create morphological features capable of surviving for more than two million years.

(ii) Even during the heyday of the 'classic' model it was obvious that the overwhelming majority of the supposed evidence for a late Pliocene incursion was confined to the London Basin. Apart from a single record of piped sands near Beachy Head on the South Downs (Edmunds 1927) there were no proven marine sediments of Plio-Pleistocene age elsewhere in the region. Attempts were made to trace the morphological platform westwards on the basis of altitude (Stevens 1959), but no prominent features could be found on the South Downs or in Wessex (Fig. 3), raising the question as to why evidence for a supposedly geomorphologically significant marine incursion should be apparently so well preserved in the London Basin and yet be unconvincingly preserved on the same rock types less than 100 km away. The false picture that had been produced by reliance on morphological evidence within a predefined altitudinal range (Fig. 3) was exposed by Green (1969, 1974) who was unable to find evidence for a late Pliocene incursion in the Wessex type area, a view subsequently supported by Small (1980, p. 59) who wrote 'Geological evidence of the extension of the Calabrian sea into the Wessex region is so slight that many would now regard the transgression as unproven. However, for the purposes of this discussion, the transgression will be assumed, though it will be argued that even if it did occur its geomorphological impact was negligible'. Whether the Red Crag incursion ever penetrated into the Hampshire Basin remains unknown and Jones (1981, p. 127) speculated that it could be represented by the 145 m level of Everard (1954). However, Everard's work was undertaken within the constraints of the Wooldridge and Linton model and at a time when British geomorphologists were infatuated with the power of the sea and therefore inclined to attribute all gently sloping platforms to marine erosion. A marine origin for the 145 m surface was advocated merely on the grounds of low west-east slope (0.2 m km^{-1}). As the mantling gravels are largely composed of sub-angular flints, Jones (1981, p. 170) subsequently dismissed the correlation in favour of a periglacial origin. Thus there is no morphological evidence for a Red Crag incursion outside the London Basin, which accords well with its downgraded status within the London Basin (see above), and only a single recorded deposit at the easternmost extremity of the South Downs: scant evidence for proposing a major formative event in the geomorphological evolution of the region.

(iii) Where evidence for the Red Crag transgression exists, discordant drainage is rare and insignificant, but where discordant drainage is flagrant and widespread, no evidence for a marine incursion exists. Thus to link the two is implausible. To be fair to Wooldridge and Linton it was Douglas Johnson who persuaded Linton that the agency required to superimpose the discordant drainage of Wessex (Linton 1932) lay in the Pliocene transgression already established in the London Basin (Wooldridge 1927), by reputedly stating (Clayton 1980, p. 11) '"why invoke a local transgression when a regional one is to hand?"' Their resulting paper (Wooldridge and Linton 1938a) was the keystone of the 'grand design' and it was only much later that the major defects in this artificial 'marriage of convenience' became apparent.

(iv) Analysis of the drainage pattern of southern England reveals it to be generally accordant with macrostructure but discordant with secondary structures (Fig. 5). It is both reasonable and logical to explain the pattern as having developed on the flexured land surface revealed following the withdrawal of the Palaeogene sea (see Fig. 4), with rivers maintaining their courses across growing folds. The rivers are, for the most part, antecedent or anteconsequents (Cotton 1948), although the discordant relationships displayed in the Central Weald and close to the axis of the Hampshire-Dieppe Syncline are probably the result of superimposition from a thick cover of Weald Clay and Palaeogene sediments, respectively.

Thus the fundamental reassessment of the Red Crag transgression begun by Jones (1974) has been supported by subsequent research, to the extent that the episode is best described as a marine incursion of limited geomorphological significance which was largely confined to the London Basin, although connected to the Atlantic Ocean by a broad seaway 'via a southern (English Channel) route' (Funnell 1995, p. 11).

The 'new' models of the early 1980s

The many and varied objections to the Wooldridge and Linton reconstruction resulted in the emergence of new models of Tertiary landscape evolution which embraced pulsed tectonism and focused on the significance of Palaeogene denudation in fashioning the gross morphology of the Chalklands (Jones 1980, 1981; Small 1980). As such they represented a rather belated conversion to the views of Pinchemel (1954), who had maintained that the major elements on the Chalk of southern Britain are inherited from an early Tertiary polycyclic surface.

Although the models of Jones (1980) and Small (1980) differ in detail, they can be amalgamated as follows. The major structural basins were initiated soon after the Chalk emerged in the latest Cretaceous and were subsequently affected by episodic deepening to the end of the Miocene, with contemporary pulsed growth of basin margins, the Isle of Wight Monocline and the Weald–Artois Anticline. As a consequence, the cyclical accumulation of Palaeogene sediments within the structural basins occurred contemporaneously with pulses of accelerated denudation on uplifted areas, some of which (e.g. the Weald) may have remained unsubmerged for most of the time. This resulted in the flanks of the basins being progressively buried by younger and younger sediments, thereby fossilizing a compound sub-Tertiary unconformity which Jones (1980, p. 37) termed a 'multi-faceted polygenetic diachronous Sub-Palaeogene surface'. The extent of Sub-Palaeocene and Sub-Eocene facets depends on the chosen position of the Palaeocene–Eocene boundary. If placed at the base of the Woolwich and Reading Beds (Cavelier & Pomerol 1986; Pomerol 1989) then the Sub-Palaeocene facet is limited to the eastern London Basin, but if the base of the Oldhaven Beds is chosen (Aubry *et al.* 1986) then it is the Sub-Eocene facet that is restricted. The existence of Sub-Oligocene and Sub-Miocene facets is more contentious.

The maximum extent of Oligocene sedimentation remains unknown so it is impossible to determine whether they overlapped previous Palaeogene deposits onto the Chalk. The limited extent of Oligocene sediments in southern England and beneath the English Channel (freshwater limestone) indicates subaerial conditions, and the widespread survival of sarsen stones (silicified sandstones or silcretes) on the Chalklands (Summerfield 1979; Summerfield & Goudie 1980) has been used to suggest the former presence of duricrusted low relief landsurfaces subject to a sub-tropical to tropical seasonal climate at a little above the present topographic surface. Three periods in the Palaeogene are favoured for silcrete formation: the Upper Palaeocene (Reading Beds), Middle Eocene (Bagshot Beds) and Upper Eocene (Barton Beds). Palaeoclimatic studies (e.g. Buchardt 1978; Collinson & Hooker 1987) have confirmed warm climatic conditions from the late Palaeocene to the early Oligocene, with evidence for markedly seasonal conditions in the late Palaeocene (Buurman 1980) and Middle Eocene (Plint 1983), indicating that sarsen formation could have occurred in any or all of the three periods. However, the detailed study of the Marlborough Downs 'type area' (Clark *et al.* 1967) favoured a Bartonian or post-Bartonian age, thereby indicating the former presence of the subaerial sub-Oligocene facet in this area. Indeed, it must be assumed that the bulk of the sub-Oligocene facet was subaerially formed and thereby subject to resculpturing by subsequent denudation, and the same is true of the Miocene facets.

The basic elements of the general model are shown in Fig. 6. Geographical location determines which facet or facets of the Sub-Paleogene surface are present, their disposition, the nature of the overlying Palaeogene cover, the timing of exhumation, the scale of resculpturing in the Neogene and Quaternary and the general extent and character of residual deposits. For example, the Chalk cuestas that form the northern and southern rims of the London Basin are generally comparable despite slight differences in structure and Palaeogene sedimentary history, but display marked differences from the South Downs and the cuestas fringing the Hampshire Basin. Thus the sequential uniformity of the Wooldridge and Linton model is replaced by diversity.

While workers in southeast England focused on the recognition of a compound Sub-Palaeogene Surface, those to the west were establishing the reality of an early Palaeogene summit surface. The tilted tablelands of east Devon and west Dorset are arguably the finest example of an uplifted and dissected erosion surface in lowland Britain. This upland plain rises from 150 m at the coast to 315 m in the north on the crest of the Black Down Hills and is developed discordantly across the Upper Greensand and outliers of Chalk. As the area lies beyond the western limits of the main outcrop of Chalk it has traditionally tended to be excluded from studies of the Chalklands, which is unfortunate.

The prevailing view in the early 1960s was that the elevational range of these tablelands 'seemingly represents the morphological extension to the west of the Mio-Pliocene peneplain of the Chalkland of Dorset and south Wiltshire'

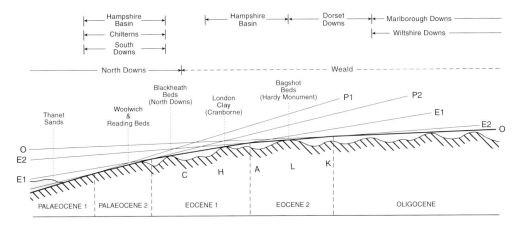

Fig. 6. The multifaceted, polygenetic Sub-Palaeogene Surface (developed from Jones 1980).

(Small 1980, p. 56), thereby indicating the continuation of this fundamental surface into Devon with the main mass of Dartmoor standing as a great monadnock some 300 m above the general level of the peneplain (Linton 1964). Indeed, Brunsden (1964) went so far as to state that the Lower Surface of Dartmoor is part of a widely developed planation surface which maintains a constant elevation from Cornwall to Kent, developed in the Neogene. This interpretation was first questioned by Waters (1960) who noted contrasting surficial deposits separated by a NW–SE line passing close to Axminster and Lyme Regis. To the west occur recognizable 'Clay-with-Flints', angular chert rubble in a sandy matrix, flints and flint breccia bound in a siliceous matrix and sarsens, all considered indicative of a subaerially developed Mio-Pliocene Peneplain. To the east, these deposits also contain local concentrations of rounded chattermarked beach cobbles of flint, interpreted as indicative of a gently flexured marine-trimmed surface that rises too high (up to 310 m) to be attributed to the late Pliocene incursion and therefore must be part of the Sub-Eocene Surface (Poole Formation = Bagshot Beds) that has survived through downfaulting. The numerous problems posed by this interpretation are discussed by Small (1980) who concluded that the tablelands are 'a complex composite "peneplain" developed by successive phases of erosion (largely subaerial but with episodes of marine trimming to the east), constituting a major early Tertiary "cycle"' (pp. 56–57).

This interpretation has been reinforced and extended by subsequent work (Isaac 1981, 1983a,b) and is excellently reviewed by Green (1985). Simply stated, the summits of both the Haldon Hills and the East Devon Plateau are now known to be mantled by residual soils indicative of tropical conditions, resting on heavily decalcified Upper Greensand. On the Haldon Hills, unrolled residual flint gravel in a well ordered kaolinite matrix (Tower Wood Gravel) is overlain by fluvial flint gravels with some Palaeozoic material (Bullers Hill Gravel), indicating formation prior to the development of the Bovey Basin in the Eocene (Hamblin 1973). On the East Devon Plateau to the north of Sidmouth, Isaac (1981, 1983a) has identified remnants of an original lateritic soil (Combpyne Soil) overlain by periglacially disturbed residual flint gravels in a kaolinitic clay matrix with fragments of laterite and siliceous breccias (Peak Hill Gravel = Tower Wood Gravel of Hamblin), dated as Danian (Lower Palaeocene) by analogy with the laterite weathering of the Intrabasaltic Formation of Northern Ireland. Thus the Summit Surface is now seen to have survived from the early Palaeogene, following the rapid removal of a significant thickness of Chalk. The mid-Eocene age of the marine shingle (Waters 1960) has also been changed, following the realization that this flint-rich deposit includes small pebbles of silicified shelly limestone which probably originated in the Purbeck Beds of southeast Dorset and could not, therefore, have been transported *eastwards* into the fluviatile Poole Formation without previously being moved *westwards* by marine agencies, i.e. by the London Clay (Lower Eocene) transgression. As similar gravels have been found on the highest parts of the Chalk in the Salisbury Plain area and in Wiltshire (Green 1969, 1974), it is now possible to state that 'an extensive erosional surface, refined by Early Eocene marine and

fluvial processes, is indicated in the broad region between Salisbury in the east and Dartmoor in the west' (Green 1985, p. 67).

The existence of such a surface, still recognizable on the highest summits, raises a number of interesting points.

1. In the western part of the region the dominant summit surface is of Palaeogene age and not the product of Neogene denudation.

2. The deeply weathered state of the Upper Greensand, together with the presence of kaolinitic clays, lateritic soils and silcretes, indicates evolution under sub-tropical or warm temperate climates of markedly seasonal character. The morphological and pedological characteristics are indicative of etchplanation.

3. The creation of the surface in Danian times implies the rapid removal of a significant thickness of Chalk (350 m) following Late Cretaceous uplift of c. 500 m (i.e. raising 350 m of Chalk from beneath 200 m of water minus the estimated Late Cretaceous eustatic fall of 40 m (Haq et al. 1987) (see Fig. 4).

4. Survival of this surface for 60 Ma, and especially the preservation of outliers of vulnerable Chalk, indicates a lengthy period of morphostasis covering most of the Tertiary. Green (1985) argued that this surface suffered only slight modification in the later Tertiary because it 'remained close to base level during the Neogene' (Green 1985, p. 71). The fluctuating eustatic curve of Haq et al. (1987) (Fig. 4) suggests that this would have been difficult, with estimated eustatic ranges of 200 m in both the Palaeogene and Neogene. The alternative explanation is that the surface was largely unaffected by base level changes, either because river lengthening during phases of eustatically lowered sea-levels cancelled the affects of base level change, or because of southward tilting, or both. Aridity might have contributed in the Palaeogene but cannot be invoked to explain the survival of this surface through the Neogene.

5. The incomplete dissection of the Summit Surface indicates the recent rejuvenation of the drainage network. Green (1985) invoked epeirogenic movements in the Pleistocene. The alternative explanation that rejuvenation was initiated by eustatic decline commencing in the Mid-Miocene (Fig. 4) seems implausible since dissection should have proceeded further. Nevertheless, the required Quaternary uplift of between 100 m and 200 m, although in line with the off-shore record (Evans & Hughes 1984), necessitated some recalibration of pre-existing interpretations of long-term landform evolution in the eastern parts of the region.

The views of Green (1985), Jones (1980, 1981) and Small (1980) on the evolution of southern England, although generally compatible, nevertheless appeared to differ in a number of important regards.

1. Green (1985) envisaged the extensive development of a duricrusted surface of low relief in the Palaeogene (Palaeocene and Eocene) which extended just above the highest summits of the region, in contrast to Jones (1980, 1981) and Small (1980) who invoked pulsed tectonism, especially for the Weald and other axes of uplift. Small (1980, p. 68) emphasized the importance of an '"early Tertiary" peneplain' but considered that it was mainly developed in the Oligocene and early Miocene.

2. The recognition of a Neogene morphostatic episode (Green 1985) appears markedly at variance with the progressive erosion envisaged by Jones and Small, especially in view of the continuing support amongst geologists for mid-Tertiary tectonism (Chadwick 1993). Small (1980, p. 53) postulated that 'the culminating Alpine movements were ... perhaps as late as "end Miocene"' thereby ensuring a 'vast amount of erosion to be assigned to the Pliocene', while Jones (1980) argued for a reduction in the emphasis placed on mid-Tertiary tectonism and, therefore, in the amount of erosion ascribed to the Neogene.

3. The authors differ as to the scale and significance of Pleistocene uplift. Green (1985) credited it with great importance in the west, implying uplift in the range 100–200 m. Small (1980) did not consider the point in detail preferring to emphasize Pliocene denudation, but appeared comfortable with his earlier view (Small & Fisher 1970) that the rivers of the western South Downs had cut down to 100–150 m O.D. by the beginning of the Quaternary, thereby implying uplift of 50–100 m. Jones (1980) also accepted the findings of Small & Fisher (1970) but subsequently (Jones 1981, p. 148) modified his position by stating 'It is clear that the whole of Southern England has experienced significant warping in the Quaternary ... the scale of which have yet to be evaluated', thereby indicating support for Worssam's (1973) suggestion of 'non-uniform uplift'.

4. All three authors indicate that the bulk of the denudation was achieved at different times: Palaeogene (Jones 1980), early Palaeogene and Quaternary (Green 1985) and Neogene (Small 1980).

The extent to which these differing interpretations represent fundamental disagreements or merely reflect local variations emphasized by

the specific geographical foci of the authors concerned, will be examined in the next section.

Advances since the early 1980s

The last decade has witnessed important developments in four areas of knowledge with a significant bearing on the long-term evolution of the region.

1. Lithostratigraphic investigations of the White Chalk (formerly the Middle and Upper Chalk) have shown it to be more variable in three dimensions than had hitherto been assumed (Mortimore 1986; Robinson 1986; Mortimore & Pomerol 1987). Although limited research has been undertaken on the relationship between lithostratigraphy and morphological detail, it is anticipated that many minor topographic features of the Chalklands will be explained in terms of lithological control. But more importantly, the greater detail provided by the new lithostratigraphic division has revealed that the Late Cretaceous sedimentary environment was spatially variable, due to the interaction of eustatic controls and local tectonics, and thereby 'finally put to rest the misconception that the White Chalk represents some uniquely uniform and consistent period of geological time' (Mortimore 1986, p. 136). Three findings from this work are of particular interest with reference to this discussion.

(i) The recognition that the pattern of sedimentation in the later Upper Cretaceous began to display a 'Tertiary Style' with signs of thinning in the vicinity of the Weald (Mortimore & Pomerol 1987), thereby confirming pre-Tertiary origins for the Wealden upwarp and the adjacent Hampshire–Dieppe and London Basins.

(ii) The recognition that an often displayed close relationship between surface tectonic structures and the distribution of sedimentary deposition centres, condensed sequences and unusual lithologies indicates contemporaneous folding and faulting (Mortimore & Pomerol 1987, 1991), so that 'structures currently interpreted as primarily Tertiary, are now seen as Cretaceous periclines with oversteepened northern limbs which subsequently became Tertiary monoclines' (Mortimore & Pomerol 1991, p. 402).

(iii) The detailed establishment of the Sub-Palaeogene overstep in the South Downs, indicating a clear younging of the Chalk subcrop towards the axis of the Hampshire–Dieppe Basin with identifiable differential erosion of at least 60 m in the Palaeocene (Mortimore & Pomerol 1991).

These findings greatly strengthen the arguments for the early establishment of the structural pattern, pulsed tectonism and the significance of Palaeogene denudation, especially on uplifted axes such as the Weald (Jones 1980, 1981).

2. Seismic investigations of deep structure in that portion of the region lying to the south of the Variscan Front have established the reality of the long-speculated link between surface folds, sub-surface faults and deeply buried thrusts in the Palaeozoic Floor/ Variscan Basement (Chadwick et al. 1983; Chadwick 1986, 1993), a linkage often referred to as 'basement control'. The evolution of surface structures is now interpreted within the framework of basin inversion, in which the subsidence patterns displayed by an original (Mesozoic) extensional sedimentary basin have been subsequently reversed in response to the change to a compressional tectonic stress regime, with the result that the basin floor has risen and the basin-fill deformed as the throw on the basin-controlling normal faults has been partially or totally reversed (Stoneley 1982; Ziegler 1987; Chadwick 1993). Two broad categories of inversion structures are recognized in the region (Fig. 7).

(i) Extensive upwarping occurred where thrust movements at depth affected weakly lithified sediments lying within grabens with poorly defined boundary faults, so that bulk shortening could only be achieved by general uplift.

(ii) Linear inversion structures, axes or zones were developed, characterized by narrow belts of en echelon faulted monoclinal or asymmetrical anticlinal folds 'which are invariably related to the partial reversal of underlying basin-controlling normal faults' (Chadwick 1993, p. 315), although others favour strike-slip movements in the late Cretaceous–Tertiary–Quaternary as an alternative, more plausible explanation for such structures (Lake & Karner 1987; Hibsch et al. 1993, 1995).

To the south of the Variscan Front, the two main upwarped areas of the Weald and the Channel High show inversions of over 1250 m and 1000 m, respectively (Chadwick 1993) and are delimited by three principal inversion axes – the Pewsey–London Platform inversion (similar to the 'Mendip Line' of Jones (1981)); the Wardour–Portsmouth inversion; and the Portland–Wight inversion – together with similar, but smaller, structures identifiable in the Weald (Butler & Pullen 1990; Chadwick 1993). The combination of major and minor inversion axes effectively compartmentalizes the region into a number of tectonic or structural units, each of which has had a different tectonic history. Thus, for example, the Mesozoic sediments

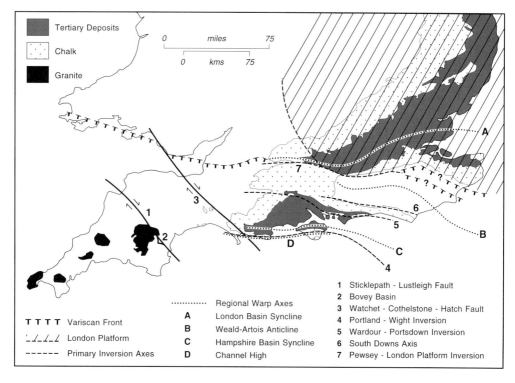

Fig. 7. A generalized portrayal of the main upwarp areas and inversion axes of southern England based on Chadwick (1993) and others, showing the compartmentalization of the area into morphotectonic regions.

deposited in the Dorset and Pewsey Basins display limited signs of upwarp compared with those laid down in the Channel and Weald Basins. The identification of this structural compartmentalization is of crucial importance because it determines morphotectonic regions, each characterized by distinct patterns of uplift, warping and erosion.

But the recognition of structural compartmentalization is important for another reason. The linkage between surface flexures and reactivated Variscan thrusts in the Palaeozoic Floor/ Variscan Basement indicates the possibility of individual inversion axes moving independently of each other in response to particular patterns of stress and thereby displaying different evolutionary histories. Advocates of region-wide profound tectonic deformation in the Miocene base their arguments on a known phase of Alpine movements and the fact that Oligocene beds are deformed by the Isle of Wight Monocline (Portland–Wight inversion axis), despite good evidence for movements of this structure in the Eocene (Plint 1982). The South Downs inversion axis similarly shows evidence of initiation in the Upper Cretaceous (Mortimore & Pomerol 1991), as does the Wardour–Portsdown axis, part of which is known to have been active in the Palaeocene (Williams-Mitchell 1956). The establishment of an inversion tectonics framework for the region therefore strengthens the argument in favour of pulsed tectonism, while at the same time reducing the need for region-wide co-ordination of fold growth. The Portland–Wight–St Valéry axis is such an important tectonic line (with some suggesting that it may even represent the Variscan Front) that it could have a unique pattern of movement in terms of magnitude, frequency and timing. Thus to argue for region-wide Miocene folding on the basis of deformation of a single major axis located on the southern border of the area is unconvincing. Clearly Miocene deformation did occur and was locally important, but there are now limited grounds for invoking Miocene movements as the formative event for the whole region.

3. Apatite fission-track studies (AFT) undertaken in the Midlands and northern England have indicated that Britain was affected by two episodes of exhumation during the Tertiary, the first commencing in the mid-Palaeocene

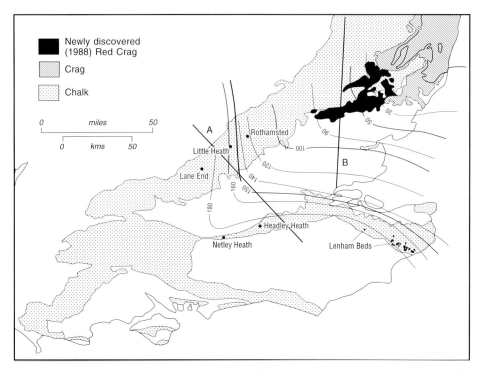

Fig. 8. The extension of known Red Crag deposits reported by Mathers & Zalasiewicz (1988) and generalized contours on the base of the Red Crag produced by combining and modifying the contours developed by Moffat & Catt (1986) and Mathers & Zalasiewicz (1988). All areas except those below sea level may be presumed to have been uplifted in the Pleistocene. A, pre-Red Crag limit of downwarping to east proposed by Wooldridge & Linton (1955); B, Braintree Line of Wooldridge & Henderson (1955).

(c. 60 Ma) and the second in the late Palaeogene (Oligocene) or early Neogene (Miocene) (Green 1986, 1989; Bray et al. 1992; Green et al. 1993; Lewis et al. 1992), with uplift and erosion amounting to 1–3 km. More recently, Japsen (1997) has used evidence from both on-shore and off-shore portions of the East Midlands Shelf to suggest that areas to the west of the present Chalk outcrop were affected equally by both phases of exhumation, while areas to the east were only affected by Neogene warping, the scale of which diminished eastwards. Despite the uncertainties associated with AFT, the results appear to provide support for those that favour a Neogene origin for the Weald upwarp (King 1981) or at least consider that its major development occurred in the Miocene (Chadwick 1993). Indeed, the very existence of uplands composed of vulnerable Chalk testifies to their recent uplift and exposure to denudation. However, the clear evidence for pulsed fold growth and for Palaeogene denudation suggests that the 'double uplift model' of Japsen (1997) is an over-simplification and cannot be translated southwards across the Variscan Front, except in the extreme west (i.e. the Southwest Peninsula, including Dartmoor). Even in the case of those 'northern' Chalklands overlying the stable London Platform, the potential significance of Neogene uplift has to be tempered by growing appreciation of the scale of Quaternary movements, as detailed below.

4. The reported discovery in the London Basin of a westward extension of *in situ* Red Crag resting on an inclined surface that cuts across Palaeocene deposits and Chalk to an elevation of over 90 m at Stansted Mountfitchet (Mathers & Zalasiewicz 1988) (Fig. 8) is important for a number of reasons.

(i) The Red Crag (Pre-Ludhamian) deposits are shallow marine (up to 25 m of water) thereby indicating a local sea-level of 120 m O.D. near Bishop's Stortford.

(ii) The Red Crag marine transgression appears to have invaded a landsurface of low relief affected by faulting oriented NNE–SSW, some of it contemporaneous (Bristow 1983; Hamblin et al. 1997).

(iii) The fauna indicate relatively cool conditions which suggests that the transgression was

probably initiated by local tectonic movements, thereby accounting for its restriction to the London Basin.

(iv) Both the Red Crag and overlying Norwich Crag show similar eastward tilt indicating warping post-2 Ma, as indicated by a widely recorded hiatus in the Crag sequence of East Anglia attributed to the Thurnian–Antian interval (Funnell 1987).

(v) The planar nature of the sub-Red Crag surface finally disproves the existence of a N–S hinge line (the 'Braintree Line' of Wooldridge & Henderson (1955); see Fig. 8) which supposedly defined the boundary between areas in the west characterized by Quaternary stability and those in the east that were downwarped towards the east due to the subsidence of the North Sea Basin.

(vi) The variations in elevation displayed by the Red Crag when compared with a late Pliocene sea-level of $c.\,20$ m (Haq et al. 1987) is firm evidence of late Neogene–Quaternary uplift and tilting, with downwarping of East Anglia (to a maximum of 50 m) and uplift of inland areas rising to a maximum of 100 m at the recently redefined western limit of the outcrop. These results verify the earlier reconstructions of the Sub-Red Crag surface (Moffat & Catt 1986) as rising westwards to $c.\,190$ m at Lane End and Netley Heath, indicating 'Quaternary' uplift of up to 170 m within the London Basin. As there are many reasons why 'Quaternary' uplift should have been even greater in the Weald (e.g. isostatic uplift due to denudational unloading, forebulge uplift generated by the weight of Pleistocene ice sheets), values well in excess of 200 m must be anticipated (Preece et al. 1990; Jones 1999).

(vii) The apparent scale of 'Quaternary' movements necessitates some recalibration of the deformation attributable to Tertiary tectonic episodes. For example, the uplift and warping attributed to Neogene movements in the 'double uplift' model of Japsen (1997) may be partly attributable to Quaternary deformation. Similarly, over 200 m of 'Quaternary' uplift in the Weald is a very significant proportion of the $c.\,1300$ m of total uplift estimated by Chadwick (1993). Thus the spatial and temporal dimensions of both Tertiary and Quaternary uplifts require further detailed investigation (Jones 1999).

A contemporary synthesis of Chalkland evolutionary geomorphology

The most important of these recent developments has been the establishment of the inversion structure framework, for this has facilitated recognition that the area is divisible into a number of morphotectonic regions, each characterized by slightly different structural and geomorphological histories. The southern margin of the London–Brabant Platform/Variscan Front has long been recognized as a fundamental boundary separating (i) a gently warped thin skin of Mesozoic strata above a shallow Palaeozoic Floor in the north from (ii) a well deformed thick Mesozoic sequence in the south. However, the strength of the three main inversion structures (the northern one of which is also claimed to be the Variscan Front (Stoneley (1982)), together with other less pronounced structures in the Weald (Fig. 7), indicates further subdivision. Thus the Weald must be seen as a complex horst structure that is largely decoupled from the North and South Downs, and the latter separated from the floor of the Hampshire–Dieppe Basin by the Wardour–Portsmouth inversion and its continuation to the east. Similarly, the heavily denuded Channel Uplands are isolated by the pronounced Portland–Wight inversion, and the main mass of the Salisbury Plain Chalklands is separated from the Dorset Downs to the south and the Marlborough Downs to the north by well developed inversion axes. As a consequence, eight distinct Chalkland regions can be recognized, indicating that the search for uniformity of Chalkland evolution must be finally abandoned in favour of a framework which includes a variable measure of differentiation.

Some indication of the anticipated variation is provided by Edmonds' (1983) study of the density of solution features (Fig. 9). The range of values from <5 per 100 km^2 in Wiltshire to >70 per 100 km^2 nearby in Dorset is surprisingly large and reveals that the greatest concentrations are in the extreme southwest and in the London Basin. Edmonds rightly points out that there are numerous inter-related factors which determine susceptibility to solution feature formation – exposed lithology, structure, erosional history, superficial cover, hydrogeology, etc. Nevertheless, it is worthy of note that relatively high concentrations occur where recent deposits are most extensive (i.e. the London Basin) and where erosion surfaces are best preserved (i.e. the London Basin and in the extreme west). By contrast, low densities of solution features are recorded where superficial deposits are generally thin and where erosion surfaces are relatively poorly preserved due to dissection. The Pewsey–London Platform inversion appears a particularly important structural divide in this context, separating as it does areas with a well

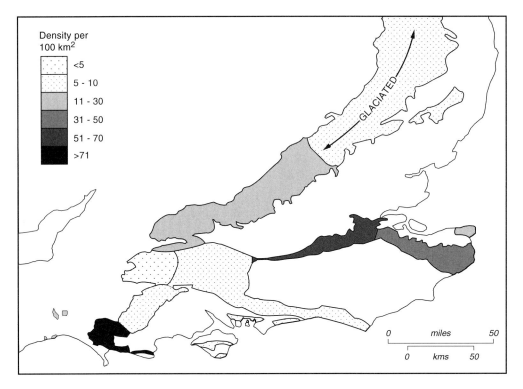

Fig. 9. The density of solution features on the Chalk (after Edmonds 1983).

developed Palaeogene legacy to the north from areas to the south which have been shown by Green (1969, 1974, 1985) to have suffered denudation in the Neogene.

This spatial partitioning also helps explain one of the great geomorphological puzzles of southern England, namely the origin of the conspicuous platforms developed on the flanks of the London Basin. That these features are only developed here suggests either that the formative events were confined to the London Basin or that conditions for preservation only existed in this area. Rejection of the link between the Red Crag transgression and the creation of these features (John 1980; John & Fisher 1984; Moffat et al. 1986) has emphasized the former explanation and they have been interpreted as elements of the Sub-Palaeogene Surface preserved through the development of monoclinal flexures (John 1980; Moffat et al. 1986). In the Chilterns, the platform has been shown to be tilted in a SW–NE direction at 0.2° (Moffat et al. 1986), roughly comparable with identified 'early Pleistocene' warping (Mathers & Zalasiewicz 1988), thereby requiring that it suffered virtually no displacement since its original formation in the Palaeocene. On the central North Downs, comparable platforms show clear evidence of dislocation in an eastward direction (Wilson 1995) and decline towards the axis of the London Basin at between $3.8 \, m \, km^{-1}$ (Wooldridge & Linton 1955) and $3.4 \, m \, km^{-1}$ (Moffat & Catt 1986), or approximately 0.2°, which is well below the regional dip of the Chalk and thereby necessitates the invoking of further monoclinal development. However, the platforms are so widely preserved as to raise doubts as to the validity of the structural explanation, especially as their prominent development on the crest of the Kentish North Downs is not associated with monoclinal development (Worssam 1973).

The alternative explanation is that the erosion surface or surfaces – for more than one 'platform' has been recognized on the central North Downs (Wooldridge & Linton 1955; Wilson 1995) – were created by non-marine agencies in the Neogene, after the completion of the majority of tectonic activity but prior to the phase of warping that occurred in the 'early Pleistocene'. The most plausible process is etchplanation. The concept of etchplanation has mainly been developed by Büdel (1982) and focuses on

landscape development through the creation of 'double surfaces of planation' or double levels of lowering: a low relief topographic surface largely shaped by wash processes over a thick regolith resting on a 'basal weathering surface' undergoing lowering through chemical attack. The combination of sub-surface lowering and surface removal creates an etchplain (the surface may sometimes be called a pediment) which proves to be a remarkably durable landform.

Etchplanation is widely recognized in the tropics where deep weathering mantles are extensively developed over irregular and active 'weathering fronts' (Thomas 1994) and there have been calls to recognize evidence for the former existence of Palaeogene etchplains in the British Isles (see Thomas & Summerfield 1987) to match their widespread identification on the continent of Europe as far north as southern Sweden. As a consequence, attention has focused on the deep weathering profiles, palaeokarst features and pediments reported from western Britain (Battiau-Queney 1984, 1987; Brunsden et al. 1976), although Migon (1997) has introduced a note of caution. This has resulted in the recognition that the 'Summit Surface' of Dorset and east Devon, with its mantle of residual soils, has many of the attributes of an etchplain.

While the undoubted warmth of the Late Palaeocene and Eocene would have proved conducive to etchplanation, the rather cooler conditions of the Neogene have often been seen to militate against the mechanism. However, modified etchplanation dating from the late Neogene has been reported from as far north as Buchan in NE Scotland (Hall 1987), so development on the highly vulnerable Chalk of southeast England is entirely feasible, if it can be explained why such well preserved etchplains are restricted to the London Basin. The simplest answer which fits the known facts is that the Red Crag transgression within the London Basin inundated the lower portions of a duricrusted low relief surface inherited from the earlier Tertiary, and in so doing, disturbed the surface layers of the weathered regolith and may have achieved minimal sculpturing of the Chalk as reported by Mathers and Zalasiewicz (1988). Regression revealed a near-coastal marine shelf developed over Chalk, Palaeogene sediments and regolith materials, beneath which developed an active weathering front due to enhanced percolation. Surface and sub-surface lowering further distinguished the formerly inundated areas from the adjacent former land areas with their still extensively preserved duricrusted carapaces. Etchplain development would have been enhanced where the late Pliocene marine shelf lay above a thin layer of Palaeogene sediments and stood the best chance of being preserved where the shelf development coincided with structurally determined relatively flat-lying portions of the Sub-Palaeogene Surface (Fig. 10).

The prospect of a new model of Tertiary landscape evolution for southern England as outlined by Jones (1981) and developed by Green (1985) can now be confirmed in a region-wide synthesis as follows.

1. Deposition of a continuous and thick (up to 550 m) sheet of Upper Cretaceous Chalk ceased in the Maastrichtian due to a combination of eustatic fall in sea-level (Hancock 1989) and tectonic deformation (Mortimore & Pomerol 1991) and 'there is little evidence of the former spread of a Late Maastrichian sea over the British Isles' (Hancock 1989, p. 584). Timing of emergence remains uncertain. Hancock (1989) indicates a rapid fall in sea-level from 68 Ma to reach Late Albian levels by 66 Ma, which must have been mainly due to tectonic movements as the eustatic curve of Haq et al. (1987) shows only a minor fall at this time. Emergence was probably complete by 65 Ma.

2. Denudation during the Palaeocene resulted in the rapid removal of up to 350 m of Chalk. Erosion was most severe on uplift axes (e.g. the Weald and the Channel High) and in the west where virtually all of a substantial layer of Chalk was quickly removed by subaerial denudation under tropical climatic conditions. A combination of eustatic fluctuations and tectonic movements resulted in the progressive encroachment of marine conditions from the east, commencing with the Thanet Sands (57 Ma: Palaeocene) and culminating in virtually complete inundation by the London Clay sea (53 Ma: Early Eocene) (Pomerol 1989). The Palaeocene and Early Eocene sediments were deposited on a multifaceted or polycyclic marine trimmed Sub-Palaeogene Surface cut in Chalk, except in the extreme west where the Upper Greensand had been exposed by the close of the Palaeocene beneath an extensive, sediment veneered, low relief surface (etchplain), the lower parts of which were easily submerged by the transgressing London Clay sea.

3. Continuing pulses of tectonic deformation through the remainder of the Palaeogene saw the further definition of the structural basins (London and Hampshire–Dieppe Basins) due to the progressive growth of the Weald–Artois Anticline and the Isle of Wight Monocline. Subaerial erosion on the axes of these upwarps led to

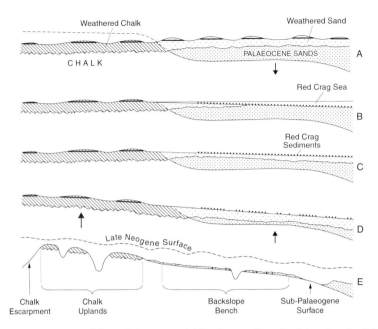

Fig. 10. Cartoon depiction of possible evolutionary model for the prominent backslope bench of the London Basin involving modified etchplanation. (**A**) Neogene duricrusted low relief surface. (**B**) Red Crag incursion following downwarping to east. (**C**) After regression. (**D**) Accentuated weathering beneath former marine platform leads to development of etchsurface. (**E**) Differential uplift in the Pleistocene leads to dissection and the removal of much regolith with the remainder greatly disturbed and mixed by periglaciation.

the development of further facets of the Sub-Palaeogene Surface, while sedimentation was progressively limited to the basin areas and eventually restricted to the Hampshire Basin in the Oligocene. Elsewhere, subaerial erosion under the hot climatic conditions of the Eocene resulted in the creation of an extensive low relief, duricrusted land surface (etchplain) over the majority the present Chalklands at an elevation a few tens of metres above the highest summits. In the west this surface had originated in the Palaeocene and slowly evolved through an essentially morphostatic episode, but to the east it had evolved through erosion of the previously deposited Palaeogene cover and Chalk, especially in areas subject to tectonic movements.

4. Rather more pronounced tectonic deformation in the mid-Tertiary (early Miocene) saw the further development of the structural pattern to virtually its present form. The uplift of upwarped areas and major inversion axes generated both relief and erosion, so that denudation accelerated on the Channel High and Weald-Artois Anticline. Elsewhere, the lesser scale of tectonic movements resulted in the development of Neogene surfaces (Miocene and Pliocene) at the expense of the late Palaeogene Summit Surface. In some areas (i.e. the West Country tablelands and the Chilterns) the limited scale of deformation resulted in the survival and continuing development of the late Palaeogene Summit Surface, sometimes to such a negligible degree as to warrant the Neogene being labelled a morphostatic episode.

5. The extent to which the lower and flatter portions of the late Neogene landsurface was invaded by Pliocene marine transgressions remains controversial. There exists evidence for a Red Crag (Pre-Ludhamian) incursion which affected the London Basin and eastern parts of the Weald but growing uncertainty as to the validity of an earlier Lenham Beds incursion. The Red Crag incursion appears the consequence of relatively localized downwarping and is considered to have caused minimal erosion, but sufficient to disrupt the surface duricrusts.

6. The subsequent marine regression revealed a gently inclined marine plain that suffered lowering through a form of etchplanation to yield the prominent 'platforms' exposed on the flanks of the London Basin.

7. At some time after 2 Ma further uplift and warping occurred, possibly as discontinuous pulses through much of the Pleistocene. While eastern East Anglia suffered subsidence, the remainder of southern England experienced

differential uplift to a maximum of at least 250 m (Jones 1999) and possibly up to 400 m (Preece *et al.* 1990). This uplift, in conjunction with oscillating sea-levels and climatic fluctuations, resulted in episodic erosion which increased in scale as relative relief developed through the Pleistocene.

This evolutionary model remains to be substantiated in a number of important regards, despite the obvious advances in knowledge over recent years. There remains uncertainty as to the true nature of the structural foundations of the area and its tectonic evolution, which results in continuing controversy as to the temporal and spatial dimensions of uplift and the relative importance of mid-Tertiary (Miocene) tectonic activity. Similarly, the recent suggestions of Pleistocene uplift and warping need to be confirmed, elaborated and accurately dated. Much work is still required on the nature and palaeoenvironmental significance of residual soils, including the varied types of silcrete, and on the number, age and geographical extent of Neogene marine incursions, most especially the continually perplexing Lenham Beds incursion. Only then, after two centuries of enquiry, will a fully detailed evolutionary geomorphology be established for this well known region.

The author gratefully acknowledges the helpful comments provided by Chris Green and Rory Mortimore on a draft of this paper.

References

AUBRY, M. P., HAILWOOD, E. A. & TOWNSEND, H. A. 1986. Magnetic and calcareous-nannofossil stratigraphy of the Lower Palaeogene formations of the Hampshire and London basins. *Journal of the Geological Society, London*, **143**, 729–735.

BATTIAU-QUENEY, Y. 1984. The pre-glacial evolution of Wales. *Earth Surface Processes and Landforms*, **9**, 229–252.

—— 1987. Tertiary inheritance in the present landscape of the British Isles. *In*: GARDINER V. (ed.) *International Geomorphology 1986 Part II*. Wiley, Chichester, 979–989.

BRAY, R. J., GREEN, P. F. & DUDDY, I. R. 1992. Thermal history reconstruction using apatite fission track analysis and vitrinite reflectance: a case study from the UK East Midlands and Southern North Sea. *In*: HARDMAN, R. F. P. (ed.) *Exploration Britain: Geological insights for the next decade*. Geological Society, London, Special Publication, **67**, 3–25.

BRISTOW, C. R. 1983. The stratigraphy and structure of the Crag of mid-Suffolk, England. *Proceedings of the Geologists' Association, London*, **94**, 1–12.

BROWN, E. H. 1960. The building of southern Britain. *Zeitscrift für Geomorphologie*, **4**, 264–274.

BRUNSDEN, D. 1964. Denudation chronology of parts of South-Western England. *Field Studies*, **2**, 115–132.

——, DOORNKAMP, J. C., GREEN, C. P. & JONES, D. K. C. 1976. Tertiary and Cretaceous sediments in solution pipes in the Devonian limestone of South Devon, England. *Geological Magazine*, **113**, 441–447.

BUCHARDT, B. 1978. Oxygen isotope palaeotemperatures from the Tertiary period in the North Sea area. *Nature*, **275**, 121–123.

BÜDEL, J. 1982. *Climatic Geomorphology* (translated by FISCHER, L. & BUSCHE, D.). Princeton University, Princeton.

BURY, H. 1910. On the denudation of the western end of the Weald. *Quarterly Journal of the Geological Society, London*, **66**, 640–692.

BUTLER, M. & PULLEN, C. P. 1990. Tertiary structures and hydrocarbon entrapment in the Weald Basin of Southern England. *In*: HARDMAN, R. P. P. & BROOKS, J. (eds), *Tectonic Events Responsible for Britain's Oil and Gas Reserves*. Geological Society, London, Special Publication, **55**, 371–391.

BUURMAN, P. 1980. Palaeosols in the Reading Beds (Palaeocene) of Alum Bay, Isle of Wight, UK. *Sedimentology*, **27**, 593–606.

CATT, J. A. 1983. Cenozoic pedogenesis and landform development in southeast England. *In*: WILSON, R. C. L. (ed.) *Residual Deposits: surface related weathering processes and materials*. Blackwell, Oxford, 251–258.

—— & HODGSON, J. M. 1976. Soils and geomorphology of the Chalk in south-east England. *Earth Surface Processes*, **1**, 181–193.

CAVELIER, C. & POMEROL, C. 1986. Stratigraphy of the Palaeogene. *Bulletin Société géologique France*, **11**, 255–265.

CHADWICK, R. A. 1986. Extension tectonics in the Wessex Basin, southern England. *Journal of the Geological Society, London*, **143**, 465–488.

—— 1993. Aspects of basin inversion in Southern Britain. *Journal of the Geological Society, London*, **150**, 311–322.

——, KENOLTY, N. & WHITTAKER, A. 1983. Crustal structure beneath southern Britain from deep seismic reflection profiles. *Journal of the Geological Society, London*, **140**, 893–911.

CHATWIN, C. P. 1927. Fossils from the ironsands on Netley Heath (Surrey). *Memoir of the Geological Survey: Summary of Progress* (1926), 154–157.

CLARK, M. J., LEWIN, J. & SMALL, R. J. 1967. The sarsen stones of the Marlborough Downs and their geomorphological implications. *Southampton Research Series in Geography*, **4**, 3–40.

CLARKE, M. R. & FISHER, P. F. 1983. The Caesar's Camp Gravel – an early Pleistocene fluvial periglacial deposit. *Proceedings of the Geologists' Association, London*, **94**, 345–355.

CLAYTON, K. M. 1980. The historical context of structure, surface and drainage in south-east England. *In*: JONES, D. K. C. (ed.) *The Shaping of Southern England*. Institute of British Geographers, Special Publication, **11**, 1–12.

COLLINSON, M. E. & HOOKER, J. J. 1987. Vegetational and mammalian formal changes in the early Tertiary of southern England. *In*: FRIIS, E. M., CHALONER, W. G. & CRANE, P. R. (eds) *The Origin of Angiosperms and their Biological Consequences*. Cambridge University, Cambridge.

COTTON, C. A. 1948. *Landscape as Developed by the Processes of Normal Erosion*. Whitcombe & Tombs, Christchurch.

DAVIS, W. M. 1895. On the origin of certain English rivers. *Geographical Journal*, **5**, 128–146.

DINES, H. G. & CHATWIN, C. P. 1930. Pliocene sandstone from Rothamsted (Hertfordshire). *Memoir of the Geological Survey – Summary of Progress (1929)*, 1–7.

EDMONDS, C. N. 1983. Towards the prediction of subsidence risk upon the Chalk outcrop. *Quarterly Journal of Engineering Geology, London*, **16**, 261–266.

EDMUNDS, F. H. 1927. Pliocene deposits on the South Downs. *Geological Magazine*, **64**, 287.

EMBLETON, C. (ed.) 1984. *Geomorphology of Europe*. Macmillan, London.

EVANS, C. D. R. & HUGHES, M. R. 1984. The Neogene succession in the South West Approaches, Great Britain. *Journal of the Geological Society, London*, **141**, 315–326.

EVERARD, C. E. 1954. The Solent River: a geomorphological study. *Transactions of the Institute of British Geographers*, **20**, 41–58.

FOSTER, C. LE NEVE & TOPLEY, W. 1865. On the superficial deposits of the valley of the Medway, with remarks on the denudation of the Weald. *Quarterly Journal of the Geological Society, London*, **21**, 443–474.

FUNNELL, B. M. 1987. Late Pliocene and early Pleistocene stages of East Anglia and the adjacent North Sea. *Quaternary Newsletter*, **52**, 1–11.

—— 1995. Global sea-level and the (pen-)insularity of late Cenozoic Britain. *In*: PREECE, R. C. (ed.) *Island Britain: a Quaternary Perspective*. Geological Society, London, Special Publication, **96**, 3–13.

GEORGE, T. N. 1974. Prologue to a geomorphology of Britain. *In*: BROWN, E. H. & WATERS, R. S. (eds) *Progress in Geomorphology*. Institute of British Geographers, Special Publication, **7**, 113–125.

GOUDIE, A. S. 1990. *The Landforms of England and Wales*. Blackwell, Oxford.

GREEN, C. P. 1969. An early Tertiary surface in Wiltshire. *Transactions of the Institute of British Geographers*, **47**, 61–72.

—— 1974. The summit surface on the Wessex Chalk. *In*: BROWN, E. H. & WATERS, R. S. (eds) *Progress in Geomorphology*. Institute of British Geographers, Special Publication, **7**, 127–138.

—— 1985. Pre-Quaternary weathering residues, sediments and landform development: examples from southern Britain. *In*: RICHARDS, K. S., ARNETT, R. R. & ELLIS, S. (eds) *Geomorphology and Soils*. Allen & Unwin, London, 58–77.

GREEN, P. F. 1986. On the thermo-tectonic evolution of Northern England: evidence from fission track analysis. *Geological Magazine*, **123**, 493–506.

—— 1989. Thermal and tectonic history of the East Midlands shelf (onshore UK) and surrounding regions assessed by apatite fission track analysis. *Journal of the Geological Society, London*, **146**, 755–773.

——, DUDDY, I. R., BRAY, R. J. & LEWIS, C. L. E. 1993. Elevated paleotemperatures prior to early Tertiary cooling throughout the UK region: Implications for hydrocarbon generation. *In*: PARKER, J. R. (ed.) *Petroleum Geology of Northwest Europe: Proceedings of the 4th Conference*. The Geological Society, London, 1067–1074.

HALL, A. M. 1987. Weathering and relief development in Buchan, Scotland. *In*: GARDINER, V. (ed.) *International Geomorphology 1986 Part II*. Wiley, Chichester, 991–1005.

HAMBLIN, R. J. O. 1973. The Haldon Gravels of south Devon. *Proceedings of the Geologists' Association, London*, **84**, 459–476.

——, MOORLOCK, B. S. P., BOOTH, S. J., JEFFERY, D. H. & MORIGI, A. N. 1997. The Red Crag and Norwich Crag formations in eastern Suffolk. *Proceedings of the Geologists' Association, London*, **108**, 11–23.

HANCOCK, J. M. 1989. Sea-level changes in the British region during the Late Cretaceous. *Proceedings of the Geologists' Association, London*, **100**, 565–594.

HAQ, B. U., HARDENBOL, J. & VAIL, P. R. 1987. Chronology of fluctuating sea levels since the Triassic. *Science*, **235**, 1156–1167.

HIBSCH, C., CUSHING, E.M., CABRERA, J., MERCIER, J., PRASIL, P. & JARRIGE, J.-J. 1993. Paleostress evolution in Great Britain from Permian to Cenozoic: a microtechnic approach to the geodynamic evolution of the southern U.K. basins. *Bull. Cent. Rech. Explor. Prod. Elf Aquitaine Production, F-31360 Boussens*, **17**, 303–330.

——, JARRIGE, J.-J., CUSHING, E. M. & MERCIER, J. 1995. Paleostress analysis, a contribution to the understanding of basin tectonics and geodynamic evolution. Example of Permian/Cenozoic tectonics of Great Britain and geodynamic implications in western Europe. *Tectonophysics*, **252**, 103–136.

HODGSON, J. M., CATT, J. A. & WEIR, A. H. 1967. The origin and development of Clay-with-Flints and associated soil horizons on the South Downs. *Journal of Soil Science*, **18**, 85–102.

——, RAYNER, J. H. & CATT, J. A. 1974. The geomorphological significance of the Clay-with-Flints on the South Downs. *Transactions of the Institute of British Geographers*, **61**, 119–129.

ISAAC, K. P. 1981. Tertiary weathering profiles in the plateau deposits of East Devon. *Proceedings of the Geologists' Association, London*, **92**, 159–168.

—— 1983a. Tertiary lateritic weathering in Devon, England, and the Palaeogene continental environment of S W England. *Proceedings of the Geologists' Association, London*, **94**, 105–114.

—— 1983b. Silica diagenesis of Palaeogene residual deposits in Devon, England. *Proceedings of the Geologists' Association, London*, **94**, 181–186.

JAPSEN, P. 1997. Regional Neogene exhumation of Britain and the western North Sea. *Journal of the Geological Society London*, **154**, 239–247.

JARVIS, M. G., ALLEN, R. H., FORDHAM, S. J., HAZELDEN, J., MOFFAT, A. J. & STURDY, R. G. 1983. *Soils of England and Wales*, Sheet 6, S.E. England. Soil Survey of England and Wales, Ordnance Survey, Southampton.

JOHN, D. T. 1974. *A study of the soils and superficial deposits of the North Downs of Surrey*. PhD thesis, University of London.

—— 1980. The soils and superficial deposits of the North Downs. *In*: JONES, D. K. C. (ed.) *The Shaping of Southern England*. Institute of British Geographers, Special Publication, **11**, Academic, London, 101–130.

—— & FISHER, P. F. 1984. The stratigraphical and geomorphological significance of the Red Crag fossils at Netley Heath, Surrey: a review and re-appraisal. *Proceedings of the Geologists' Association, London*, **95**, 235–247.

JONES, D. K. C. 1974. The influence of the Calabrian transgression on the drainage evolution of south-east England. *In*: BROWN, E. H. & WATERS, R. S. (eds) *Progress in Geomorphology*. Institute of British Geographers, Special Publication, **7**, 139–158.

—— 1980. The Tertiary evolution of south-east England with particular reference to the Weald. *In*: JONES, D. K. C. (ed.) *The Shaping of Southern England*. Institute of British Geographers, Special Publication, **11**, Academic, London, 13–47.

—— 1981. *The Geomorphology of the British Isles: Southeast and Southern England*. Methuen, London.

—— 1999. On the uplift and denudation of the Weald. This volume.

KING, C. 1981. *The stratigraphy of the London Clay*. Tertiary Research Special Paper, **6**, Backhuys, Rotterdam.

LAKE, S. D. & KARNER, G. D. 1987. The structure and evolution of the Wessex Basin, southern England: an example of inversion tectonics. *Tectonophysics*, **137**, 347–348.

LEWIS, C. L. E, GREEN, P. F., CARTER, A. & HURFORD, A. J. 1992. Elevated K/T palaeotemperatures throughout Northwest England: three kilometres of Tertiary erosion? *Earth and Planetary Science Letters*, **112**, 131–145.

LINTON, D. L. 1932. The origin of the Wessex Rivers. *Scottish Geographical Magazine*, **48**, 149–166.

—— 1964. Tertiary landscape evolution. *In*: WREFORD-WATSON, J. & SISSONS, J. B. (eds) *The British Isles: a systematic geography*. Nelson, London, 110–130.

LOVEDAY, J. A. 1962. Plateau deposits of the southern Chiltern Hills. *Proceedings of the Geologists' Association, London*, **73**, 83–101.

MATHERS, S. J. & ZALASIEWICZ, J. A. 1988. The Red Crag and Norwich Crag formations of southern East Anglia. *Proceedings of the Geologists' Association, London*, **99**, 261–278.

MIGON, P. 1997. Palaeoenvironmental significance of grus weathering profiles: a review with special reference to northern and central Europe. *Proceedings of the Geologists' Association, London*, **108**, 57–70.

MOFFAT, A. J. & CATT, J. A. 1986. A re-examination of the evidence for a Plio-Pleistocene marine transgression on the Chiltern Hills III. Deposits. *Earth Surface Processes and Landforms*, **11**, 233–247.

——, ——, WEBSTER, R. & BROWN, E. H. 1986. A re-examination of the evidence for a Plio-Pleistocene marine transgression on the Chiltern Hills. I. Structures and Surfaces. *Earth Surface Processes and Landforms*, **11**, 95–106.

MORTIMORE, R. N. 1986. Stratigraphy of the Upper Cretaceous White Chalk of Sussex. *Proceedings of the Geologists' Association, London*, **97**, 97–139.

—— & POMEROL, B. 1987. Correlation of the Upper Cretaceous White Chalk (Turonian to Campanian) in the Anglo-Paris Basin. *Proceedings of the Geologists' Association, London*, **98**, 97–143.

—— & —— 1991. Upper Cretaceous tectonic disruptions in a placid Chalk sequence in the Anglo-Paris Basin. *Journal of the Geological Society, London*, **148**, 391–404.

PINCHEMEL, P. 1954. *Les plaines de Craie du nord-ouest du Bassin Parisien et du sud-est du Bassin de Londres et leur bordures*. Armand Colin, Paris.

PLINT, A. G. 1982. Eocene sedimentation and tectonics in the Hampshire Basin. *Journal of the Geological Society, London*, **139**, 249–254.

—— 1983. Sandy fluvial point-bar sediments from the Middle Eocene of Dorset, England. International Association of Sedimentologists, Special Publication, **6**, 355-368.

POMEROL, C. 1989. Stratigraphy of the Palaeogene: hiatuses and transitions. *Proceedings of the Geologists' Association*, **100**, 313–324.

PREECE, R. C., SCOURSE, J. D., HOUGHTON, S. D., KNUDSEN, K. L. & PENNY, D. N. 1990. The Pleistocene sea-level and neotectonic history of the eastern Solent, southern England. *Philosophical Transactions of the Royal Society of London*, **B328**, 425–447.

RAMSEY, A. E. 1864. *The Physical Geology and Geography of Britain*. Stanford, London.

ROBINSON, N. D. 1986. Lithostratigraphy of the Chalk Group of the North Downs, southeast England. *Proceedings of the Geologists' Association, London*, **97**, 141–170.

SELLWOOD, B. W., SCOTT, J. & LUNN, G. 1986. Mesozoic basin evolution in Southern England. *Proceedings of the Geologists' Association, London*, **97**, 259–289.

SMALL, R. J. 1964. Geomorphology. *In*: MONKHOUSE, F. J. (ed.) *A Survey of Southampton and its Region*. Camelot, Southampton, 37–50.

—— 1980. The Tertiary geomorphological evolution of south-east England: an alternative interpretation. *In*: JONES, D. K. C. (ed.) *The Shaping of Southern England*. Institute of British Geographers, Special Publication, **11**, Academic, London, 49–70.

—— & FISHER, G. C. 1970. The origin of the secondary escarpment of the South Downs. *Transactions of the Institute of British Geographers*, **49**, 97–107.

STEVENS, A. J. 1959. Surfaces, soils and land use in north-east Hampshire. *Transactions of the Institute of British Geographers*, **26**, 51–66.

STONELEY, R. 1982. The structural development of the Wessex Basin. *Journal of the Geological Society, London*, **139**, 543–554.

SUMMERFIELD, M. A. 1979. Origin and palaeoenvironmental interpretation of sarsens. *Nature*, **281**, 137–139.

—— & GOUDIE, A. S. 1980. The sarsens of southern England: their palaeoenvironmental interpretation with reference to other silcretes. *In*: JONES, D. K. C. (ed.) *The Shaping of Southern England*. Institute of British Geographers, Special Publication, **11**, Academic, London, 71–100.

THOMAS, M. F. 1994. *Geomorphology in the Tropics: a study of Weathering and Denudation in Low Latitudes*. Wiley, Chichester.

—— & SUMMERFIELD, M. A. 1987. Long-term landform development: key themes and research problems. *In*: GARDINER, V. (ed.) *International Geomorphology 1986 Part II*. Wiley, Chichester, 935–956.

TOPLEY, W. 1875. *The Geology of the Weald*. Memoir of the Geological Survey, UK.

WATERS, R. S. 1960. The bearing of superficial deposits on the age and origin of the upland plain of east Devon, west Dorset and south Somerset. *Transactions of the Institute of British Geographers*, **28**, 89–95.

WILLIAMS-MITCHELL, E. 1956. The stratigraphy and structure of the chalk of Dean Hill anticline, Wiltshire. *Proceedings of the Geologists' Association, London*, **26**, 221–227.

WILSON, D. I. 1985. *Geomorphological Significance of the Soils and Superficial Deposits of the North Downs*. PhD thesis, University of London.

—— 1995. Soils of the central North Downs: their distribution, derivation and geomorphological significance. *Zeitschrift für Geomorphologie*, **39**, 433–460.

WOOLDRIDGE, S. W. 1927. The Pliocene period in the London Basin. *Proceedings of the Geologists' Association, London*, **38**, 49–132.

—— 1952. The changing physical landscape of Britain. *Geographical Journal*, **118**, 297–308.

—— & HENDERSON, H. C. K. 1955. Some aspects of the physiography of the eastern part of the London Basin. *Transactions of the Institute of British Geographers*, **21**, 19–31.

—— & LINTON, D. L. 1938a. Influence of the Pliocene transgression on the geomorphology of south-east England. *Journal of Geomorphology*, **1**, 40–54.

—— & —— 1938b. Some episodes in the structural evolution of south-east England considered in relation to the concealed boundary of Meso-Europe. *Proceedings of the Geologists' Association, London*, **49**, 264–291.

—— & —— 1939. *Structure, Surface and Drainage in South-East England*. Institute of British Geographers, Publication **10**.

—— & —— 1955. *Structure, Surface and Drainage in South-East England*. Philip, London.

WORSSAM, B. C. 1963. *The Geology of the Country around Maidstone*. Memoir of the Geological Survey, UK.

—— 1973. *A new look at river capture and at the denudation history of the Weald*. Institute of Geological Sciences, Report, **73/17**.

ZIEGLER, P. A. 1987. Compressional intraplate deformations in the Alpine foreland – an introduction. *Tectonophysics*, **137**, 1–5.

On the uplift and denudation of the Weald

DAVID K. C. JONES

Department of Geography, London School of Economics, Houghton Street, London WC2A 2AE, UK

Abstract: The evolution of the Weald has been the subject of continuing controversy over nearly two centuries of investigation. This paper reviews current knowledge of the area's development during the Tertiary against this background and produces an evolutionary geomorphology focused on a cross-section reconstruction of the Weald stretching from the south coast to the Thames. Estimations of the original thickness of Chalk (400–460 m) and other Mesozoic strata indicate a complex fold structure which, if ever fully developed, would have had a crestal elevation of 1400 m. The additional accumulation of an estimated 60–220 m of Palaeogene sediments reveals gross denudation of up to 1350 m in 66 Ma or 20.4 m Ma^{-1}, a low rate on relatively erodible strata indicating periods of relative morphostasis in the Tertiary. Four models of evolution are identified – 'Neogene flexuring and erosion', 'pulsed tectonism focusing on Palaeogene denudation', the classic Wooldridge and Linton interpretation, and the model detailed in this paper – and applied to the cross-sectional reconstruction to produce spatial and temporal patterns of gross denudation and denudation rates. The analysis reveals that all models are possible theoretically and that denudation rates lie in the lower half of the range attributed to 'normal denudation'. The model detailed in this paper has a proposed pattern of evolution for the Central Weald that involves early Palaeogene uplift, warping and denudation, followed by slower denudation in the later Palaeogene, stimulated erosion in the Neogene due to tectonic activity, culminating with severe denudation and relief generation in the Quaternary due to uplift of up to 260 m. The Chalk downlands, by contrast, are considered to have developed from the Pleistocene exhumation and dissection of areas that evolved more slowly in the Tertiary and were dominated by low relief landsurfaces at no great elevation above contemporary summits that experienced lengthy periods of morphostasis.

The Weald is one of the best known examples of denuded anticline landscapes yet its evolution has remained controversial despite numerous studies. The purpose of this paper is to review current knowledge of the long-term development of the area so as to propose an evolutionary geomorphology and bring into focus remaining uncertainties. This is to be achieved by applying a number of evolutionary interpretations to a reconstructed cross-section of the Weald in order to assess their comparative feasibility.

Background to the evolutionary geomorphology of the Weald

The Weald represents the heavily eroded western half of the broad Weald–Artois Anticline, its eastern extremity in the Boulonnais of northern France separated by the 42–95 km expanse of the Dover Straits (Fig. 1A). Superimposed on this 210 km long upwarp are short, roughly E–W oriented asymmetrical periclines, often arranged en echelon. The whole complex of structure is considered to have developed mainly during the Tertiary. Denudation has removed the original cover of Chalk over an area of 4500 km^2 of southern England to reveal concentric outcrops of Lower Cretaceous sandstones and clays (Fig. 1B), truncated in the east by the discordant Channel coast, and rimmed by imposing infacing Chalk escarpments on the other three sides that generally rise to over 150 m and reach 266 m on the North Downs, 246 m on the Hampshire Downs and 271 m on the South Downs. At the core of the structure are the predominantly arenaceous Hastings Beds which form the elevated ground of the Central or 'High' Weald, rising to a maximum of 240 m (Fig. 2). This elevated central area is bounded on three sides by a broad horse-shoe-shaped lowland (The Low Weald), generally below 60 m, developed predominantly on the thick Weald Clay. In the southeast this lowland extends right up to the foot of the Chalk escarpment but elsewhere up to three intervening cuestas occur on the Upper and Lower Greensands. The most impressive of these is developed on the lithologically variable Lower Greensand Hythe Beds and results in high ramparts in the northwest that rise 200 m above the adjacent Low Weald to elevations of 273 m (Hindhead), 280 m (Blackdown) and 294 m (Leith Hill) (Fig. 2).

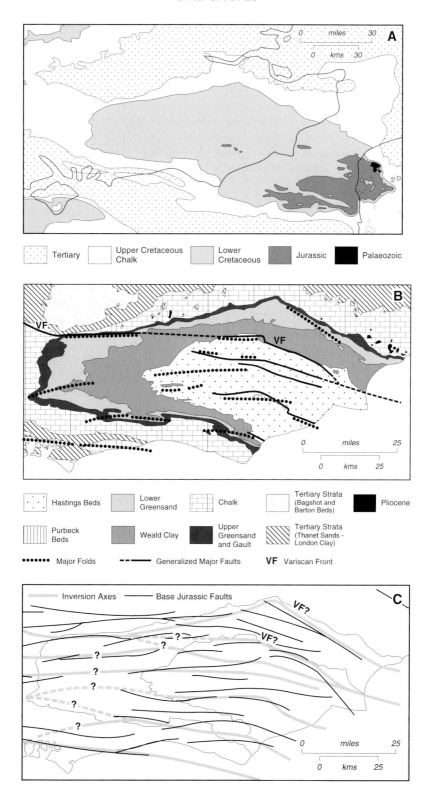

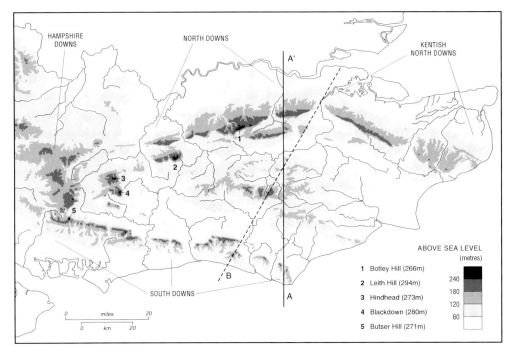

Fig. 2. The topography and drainage of the Weald. A–A′, Approximate line of generalized cross-section used for geological reconstruction. B, 'Lewes–Medway Line' of Jones (1980).

The radial drainage pattern is concordant with macrostructure but displays marked discordance with the secondary structures in the southern half of the region (Fig. 2). This discordance was formerly explained as the result of superimposition from a high-level marine erosion surface (Wooldridge & Linton 1955) but more recently has been interpreted as the product of consequent drainage lines developed in the Palaeogene maintaining their courses across growing folds (Jones 1974, 1981, 1999). The most dramatic discordance is displayed by the Straits of Dover which cut across the eastern part of the Weald–Artois Anticlinorium. There are three current explanations for this remarkable coincidence: the escape southwards of meltwaters from Pleistocene pro-glacial lakes impounded in the southern North Sea area and fed by the Thames–Rhine systems (see Jones 1981; Gibbard 1988, 1995); a single formative event involving the catastrophic escape of meltwaters in the late Pleistocene (Smith 1989); and downfaulting of the NE–SW oriented Faille du Pas de Calais 800 000–900 000 years ago (Colbeaux et al. 1980; Amédro & Robaszyinski 1987). A combination of fault movements and meltwater overflows appears the most likely explanation for the present seaway, with much of the initial excavation achieved in the Anglian (c. 450 000) and with subsequent fashioning by interglacial high sea-levels. Prior to this the Weald–Artois Anticlinorium is considered to have been dry land since uplift about 2 Ma led to the closure of a previous seaway (Funnell 1995), with much of the presently drowned portion draining northwards via the now obliterated Lobourg River (Stamp 1927).

A detailed history of investigations into the evolution of the Weald is beyond the scope of this paper and partial reviews are to be found in Gallois (1965), Clayton (1969, 1980), Linton (1969), Kirkaldy (1975) and Jones (1981). However, it is instructive to note that nearly two centuries have elapsed since John Farey referred to the Weald as a 'great southern denudation' in 1805. Early interpretations of Wealden landscape evolution were 'catastrophist' by nature with the rivers following faulted rifts through

---

Fig. 1. (**A**) The Weald–Artois Anticline; (**B**), simplified geology and main structural elements of the Weald; and (**C**) possible structural partitioning of the Weald based on known faults in the base Jurassic as reported by Butler & Pullen (1990).

the Chalklands (Martin 1828; Lyell 1833) and escarpments representing abandoned marine cliffs (Lyell 1833); it was not until the mid-nineteenth century that supporters of fluvial erosional development gained ascendancy. Although many early authors had a partial appreciation of the potential of fluvial erosion (Conybeare & Phillips 1822; Buckland 1826), the breakthrough is accredited to Greenwood (1857) who rapidly converted Ramsey (1864) away from the marine hypothesis. Foster and Topley (1865) reviewed and rejected the earlier catastrophic explanations and the combined efforts of Whitaker (1867) and Topley (1875) finally established the role of fluvial erosion acting on different bedrocks as the sculpturing agency in generating the physiography of the Weald, although glaciation has also been invoked (Fisher 1866; Prestwich 1890; Martin 1920; Sherlock 1929; Kellaway 1971; Kellaway et al. 1975) and dismissed (Jones 1981).

Despite the topographic evidence for recent denudation it is the apparent accordance of summits that has attracted most attention since first commented upon by Ramsey (1864). Foster and Topley (1865) regarded the Weald as having been excavated from beneath a 'plane of marine denudation' which Topley (1875) extended over the whole of lowland Britain south and east of a line running from the Yorkshire coast near Whitby to the SE corner of Dartmoor. Analysis of drainage–structure relationships led Bury (1910) to support the 'marine hypothesis' but Davis (1895) argued that the summit accordance represented the remnants of a subaerial peneplain. This debate as to the origin of the 'Summit Surface' or 'Hill-top Plain' was to continue until apparently resolved by Wooldridge and Linton (1938a, 1939, 1955) who favoured a subaerial origin in the Neogene to form the Mio-Pliocene Peneplain or Pliocene landsurface, surviving above 210 m, the lowest parts of which were trimmed by the Plio-Pleistocene sea, resulting in the discordant drainage pattern commented upon earlier. However, the reality of this surface has subsequently been challenged on numerous grounds (see Jones 1981, 1999). Although a low relief landsurface is considered to have existed over most of southern England in the Pliocene, it is no longer interpreted as the product of Neogene peneplanation but rather as inherited in large part from the Palaeogene. Little of this surface is considered to have survived the ravages of Pleistocene denudation in SE England although it is extensively preserved in western counties (Green 1985). In the case of the Weald and other similar axes of uplift, the scale of Tertiary tectonic activity undoubtedly resulted in significant erosion during the Neogene so that the nature of the end Neogene landsurface in these areas has remained the subject of controversy.

Interpretations of the style and age of tectonic evolution have also changed over time. The anticlinal structure was probably first recognized by John Farey in an unpublished map of 1806 and Topley (1875) was able to describe the Weald as a truncated dome with a number of minor flexures approximately parallel to the central axis. The established description of the Weald as 'an anticlinorium with subsidiary flexures en echelon' (Lake 1975, p. 551) was therefore in existence by the turn of the century, as was the mid-Tertiary date of flexuring (see Bury 1910), although recent generations of geologists have tended to favour placing this episode in the Miocene (e.g. King 1981; Chadwick 1993). The opposing view of 'pulsed tectonism' (George 1974), involving the long-term incremental development of the flexure pattern with a possible peak in tectonic activity in the Miocene, has been supported by stratigraphic and geomorphological investigations (Jones 1980, 1981, 1999; Small 1980) and further extended by the possibility of uplift, warping and faulting in the later Neogene and Pleistocene (Worssam 1973; Jones 1981; Amédro & Robaszyinksi 1987; Preece et al. 1990; Bergerat & Vandyke 1994). However, the most dramatic change to date has been the abandonment of the classic 'tensional-Alpine compressional' model of fold development (Wooldridge & Linton 1938b, 1955), in which the sedimentary infill of the Mesozoic Wealden Basin was envisaged to have been buckled against the unyielding London–Brabant Platform by compressional movements from the south, and its replacement by inversion tectonics (Butler & Pullen 1990; Chadwick 1993). The linkage between certain surface folds and reactivated faults/ thrusts in the Palaeozoic Floor/Variscan Basement was recognized by Wooldridge & Linton (1939, 1955) and emphasized by Lake (1975). The extent of this coupling or 'basement control' was subsequently developed by Stoneley (1982) for the adjacent Wessex Basin and then for the whole of southern England by Chadwick (1986, 1993), so that the lines of small surface folds en echelon are now interpreted as drape structures developed over major faults in the Mesozoic strata which, in turn, pass downwards into southward inclined thrusts in the Variscan Basement.

Inversion tectonics involves the creation of two distinct scales of structure when extensional sedimentary basins are subjected to compression:

regional upwarps to accommodate crustal shortening, and linear inversion axes created by the reversal of faults whose movements originally defined the sedimentary basins (Chadwick 1993). The extent to which the latter and their associated small surface periclinal structures are the product of vertical (Chadwick 1993) as against strike-slip movements (Lake & Karner 1987) remains controversial. Irrespective of origin, the relationship between regional upwarp and linear inversion axes is important in the Weald. The two dominant inversion axes identified to date are the Dorking–Penshurst–Tonbridge–Biddenden axis (Lake 1975) which forms the eastern end of the Pewsey–London Platform axis of Chadwick (1993) and may represent the Variscan Front (Lake 1975; Chadwick 1986), and the continuation of the Wardour–Portsdown axis off the Sussex coast (Fig. 1C). Four other secondary axes can also be identified: a monoclinal flexure running parallel to, and to the south of, the North Downs escarpment (Stoneley 1982), corresponding in part with the Maidstone lineament of Lake (1975) and originally claimed by some to overlie the Variscan Front (Figs 1B and 1C); the Groombridge–Benenden axis (Lake 1975);

the Ashdown Forest/Crowborough–Burwash–Mountfield–Fairlight axis (Lake 1975) with its well developed faulted periclines arranged en echelon and associated Purbeck (Upper Jurassic) inliers (Fig. 1A); and the line of faulted periclines running from Fernhurst in the west to Pevensey in the east (Jones 1981; Stoneley 1982) that has been called the 'South Downs Axis' (Jones 1999). These axes not only effectively partition or compartmentalize the area but also largely decouple the Weald from the less disturbed provinces to north (above the London Platform) and to south (the relatively undeformed Dieppe Basin).

The concept of a separate structural evolutionary history for the Weald–Artois Anticlinorium compared with adjacent areas has not been pursued with any vigour in the British literature but is well established in the French, possibly due to the exposure of Palaeozoic rocks in the Boulonnais (Fig. 1A). The French interpretation of events, well summarized in Robaszynski and Amédro (1986), envisages that reactivation of WNW–ESE oriented Variscan faults in the late Cretaceous to Eocene and Miocene led to the uplift of the core of the structure, sometimes referred to as the Weald–Boulonnais–Artois

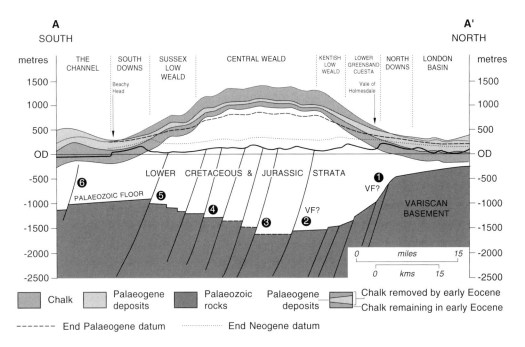

Fig. 3. Reconstructed cross-section of the Weald on an approximately N–S line running from the easternmost South Downs near Beachy Head in the south, to the River Thames near Dartford in the north. The subsurface geology is highly generalized. VF = Variscan Front. Numbers refer to tectonic axes as follows: 1, Maidstone lineament; 2, Penshurst–Biddenden axis; 3, Groombridge–Benenden axis; 4, Ashdown–Fairlight axis; 5, South Downs axis; 6, Wardour–Portsdown axis.

horst. A further set of transverse faults, oriented N30°, was established in the Lower Cretaceous and reactivated in the Pleistocene, thereby contributing to the formation of the Dover Straits and possibly having more widespread effects as this orientation coincides with the Lewes–Medway Line identified by Jones (1980) (Fig. 2) which marks the changing trend of the Weald structure. Thus the complex structure of the Central Weald, where the fault-bounded 'structural highs' have a horst-like form separated by faulted synclines (Lake 1975), may be indicative of the westward extension of a much larger, horst-like structure recognized in northern France.

Despite recent advances in knowledge, numerous uncertainties still exist concerning the evolutionary geomorphology of the Weald because it has been the focus of denudation in the later Tertiary and Quaternary. Some of these uncertainties will be addressed in the remainder of this paper which seeks to review four models of evolution by relating them to a geological reconstruction (Fig. 3) along a line of section used by Wooldridge & Linton (1955, p. 19) that runs northwards across the Weald from 13 km south of the Channel (Sussex) coast near Beachy Head, via the Ashdown Forest and Sevenoaks, to the River Thames near Dartford; this line of section was used in a previous study (Jones 1980) thereby allowing direct comparison with an earlier evolutionary reconstruction.

Geological reconstruction

The cross-sectional reconstruction of the Weald (Fig. 3) shows the relationship between the present topographic profile, the Palaeozoic Floor/Variscan Basement and the estimated form of the fully developed Weald–Artois Anticlinorium based on the thickness of Mesozoic strata that existed in the late Cretaceous but has subsequently been removed. Sub-surface reconstruction is diagrammatic with the varying level of the Variscan Basement following the depths indicated by Chadwick (1993, fig. 1b). Along the line of section this fundamental surface lies at a little below −1000 m at the South Coast and declines to −1600 m beneath the Central Weald before rising abruptly beneath the northern margin of the Weald to form the gently northward rising surface of the London Platform to reach −350 m beneath the River Thames.

Reconstructing the theoretical cross-sectional form of a fully developed Wealden anticlinorium is an even more speculative exercise. Chadwick (1993, p. 314) estimates 'the total (Late Cretaceous to Miocene) relative uplift' to be in excess of 1250 m (visual estimation of his Figure 4) suggests a maximum of c. 1400 m) based on mapping of the base of the Chalk and sonic velocity estimations of Weald Clay compaction. His reconstruction, although very generalized, is instructive in that it shows the Weald upwarp to be slightly asymmetrical with a relatively steep northern limb, wholly in line with the form anticipated from the basin inversion model now widely accepted.

The alternative approach involves using published isopach maps to reconstruct the anticlinorium as has been attempted by Jones (1980) and Butler & Pullen (1990). The latter computed that 'a regional map at base Upper Chalk level or near base Senonian' (p. 375) revealed a crestal elevation of c. 1220 m, 'giving a probable overall Tertiary uplift of the eastern part of the basin in excess of 1525 metres' (p. 380). However, uncertainty as to the thickness of Upper Chalk and the growing awareness that significant uplift may have continued into the Quaternary require that their conclusions are reassessed.

Data are available on Lower Cretaceous strata (Sellwood *et al.* 1986) but estimating the original thickness of Upper Cretaceous Chalk remains controversial because of variable denudation in the Palaeogene. Hancock (1975) indicated 400–500 m for the Wessex–Paris Basin but less over the stable London Platform, while Lake (1975) suggested a maximum of 350–400 m for the Weald. More recent lithostratigraphic investigations have revealed the true complexity of Chalk sedimentation and the scale of local variations due to contemporaneous tectonic activity (Mortimore & Pomerol 1991) and have shown broad but uneven changes in the thickness 'of all the White Chalk onto the North Downs (London–Brabant Platform) compared with the standard sections in Sussex' (Mortimore & Pomerol 1987, p. 136). The detailed sections of Mortimore (1986) indicate the preservation of about 290 m of Chalk in the easternmost South Downs near Eastbourne. However, the sequence is truncated in the Culver Chalk (Lower Campanian, c. 80 Ma) indicating the removal of sediments accumulated over a 12 million year period of raised sea-level in the Upper Cretaceous prior to the onset of regression at c. 68 Ma (Hancock 1989). As this represents 52% of the duration of White Chalk sedimentation in this area, the amount removed must have been substantial. An estimated total thickness of 460 m therefore appears reasonable in the light of current knowledge regarding Upper Cretaceous

sedimentation patterns (Mortimore & Pomerol 1987, 1991; Hancock 1989), thickening to 480 m off the coast at the southern end of the cross-section (Fig. 3).

Estimating the original thickness of Chalk in the London Basin is even more problematic because of the scale of early Palaeogene denudation (Curry 1965). A proven Chalk thickness of 197 m near the northern end of the cross-section (Sellwood *et al.* 1986), combined with the lithostratigraphic investigations of Robinson (1986) which indicate *c.* 170 m preserved in the North Downs, represent severe underestimations as many higher zones are missing. The preservation of 247 m of Chalk in the western London Basin and 289 m in Thanet suggests original deposition well in excess of 350 m and a figure of 400 m will be used in this study. These values suggest that the Chalk was about 420 m thick over the crest of the Weald. Combining these thicknesses with the isopach data of Sellwood *et al.* (1986) indicates that removal of Mesozoic strata from over the Central Weald reaches an estimated 1205 m at Ashdown, 1273 m at Britling and 1285 m at Bolney, thereby indicating a general figure of 1250 m to be acceptable. This numerical reconstruction reveals a broad, slightly asymmetrical upwarp with a crestal elevation of 1400 m and an amplitude of *c.* 1200 m, slightly less than the general estimate of Chadwick (1993), but a significant increase on the 1180 m and 990 m respectively reported in Jones (1980).

In order to compute total or gross denudation it is necessary also to estimate the thickness of Palaeogene sediments that were deposited and then removed. Some 730 m of Palaeogene sediments are preserved in the Hampshire Basin, 380 m in the Dieppe Basin (Curry & Smith 1975) to the south of the southern end of the line of section and 240 m in the core of the London Basin. However, none of these areas contains a complete record of deposition, which is known to have been spatially very variable over southern England and France due to the combination of eustatic fluctuations and contemporary localized tectonic activity (Pomerol 1989). The lack of sediments of Oligocene and Miocene age is generally taken to indicate subaerial erosion and non-deposition (Jones 1981), although other authors argue that the absence of sediments should not be interpreted in this way as the deposits may have subsequently been removed (Small 1980; Curry 1989). This paper adopts a conservative, 'subaerial' standpoint in arguing that Oligocene and Miocene sediments were never laid down over the Weald and its lower flanks. It is also assumed that the Palaeogene sequences preserved in the London and Hampshire Basins represent a reasonable estimation of total net Palaeogene sedimentation in these areas (i.e. excluding sediments removed during the Palaeogene as recorded by the well established hiatuses; see Pomerol 1989), although subaerial and submarine erosion may have truncated the sequence preserved beneath the Channel in the Dieppe Basin (Curry 1989).

Further uncertainty is provided by disagreement regarding the degree to which the early development of the Weald upwarp limited the extent to which it was submerged by the Palaeogene transgressions and thereby influenced the thickness of Palaeogene deposits. Sedimentological studies of the London Clay (early Eocene) have repeatedly failed to find evidence for a Weald island or shoal to indicate the early existence of the Weald upwarp (Wrigley 1940; Davis & Elliot 1957; Curry 1965; King 1981) and King (1981) goes so far as to favour a Miocene date for the folding of the Weald. Structural geologists and sedimentologists like Guy Plint, on the other hand, appear willing to accept some tectonic development in the late Cretaceous and early Palaeogene (Palaeocene and Eocene) but insist on the dominant role of Miocene deformation because (i) this corresponds with an important phase of Alpine deformation and (ii) the well known fact that Oligocene beds on the Isle of Wight are greatly affected by the Isle of Wight Monocline/Portland–Wight Inversion (Stoneley 1982; Chadwick 1993). However, the largest and most diverse group continues to advocate significant uplift with some flexuring in the Palaeogene, at both small and large scales, based on stratigraphic studies and tectonic consideration (Wooldridge & Linton 1955; Jones 1980, 1981, 1999; Small 1980), although notions of a 'Weald island' or 'shoal' in the London Clay Sea (Prestwich 1852; Stamp 1921; Wooldridge & Linton 1955) are now rejected. Here the argument is between those who consider pre-Eocene flexuring to have been insufficient to create an island in the London Clay sea and those who argue that the topographic expression of significant tectonic movements had been minimized by denudation.

The crucial evidence for the early development of a Weald upwarp is to be found in the form of the sub-Tertiary unconformity or Sub-Palaeogene Surface as it is now known (Jones 1980). This deformed, marine-trimmed erosion surface has been widely exhumed from beneath covering Palaeogene sediments in the London Basin, Hampshire Basin and the South Downs, and displays long known overstep and overlap relationships as summarized in Jones (1980, 1999). Wooldridge & Linton (1955) emphasized that if

the rate of overstep displayed on the North Downs (1:175 or 5.7 m km^{-1}) was continued to the south it would result in the complete overstep of the Chalk in 16–24 km, implying unroofing of the Chalk over the Central Weald by the early Eocene. While some may argue that local deformation of the Sub-Palaeogene surface or atypical localized erosion may have exaggerated the apparent scale of early Palaeogene denudation over the Weald, the presence of Palaeogene residuals near the crest of the North Downs (Wooldridge & Linton 1955; Jones 1980, 1999; Wilson 1995) has the reverse effect of reinforcing the apparent scale of Palaeogene denudation. Further, it has to be noted that lithostratigraphic studies of the South Downs Chalk indicate the late Cretaceous development of local anticlines and the onset of regional warping (Mortimore & Pomerol 1991) and have conclusively shown that the Woolwich and Reading Beds overstep successively lower beds in the Campanian from Portsmouth in the west to Brighton in the east, indicating a systematic variation in Palaeocene erosion of between 60 and 100 m. Projecting these values northwards over the Weald is highly speculative because of the known scale of secondary structure growth in the Palaeocene (Mortimore & Pomerol 1991) and uncertainty regarding the form of early Tertiary uplift. However, even a conservative estimate indicates a further 135 m of overstep by the crest of the Weald upwarp, making a total of 235–250 m or approximately equivalent to that known to have been removed from over the North Downs by the same time. But this figure must be recognized as an underestimation for two reasons. First, the already mentioned sub-Tertiary overstep on the North Downs indicates greater denudation over the Weald. Second, the projection of the unconformity northwards from the South Downs fails to take account of the possibility that the central parts of the Weald may have been uplifted more than adjacent areas due to the initial development of the Weald–Boulonnais–Artois horst (Robaszynski & Amédro 1986). As a consequence, a figure in the range 300 to 350 m seems more likely and 320 m will be used in this study, indicating that three-quarters of the Chalk had been removed from the axis of uplift by the end of the Palaeocene.

These various lines of evidence point to the existence of a well developed Weald upwarp in the early Palaeogene and significant, if not total, removal of the Chalk cover from the crest of the Weald upwarp by the Eocene. This implies that the central Weald may have been the focus of subaerial denudation for most of the Palaeogene, except for a period of submergence beneath the London Clay sea (c. 53.5–52 Ma; Pomerol 1989). As a consequence, Palaeogene sediments would have thinned towards the axis of the upwarp with only the London Clay deposited on the crest. Up to 200 m of London Clay is thought to have been deposited in southern England (Davis & Elliot 1957) but as shallower water conditions would have existed over the Central Weald, deposition would have been reduced in this area and 60 m is used in this study. Thus the thicknesses of Palaeogene sediments used in this cross-sectional reconstruction are 420 m off the south coast reducing to 60 m over the Central Weald before expanding to 280 m in the London Basin. These deposits were laid down everywhere on a surface eroded in Chalk that progressively overstepped younger zones towards the crest of the Weald upwarp where it lay either in the Turonian or in the Cenomanian, indicating the almost complete removal of the White Chalk by 53 Ma.

The deduced position of the Sub-Palaeogene surface in relation to the Chalk as it existed in the early Eocene (c. 53 Ma) is shown in Fig. 3, together with the estimated thickness of overlying Palaeogene sediments. The extent of Chalk that had been removed by that time is apparent (c. 320 m), as are the amounts of Chalk and other strata remaining to be removed and the subsequent scale of uplift and deformation.

Gross denudation

The cross-sectional reconstruction reveals dramatic variations in Tertiary gross denudation across southern England since the latest Cretaceous (Fig. 4). Over the Central Weald estimated gross denudation of Mesozoic and Tertiary deposits varies between 1140 m and 1350 m, with values falling away asymmetrically to north and to south. Over the crest of the North Downs the value has fallen to 425 m and it fluctuates between 425 m and 500 m over the Chalk backslopes (excluding locally high values produced by incised valleys) before slowly rising northwards to 540 m near the River Thames. Values over the South Downs are similar, ranging from 390 m to 460 m before rising to about 740 m over a submerged anticline beneath the Channel.

The period of time over which this denudation was achieved is also open to debate, as no precise date for the late Cretaceous emergence has been specified. Hancock (1989) indicates the onset of rapid regression at 68 Ma, suggesting that erosion may have commenced by 66 Ma. Using this commencement date results in mean

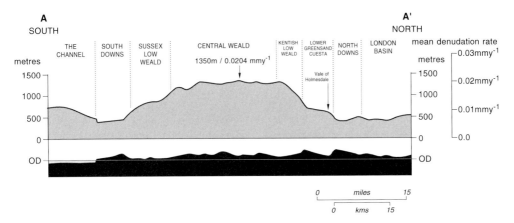

Fig. 4. Estimated gross denudation for the last 66 Ma along the line of section.

denudation rates of 17.3–20.45 m Ma^{-1} for the Central Weald, c. 8.2 m Ma^{-1} for the London Basin, 6.4–7.6 m Ma^{-1} for the North Downs, 5.9–7.0 m Ma^{-1} for the South Downs rising to c. 11.2 m Ma^{-1} in the Channel. These are all exceptionally low mean denudation rates indicating the plausibility of the reconstruction. Clearly such long-term rates can conceal episodes of greater activity; for example, the consequences of a postulated mid-Tertiary folding episode, and this aspect will be considered later. But the reverse is also true and any phases or pulses of accelerated activity must be compensated for by periods of limited denudation due to submergence, approximation of land surface to base level, low topography or reduced rates of erosion. Morphostatic episodes must, therefore, have characterized the Tertiary evolutionary geomorphology of the region (Green 1985) and such conditions were undoubtedly most long-lived on the Chalklands least affected by the growth of linear inversion axes (Jones 1999).

Consideration must also be given to the alternative scenario that the Weald upwarp either did not exist in the early Palaeogene or was of limited development, thereby resulting in the deposition of a thicker layer of Palaeogene sediments over the Weald. In this scenario 200 m of Palaeogene cover is assumed for the Weald, thickening to 300 m in the north and over the South Downs and reaching 450 m in the Channel. The resulting pattern of gross denudation (Fig. 5A) is similar to that discussed previously (Fig. 4) except that it is increased to between 1280 m and 1480 m over the Central Weald, 480–550 m over the North Downs and 530–600 m over the South Downs. The resultant changes to gross denudation rates are minimal: 19.4–22.4 m Ma^{-1} over the Central Weald, 7.3–8.3 m Ma^{-1} over the North Downs and 8.0–9.1 m Ma^{-1} over the South Downs.

The geological reconstruction (Fig. 3) also facilitates the computation of net denudation (i.e. surface lowering) achieved since emergence at the close of the Cretaceous. The estimated values are 1080–1280 m over the Central Weald, 840–1240 m over the northern Low Weald, 620 m over the Lower Greensand Escarpment, 480 m for the Vale of Holmesdale, 240–280 m over the North Downs and 200–370 m over the Tertiary deposits in the London Basin. Southwards from the Central Weald the values fall to 480–750 m over the southern Low Weald and 215–300 m over the South Downs before rising to 380 m in the Channel. The striking feature of these data is the limited net denudation recorded for the North and South Downs, indicative of relatively slow rates of landscape evolution for lengthy periods of the Tertiary (see Jones 1999).

Evolution in the later Palaeogene

Construction of the cross-section requires consideration of three further elements of evolution:

(1) the scale and spatial variability of late Palaeogene denudation so that the final position of the partially evolving Sub-Palaeogene Surface can be established (i.e. the end-Palaeogene datum);
(2) the impact of 'mid-Tertiary' tectonic deformation;
(3) the position and disposition of the end-Tertiary surface (late Neogene landsurface). Because of uncertainties regarding the scale

and timing of tectonic activity in the Tertiary it is proposed to consider this aspect last.

The evolution of the Weald in the late Eocene and Oligocene has always proved problematic due to lack of evidence. Thus while Wooldridge & Linton (1955) felt able to state '... that South-East England had taken on a strong semblance of its present structural plan as a result of pre-Eocene movements' (p. 10) and referred to '... the severe but indeterminate denudation of the Wealden crest in pre-Eocene times' (p. 21), they could offer little on subsequent development in the later Palaeogene except for remarking on '... a phase of widespread if gentle, uplift and warping, which, in detail, may have been premonitory of the more violent movements which followed in Oligo-Miocene times' (p. 11) and merely concluded that 'what happened in the major upwarped areas between the Eocene depressions, must for ever remain a subject for surmise, or, at best, for argument by analogy' (p. 11). Sadly, this largely remains true for the whole of the 28 million years between the withdrawal of the London Clay sea and the close of the Oligocene.

Subsequent evolutionary interpretations (Jones 1980, 1981; Small 1980) embraced pulsed tectonism and envisaged that upwarped areas would have continued to experience uplift episodes throughout the lengthy later Palaeogene. Thus the Weald was considered to have experienced pulses of stimulated erosion so that the early Palaeogene cover was progressively stripped from crestal areas to reveal Mesozoic outcrops that were then subjected to denudation (Jones 1980). However, Green (1985), in a paper largely focused on areas well to the west of the Weald, argued that the later Palaeogene was characterized by limited erosion because southern England was predominantly a low relief, duricrusted landsurface developed under the sub-tropical seasonal climate of the Eocene. This conclusion was based on several lines of reasoning: (i) the widespread evidence of residual soils and remnant silcretes, dating from the Palaeogene; (ii) the survival of early Tertiary tablelands; (iii) the fact that the offshore record of Middle and Upper Eocene deposits beneath the Channel consists predominantly of carbonates (Curry et al. 1970); and (iv) that the volume of Oligocene deposits is small, with the offshore record limited to a freshwater limestone.

The lack of an offshore record of terrigenous sediments has always proved problematic for proponents of the 'denudation model' (Jones 1980, 1981; Small 1980) until the recognition of a major eustatic fall in sea-level of about 130 m in the Middle Oligocene (30 Ma) (Haq et al. 1987), which would have rendered the Tertiary sedimentary basins (i.e. the Hampshire–Dieppe Basin, etc.) dry land. The lack of Middle and Upper Eocene terrigenous sediments may also reflect the significant scale of submarine erosion that has affected the floor of the Channel and the fact that sediments have tended to have a relatively short residence time in the Channel before being transported either westwards to the continental slope or eastwards to the subsiding North Sea Basin (Curry 1989). Thus the lack of sediments beneath the Channel is insufficient grounds for arguing for lack of erosion over the whole of southern England during the later Palaeogene.

In a recent review (Jones 1999) a distinction has been drawn between relatively stable areas in the later Palaeogene, upon which were developed extensive low-relief surfaces (etchplains) mantled by thick residual soils and duricrusts (silcretes), and adjacent areas experiencing pulsed uplift and characterized by progressive denudation. This division (structural compartmentalization or structured partitioning) into units with distinctive evolutionary histories (morphotectonic regions) has been facilitated by the establishment of an inversion tectonics framework for southern England (Chadwick 1993). This recognizes that the major structures of the Weald upwarp and Hampshire Basin are crossed by a series of linear inversion axes, predominantly oriented E–W, where surface patterns of asymmetrical faulted periclines arranged en echelon reflect reactivated thrusts in the Variscan Basement. These belts of pronounced tectonic disturbance define blocks whose disposition reflects both the scale of inversion axis movement and their location with regard to the pattern of major structures. As there is no reason why all the inversion axes should have displayed co-ordinated uniformity of movement, it is realistic to envisage that areas of relative stability may have experienced morphostasis while adjacent tracts, such as the Weald, were undergoing uplift and denudation.

There thus appear good grounds for arguing that the Weald continued to experience uplift and denudation throughout the Eocene and Oligocene. The crucial sedimentological evidence for this interpretation is the reported presence of pebbles of Upper Greensand chert in the Barton Basement Bed (Upper Eocene) of the London Basin (Dewey & Bromehead 1915; Dines & Edmunds 1929; Oakley 1939; Wooldridge & Goldring 1953; Wooldridge & Linton 1955) indicating that denudation had cut deep

into the Mesozoic sequence within the Weald upwarp by 42 Ma, thereby confirming the well developed nature of the structure by that time. Indeed, Jones (1981) deduced that this evidence required that the crest of the Weald had been uplifted by 450 m relative to the London Basin and that 550 m of Mesozoic strata had been removed by 42 Ma, equivalent to 50% of the present-day upwarp amplitude in the western Weald; these figures must be increased by c. 80 m because of new estimations of Chalk thickness (i.e. to 530 m and 630 m respectively). It was this reasoning that led Jones (1980) to postulate equivalent denudation on other parts of the uplift axis so that the end Palaeogene erosional datum was shown to lie at 450 m over the Central Weald at the line of cross-section. This conclusion is now considered problematic because it leaves little scope for Miocene flexuring and Pleistocene uplift (Jones 1999). Assuming the sedimentological evidence to be sound then an explanation has to be found as to why the western end of the Weald upwarp should have experienced greater denudation than has been deduced from studies of the Sub-Palaeogene surface, which suggest 300–350 m of erosion by 53 Ma (see earlier). The difference of 200–250 m may be due to the fact that the western Weald experienced more pronounced uplift in the early Palaeogene, possibly due to the convergence of a number of structural lineaments, including the Variscan Front (Fig. 1C), but was less affected by later (Neogene) movements than other parts of the Weald. A more probable alternative is that uplift and denudation continued through the Eocene and into the Oligocene so that the sedimentological evidence is merely indicative of a later stage in the evolutionary process.

As a consequence of these various lines of reasoning it is concluded that the Weald did suffer continuing uplift and denudation in the Eocene and Oligocene. It is also concluded that the earlier reconstruction (Jones 1980) shows the maximum denudation attributable to this lengthy period of evolution and that a more cautious or conservative estimation should be used on this occasion. The position of the end Palaeogene surface shown in Fig. 3 is, therefore, a *minimal* estimate and shows it passing above the present Central Weald at 750–850 m, indicating the removal of Palaeogene sediments and a further 210 m of Mesozoic strata in the 27 million years since the withdrawal of the London Clay sea. By this time erosion would have penetrated over 500 m into the Mesozoic sequence of the Weald upwarp so as to expose extensive outcrops of Lower Greensand.

The position of the end Neogene datum

Wooldridge & Linton (1955) believed that the end Neogene (end Pliocene) landsurface was a dissected unwarped subaerial peneplain preserved above 210 m, for they wrote (p. 149) '... the south country has been sculptured from an uplifted plain of which relics are preserved in our highest hill summits. A plane, 800 feet above sea-level, would, indeed, pass close to all these summits'. The existence of this 'Mio-Pliocene Peneplain' has been extensively criticized and the concept abandoned (see Jones 1981) although an ancient polygenetic summit surface, to varying degrees inherited from the Palaeogene, is recognized as surviving on the highest summits (Green 1985; Jones 1999). In some areas, such as Wessex, this early surface is known to have been dissected in the Neogene (Green 1974) with the development of inclined surfaces dated as Miocene and Pliocene, the latter declining to 150 m. Indeed, Green (1973) has stated that the 'end Tertiary surface' may have been at 110 m to the south of Salisbury, while Small & Fisher (1970) argued that rivers in the western South Downs had cut down to 100–150 m by the close of the Pliocene.

These variations in level have led to the questioning of Quaternary stability (Worssam 1973; Jones 1981) and the possibility of differential uplift. As such, they represent a return to the ideas of Jukes-Browne (1911) who argued that arching of the Weald following the deposition of the Lenham Beds (Pliocene?) had caused the Central Weald to rise by 172 m with reference to the North Downs, a conclusion generally supported by Worssam (1973).

Recent investigations of the Late Pliocene Red Crag transgression (>2.4 Ma) in the London Basin have clearly shown that these shallow marine sediments (up to 25 m of water) were deposited on a marine-trimmed surface developed from a pre-existing low relief subaerial landscape, which was then warped at some time post −2 Ma (Mathers & Zalasiewicz 1988). Wooldridge & Linton (1955) argued that only the northeasternmost portion of the London Basin had been downwarped to the east of a hingeline subsequently identified as 'the Braintree Line' (Wooldridge & Henderson 1955), but several lines of evidence now exist to show that there is no such hinge and that warping affected the whole of the London Basin (Jones 1999). As most investigations of the Chalkland rim of the London Basin still maintain that there is some evidence for a Red Crag marine incursion up to about 180–200 m (John & Fisher 1984; Moffat & Catt 1986; Wilson 1995) but that it

was a cool sea transgression induced by downwarping (Mathers & Zalasiewicz 1988), the clear implication is for differential uplift of up to 170 m in the Quaternary (Jones 1999). Extrapolation of the generalized degree of warping into the Weald indicates minimal estimates of uplift at the structural crest of 230–260 m, implying that the end Neogene land surface lay at 300–340 m over the Central Weald (assuming a Red Crag sea-level of 25 m and Neogene topography of about 50 m) before descending again over the South Downs. As the existing topography rises to 294 m at Leith Hill and the Chalk downlands rise to 271 m at Butser Hill (Fig. 2), both of which would have suffered lowering due to scarp retreat in the Quaternary, together with the effects of solutional lowering in the case of the latter which may be conservatively estimated at 15 m (i.e. the minimum rate of solutional lowering for half of the Quaternary, assuming periglacial conditions for the remainder, i.e. 17 mm ka^{-1} for 0.9 million years), then these estimates appear acceptable.

Indeed, some may consider these estimates conservative as they fail to take account of the possibility that central parts of the Weald may have suffered greater uplift than surrounding Chalklands. Greater uplift has already been postulated. For example, the arguments advanced by Jukes-Brown (1911) point to the Central Weald having been uplifted by 330 m since Pliocene times and Preece et al. (1990) have suggested local updoming of over 400 m since the late Neogene, both of which indicate an end Pliocene datum rising to 380–420 m.

However, it has to be recognized that the evidence for elevating the end Neogene surface is located in the western Weald and it is here that the relative relief is greatest and the topography most accentuated (Fig. 2). By contrast, to the east of the Lewes–Medway Line (Jones 1980) the topography is more subdued and elevations on all main topographic units are lower. The evidence for the faulted development of the Dover Straits (Colbeaux et al. 1980; Amédro & Robaszyinski 1987), together with the eastward slope of the Sussex Low Weald (Thornes & Jones 1969) and other lines of evidence which will be discussed elsewhere, all point to the fact that the Weald may have experienced both variable uplift at some time after 2 Ma and downwarping to the east at a later stage, perhaps coincident with the faulting at 800–900 ka. As the line of section crosses the Central Weald coincident with the Lewes–Medway Line (Fig. 2), the conservative elevations of the late Neogene datum (300–340 m) will be used in this study even though they may prove to be slightly high for the area to the east. A surface at this elevation would have been mainly developed in Weald Clay with low ridges of Hastings Beds created by the emerging crests of the Ashdown Forest/Crowborough and Battle anticlines.

The Neogene

The relatively short span of the Neogene (20.7 Ma) has traditionally been of supreme importance in the Tertiary evolution of southern England because of the supposed occurrence of an intense phase of folding in the 'mid-Tertiary', Oligo-Miocene or Miocene, followed by severe erosion to create a low relief surface by the close of the Pliocene. The latter is no problem because the rocks exposed in the Weald during the Neogene were not resistant, except for the Lower Greensand Hythe Beds, as the tougher sandstones of the Hastings Beds had yet to be exposed.

The scale and impact of folding in the Neogene has always been controversial because of the lack of stratigraphic control. The disturbance of Oligocene beds by the Isle of Wight Monocline is the crucial evidence for an Oligo-Miocene age for the folding, although Jones (1999) has questioned whether movements of a primary inversion axis on the southern margin of southern England necessarily means that all major movements have to be attributed to this period of tectonism. The cross-sectional analysis described in this paper attributes between 80 m and 530 m of surface lowering to Neogene erosion (i.e. the difference in elevation between the end Palaeogene and end Neogene datums shown in Fig. 3), or a little more than 40% of the total denudation of the Weald. These figures could be raised slightly if the proposed magnitude of Pleistocene uplift were reduced, but even so the amount of uplift attributable to the Neogene would still be only 50% of the total, a radically different conclusion from that conveyed by most authors. The suggested increase in doming since the late Neogene (Preece et al. 1990) merely serves to reduce the significance of early Neogene tectonism yet further.

The reduction in significance accredited to early Neogene flexuring also reflects greater emphasis on Palaeogene warping and denudation, with the latter crucially dependent on the reported presence of Lower Greensand Hythe Beds material in the Eocene deposits of the London Basin (see earlier discussion). Although some of this evidence may be considered speculative, it has to be noted that the Weald is a

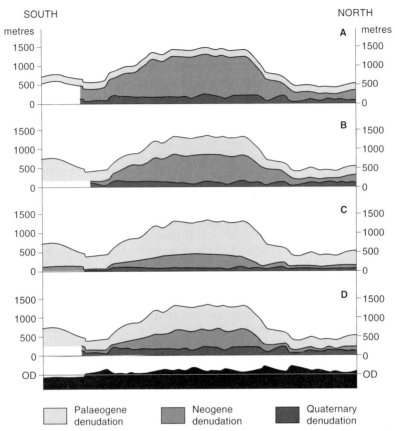

Fig. 5. Compared patterns of gross denudation for four evolutionary models. (**A**) The 'Miocene Deformation Model' as suggested by King (1981) and others, who envisage limited development of the macroflexure in the Palaeogene. (**B**) The 'classic' Wooldridge & Linton (1955) interpretation. (**C**) The 'Palaeogene Denudation Model' proposed by Jones (1980). (**D**) The evolutionary model proposed in this paper.

complex horst-like structure with strong W–E *and* NW–SE lineations (see Fig. 1A) and other fracture orientations (Bergerat & Vandyke 1994), located in proximity to a number of stress sources known to have been active in the Tertiary (e.g. rifting to form the northern North Atlantic, subsidence of the North Sea Basin, rifting of the Lower Rhine embayment, the formation of the Alps, etc.). Emphasis on a 'mid-Tertiary' or early Neogene tectonic episode linked to one specific stage in the evolution of the Alps has for long been considered highly selective (George 1974; Jones 1981) and seems to have almost attained the status of unquestioned dogma. The interpretation of the Wealden core as a horst blocked-out in the Cretaceous, uplifted in the late Cretaceous to Eocene and rejuvenated in the Miocene (see Robaszynski & Amédro 1986), together with the new evidence for uplift in the end Neogene–early Pleistocene, seem strong grounds for a radical downgrading in significance of the Miocene tectonic episode with reference to the Weald. Indeed, the convergence of various lines of evidence now suggests tectonic movements from the late Cretaceous to the late Pleistocene, with an episode of more intense local deformation that 'may have been associated with a pulse of Alpine movements in the Miocene' (Butler & Pullen 1990, p. 382).

Models of evolution

The pattern of evolution for the Weald advanced in this paper is at variance with previous reconstructions, most notably with regard to the increased scale of Pleistocene movements and the reduction in emphasis placed on Neogene events. To show how it differs from previously described or implied evolutionary reconstructions, cross-sections have been prepared of gross denudation attributable to specific periods

of evolution for four evolutionary models (Fig. 5): (a) minimal development of the Weald upwarp during the end Cretaceous–early Palaeogene with resulting thick Palaeogene sedimentary sequence and profound Miocene deformation, as proposed by King (1981); (b) the Wooldridge & Linton (1955) interpretation involving the early development of a Weald upwarp but with the main tectonic movements in the Miocene followed by the creation of the Mio-Pliocene peneplain; (c) the pulsed tectonism model of Jones (1980) which resulted in a minimalistic view of Neogene events; and (d) the model advanced in this paper. Each model has been applied to the cross-sectional reconstruction described earlier (Fig. 3) to yield patterns of denudation attributable to the Palaeogene, Neogene and Quaternary (Fig. 5).

The 'Miocene deformation model' (Fig. 5A) clearly shows the large volume of denudation that has to be attributed to the Neogene. Such intense erosion does not fit well with Green's (1985) reconstruction of the Neogene as characterized by morphostasis but with relief development that '...is most readily explained as a result of localized structural activity' (p. 74). While the 'structural compartmentalization model' of Jones (1999) offers a potential solution to this dilemma, there appears too much evidence for tectonic activity in the Late Cretaceous and Palaeogene for this interpretation to warrant further consideration.

The 'classic' Wooldridge & Linton (1955) 'grand design' for southern England (Fig. 5B) has been interpreted for the Weald employing a compromise between their emphasis on the importance of 'mid-Tertiary' (Oligo-Miocene) folding and their observation of 'the severe but indeterminate denudation of the Wealden crest in pre-Eocene times' (Woodridge & Linton 1955, p. 21). As a consequence, moderate denudation has been attributed to the Palaeogene but less than that required by the evidence of Lower Greensand debris from the Weald being deposited in Eocene sediments in the London Basin. The limited scale of Pleistocene denudation is a feature of this reconstruction.

The 'Palaeogene denudation model' (Fig. 5C) of Jones (1980) is problematic for two main reasons. First, it employs the evidence for the exposure of the Lower Greensand by the Upper Eocene to argue that the end-Palaeogene surface lay at relatively low elevation over the Central Weald ($c.\,450$ m) thereby reducing Miocene deformation and Neogene denudation to minimal levels. Second, Small & Fisher's (1970) observation that drainage lines had cut down to 100–150 m by the end of the Pliocene was accepted and applied uniformly throughout the Weald, resulting in minimal assessments of Quaternary denudation. This interpretation was clearly a product of a time when efforts were being made to find an alternative evolutionary model to replace the long accepted Wooldridge & Linton reconstruction (Jones 1999) and must, therefore, be recognized as an extreme position. The model advanced in this paper (Fig. 5D) represents a reassessment and has resulted in increased denudation attributed to both the Neogene and the Quaternary.

To facilitate further comparison, generalized graphs of erosion through time for each of the four evolutionary models have been prepared for three locations on the cross-section: the Central Weald (Fig. 6), the crest of the South Downs (Fig. 7A) and on the backslope bench of the North Downs (Fig. 7B). As expected, the highest estimated mean erosion rates are recorded in the Weald (Fig. 6) for early Palaeogene denudation in Jones' (1980) model (600 m over 66–53.5 Ma or $48.0\,\text{m}\,\text{Ma}^{-1}$) and Neogene to present denudation in the 'Miocene deformation model' (Fig. 5A) which works out at 1200–1320 m in $c.\,20$ Ma or 60.0–$66.0\,\text{m}\,\text{Ma}^{-1}$. Both lie in the central portion of the range of values attributable to 'normal denudation' (Summerfield 1991), although the latter imply lengthy periods with a mean relative relief of over 500 m (Ahnert 1970) which seems unlikely. Moderate erosion rates are also implied by one interpretation of the Wooldridge & Linton (1955) model which envisages limited uplift in the early Palaeogene followed by exposure of the Lower Greensand Hythe Beds by the Upper Eocene, indicating 350 m of erosion in 10 Ma or $35\,\text{m}\,\text{Ma}^{-1}$. All other evolutionary sequences involve lower erosion rates and are therefore feasible. On the basis of this analysis only the 'Miocene deformation model' appears doubtful.

Examination of Fig. 6 reveals the range of interpretations that can be deduced from the Wooldridge & Linton (1955) reconstruction. As some of their comments were contradictory and others have subsequently been rejected (e.g. the 'Wealden Island' in the London Clay sea) it is only possible to construct a series of possible denudational sequences. The evidence that they presented indicates the lower line shown in Fig. 6 (with and without a 'Wealden Island') but their conclusions suggest the upper line. The arguments presented in this paper have been employed to create a conservative reconstruction of Palaeogene denudation, as compared with Jones (1980), and a second variation is included to show the denudation required to expose the Hythe Beds by the Upper Eocene. Also shown

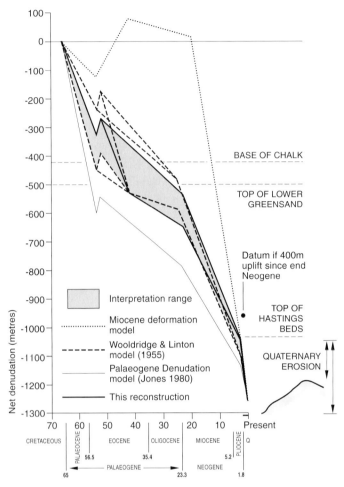

Fig. 6. Patterns of denudation for the Central Weald. The graph for the 'Miocene Deformation Model' differs from the others because the lack of early Tertiary deformation is assumed to have caused a thicker accumulation of Palaeogene sediments and later emergence. The series of graphs for the Wooldridge & Linton (1955) interpretation reflect varying possibilities arising from conflicting statements. As the reconstruction advanced in this paper adopts a conservative view of Palaeogene denudation but emphasizes Quaternary uplift, the true pattern of denudation is likely to fall within the shaded zone.

is the adjustment required by 400 m of uplift since the late Neogene (Preece *et al.* 1990), which is considered unlikely (see later). The evolutionary sequence proposed in this paper envisages 320 m of erosion by 53.5 Ma (i.e. 320 m in 12.5 Ma or 25.6 m Ma^{-1}) followed by 60 m of sedimentation with a further 270 m of erosion by the close of the Palaeogene (i.e. 270 m over the period 52.0 Ma to 35.4 Ma or 16.3 m Ma^{-1}). Uplift in the early Neogene then heralded a phase of renewed denudation resulting in further erosion of 510 m of strata by 2 Ma (i.e. 15.3 m Ma^{-1}), a figure that is reduced to 430 m or 12.9 m Ma^{-1} if the uplift estimates of Preece *et al.* (1990) are to be believed. However, uncertainty still provides for a range of possibilities and the true evolutionary sequence probably lies within the shaded area in Fig. 6.

The evolutionary graphs for the South Downs crest (Fig. 7A) and the North Downs backslope platform (Fig. 7B) reveal that denudation rates resulting from all postulated evolutionary models have generally been much lower over the Chalklands, with the highest rates recorded for early Palaeogene denudation in the London Basin (200 m in 8.5 Ma or 23.5 m Ma^{-1}) and on the South Downs (220 m in 11 Ma or 20.0 m Ma^{-1}), and for early Neogene to present denudation in the 'Miocene deformation model' for the South Downs (320–600 m in 20 Ma or

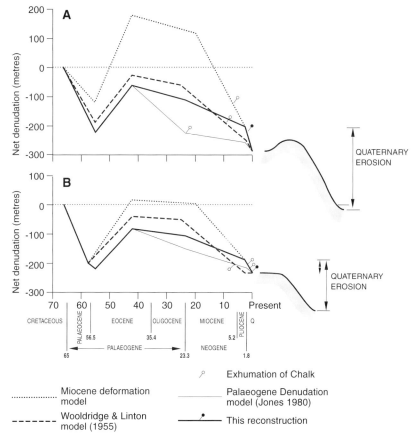

Fig. 7. Patterns of denudation for (**A**) the crest of the South Downs and (**B**) the backslope bench of the North Downs. Variations in the extent of early Palaeogene erosion, thickness of Palaeogene cover and timing of cessation of Palaeogene sedimentation should be noted.

$16-30 \, \text{m} \, \text{Ma}^{-1}$). Thus all reconstructions lie within the lower range of 'normal denudation'. The evolutionary sequence advanced in this paper indicates late Palaeogene denudation rates of $1.3-2.7 \, \text{m} \, \text{Ma}^{-1}$ and Neogene rates of $3.9-4.3 \, \text{m} \, \text{Ma}^{-1}$, wholly in line with suggestions that areas away from uplift axes experienced lengthy periods of morphostasis (Green 1985). The graphs also clearly show two important points. First, that landsurfaces in the late Palaeogene and Neogene lay at no great elevation above the highest elements of the existing Chalkland topography, as has often been advocated in order to account for the widespread survival of Palaeogene silcretes. Second, they reveal varying dates for the exhumation of the Chalk from beneath the Palaeogene cover. The sequence proposed in this paper indicates exhumation in the Pleistocene and is wholly consistent with the surprisingly widespread survival of well developed Chalkland topography and the evidence for Pleistocene uplift. The fact that certain evolutionary sequences indicate exposure in the Pliocene or Miocene raises serious questions as to their validity unless supported by sedimentological evidence for Neogene residual soils. As none are known from these specific areas (see Jones 1999), such models must be abandoned in favour of the sequence outlined in this paper.

Finally, brief mention must be made of the deduced rates of Quaternary denudation. Assuming that uplift occurred at 2 Ma then the estimated ranges of denudation for the Central Weald (140–258 m, Fig. 6), South Downs (40–290 m, Fig. 7A) and North Downs (46–176 m, Fig. 7B) yield rates of $70.0-129.0 \, \text{m} \, \text{Ma}^{-1}$, $20-145 \, \text{m} \, \text{Ma}^{-1}$ and $23-88 \, \text{m} \, \text{Ma}^{-1}$ respectively. The lower values fall within the lower end of the range of values for normal denudation with low relative relief but the higher figures fall in the higher part of the range and are more typical

of high local relief. The repeated oscillations from periglacial to temperate conditions in the Pleistocene produced numerous episodes of intense denudation, so all the values are considered feasible. However, increasing the magnitude of late Neogene–early Pleistocene uplift (Preece *et al.* 1990) significantly raises the values for the Central Weald, thereby suggesting that 400 m of uplift may be excessive for the section of the Weald investigated in this paper.

Conclusion

Substantial progress has been made towards establishing an objective evolutionary geomorphology for the Weald. Increased knowledge of deep structure and inversion tectonics has proved crucial to this achievement. The long-standing model of structural development as largely confined to a brief 'mid-Tertiary' compressional episode has been abandoned in favour of longer-term or 'pulsed' evolution characterized by the progressive development of the structural pattern from the late Cretaceous to the Pleistocene, with prominent episodes in the early Palaeogene, early Neogene and Plio-Pleistocene. Uplift of inversion axes to a large extent decoupled the Weald–Artois upwarp from adjacent areas of greater stability, thereby resulting in contrasting evolutionary sequences for the Weald as compared with the bounding Chalklands. The scale of movements was so much greater in the Weald that erosion appears to have characterized the majority of the Tertiary. As a consequence, it would always have displayed relief although this may have become relatively subdued in the later Palaeogene as the Lower Greensand was progressively stripped away to reveal the Weald Clay. On the Chalklands, by contrast, morphostatis characterized the later Palaeogene and Neogene so that low relief landsurfaces dominated until late Pliocene–early Pleistocene uplift initiated incision and escarpment development. This last uplift of the Weald stimulated erosion, leading to the exposure of the Hastings Beds and the development of the contemporary landscape. Recognition that the Weald was relatively uplifted at this late stage represents a radical change from the 'classic' Wooldridge & Linton interpretation which envisaged stability in the Quaternary with progressive incision due to eustatic decline.

The author gratefully acknowledges the helpful comments of Chris Green and Alastair Ruffell on a draft version of this paper.

References

AHNERT, F. 1970. Functional relationship between denudation, relief and uplift in large mid-latitude drainage basins. *American Journal of Science*, **284**, 1035–1055.

AMÉDRO, F. & ROBASZYINSKI, F. 1987. Influences eustatiques et contrôle tectonique de la sedimentation dans la partie moyenne du Crétacé du Nord de la France. *Mémoires geologiques de l'Université de Dijon*, **11**, 57–66.

BERGERAT, F. & VANDYKE, S. 1994. Palaeostress analysis and geodynamical implications of Cretaceous–Tertiary faulting in Kent and Boulonnais. *Journal of the Geological Society, London*, **151**, 439–448.

BUCKLAND, W. 1826. On the formation of the valley of Kingclere and other valleys by the elevation of strata that enclose them. *Transactions Geological Society* (second series), **2**, 119–130.

BURY, H. 1910. On the denudation of the western end of the Weald. *Quarterly Journal of the Geological Society, London*, **66**, 640–692.

BUTLER, M. & PULLEN, C. P. 1990. Tertiary structures and hydrocarbon entrapment in the Weald Basin of southern England. *In*: HARDMAN, R. P. P. & BROOKS, J. (eds) *Tectonic Events Responsible for Britain's Oil and Gas Reserves*. Geological Society, London, Special Publications, **55**, 371–391.

CHADWICK, R. A. 1986. Extension tectonics in the Wessex Basin, southern England. *Journal of the Geological Society, London*, **143**, 465–488.

—— 1993. Aspects of basin inversion in southern Britain. *Journal of the Geological Society, London*, **150**, 311–322.

CLAYTON, K. M. 1969. Post-war research on the geomorphology of south-east England. *Area*, **1**(2), 9–12.

—— 1980. The historical context of structure, surface and drainage in south-east England. *In*: JONES, D. K. C. (ed.) *The Shaping of Southern England*. Institute of British Geographers, Special Publication, **11**, 1–12.

COLBEAUX, J. P., DUPUIS, C., ROBASZYNSKI, F., AUFFRET, J. P., HAESAERTS, P. & SOMMÉ, J. 1980. Le detroit du Pas de Calais: un élément dans la tectonique de l'Europe Nord-Occidentale. *Bulletin d'Information des Geologues du Bassin de Paris*, **17**, 41–54.

CONYBEARE, W. D. & PHILLIPS, W. 1822. *Outline of the Geology of England and Wales*. Phillips, London.

CURRY, D. 1965. The Palaeogene beds of South-east England. *Proceedings of the Geologists' Association, London*, **76**, 151–173.

—— 1989. The rock floor of the English Channel and its significance for the interpretation of marine unconformities. *Proceedings of the Geologists' Association, London*, **100**, 339–352.

—— & SMITH, A. J. New discoveries concerning the geology of the central and eastern parts of the English Channel. *Philosophical Transactions of the Royal Society, London*, **A279**, 155–167.

——, HAMILTON, D. & SMITH, A. J. 1970. *Geological and shallow subsurface geophysical investigations in the Western Approaches to the English Channel*. Institute of Geological Sciences Scientific, Report **70/3**.

DAVIS, A. G. & ELLIOT, G. F. 1957. The palaeogeography of the London Clay Sea. *Proceedings of the Geologists' Association, London*, **68**, 255–277.

DAVIS, W. M. 1895. On the origin of certain English rivers. *Geographical Journal*, **5**, 128–146.

DEWEY, H. G. & BROMEHEAD, F. H. 1915. *The geology of the country around Windsor and Chertsey.* Memoir of the Geological Survey, UK.

DINES, H. G. & EDMUNDS, F. H. 1929. *The geology of the country around Aldershot and Guildford.* Memoir of the Geological Survey, UK.

FISHER, O. 1866. On the probably glacial origin of certain phenomena of denudation. *Geological Magazine*, **3**, 483–487.

FOSTER, C. LE NEVE & TOPLEY, W. 1865. On the superficial deposits of the valley of the Medway, with remarks on the denudation of the Weald. *Quarterly Journal of the Geological Society, London*, **21**, 443–474.

FUNNELL, B. M. 1995. Global sea-level and the (pen-)-insularity of late Cenozoic Britain. *In*: PREECE, R. C. (ed.) *Island Britain: A Quaternary Perspective.* Geological Society, London, Special Publications, **96**, 3–13.

GALLOIS, R.W. 1965. *The Wealden District.* Memoir of the Geological Survey, UK.

GEORGE, T. N. 1974. Prologue to a geomorphology of Britain. *In*: BROWN, E. H. & WATERS, R. S. (eds) Progress in Geomorphology. Institute of British Geographers, Special Publication, **7**, 113–125.

GIBBARD, P. L. 1988. The history of the great northwest European rivers during the past three million years. *Philosophical Transactions of the Royal Society of London*, **B318**, 559–602.

—— 1995. The formation of the Strait of Dover. *In*: PREECE, R. C. (ed.) *Island Britain: A Quaternary Perspective.* Geological Society, London, Special Publications, **96**, 15–26.

GREEN, C. P. 1973. Pleistocene River Gravels and the Stonehenge Problem. *Nature*, **243**, 214–216.

—— 1974. The summit surface on the Wessex Chalk. *In*: BROWN, E. H. & WATERS, R. S. (eds) Progress in Geomorphology. Institute of British Geographers, Special Publication, **7**, 127–138.

—— 1985. Pre-Quaternary weathering residues, sediments and landform development: examples from southern Britain. *In*: RICHARDS, K. S., ARNETT, R. R. & ELLIS, S. (eds) *Geomorphology and Soils.* Allen & Unwin, London, 58–77.

GREENWOOD, G. 1857. *Rain and Rivers; or Hutton and Playfair against Lyell and all comers.* London.

HANCOCK, J. M. 1975. The petrology of the Chalk. *Proceedings of the Geologists' Association, London*, **86**, 499–535.

—— 1989. Sea-level changes in the British region during the Late Cretaceous. *Proceedings of the Geologists' Association, London*, **100**, 565–594.

HAQ, B. U., HARDENBOL, J. & VAIL, P. R. 1987. Chronology of fluctuating sea levels since the Triassic. *Science*, **235**, 1156–1167.

JOHN, D. T. & FISHER, P. F. 1984. The stratigraphical and geomorphological significance of the Red Crag fossils at Netley Heath, Surrey: a review and re-appraisal. *Proceedings of the Geologists' Association, London*, **95**, 235–247.

JONES, D. K. C. 1974. The influence of the Calabrian transgression on the drainage evolution of southeast England. *In*: BROWN, E. H. & WATERS, R. S. (eds) *Progress in Geomorphology.* Institute of British Geographers, Special Publication, **7**, 139–158.

—— 1980. The Tertiary evolution of south-east England with particular reference to the Weald. *In*: JONES, D. K. C. (ed.) *The Shaping of Southern England.* Institute of British Geographers, Special Publication, **11**, Academic, London, 13–47.

—— 1981. *The Geomorphology of the British Isles: Southeast and Southern England.* Methuen, London.

—— 1999. Evolving models of the Tertiary evolutionary geomorphology of southern England, with special reference to the Chalklands. *This volume.*

JUKES-BROWN, A. J. 1911. *The Building of the British Isles.* Stanford, London.

KELLAWAY, G. A. 1971. Glaciation and the stones of Stonehenge. *Nature* **232**, 30–35.

——, REDDING, J. H., SHEPHARD-THORN, E. R. & DESTOMBES, J-P. 1975. The Quaternary history of the English Channel. *Philosophical Transactions of the Royal Society of London*, **A279**, 189–218.

KING, C. 1981. *The Stratigraphy of the London Clay and Associated Deposits.* Tertiary Research Special Paper, **6**, Backhuys, Rotterdam.

KIRKALDY, J. F. 1975. William Topley and 'The Geology of the Weald.' *Proceedings of the Geologists' Association, London*, **86**, 373–388.

LAKE, R. D. 1975. The structure of the Weald – a review. *Proceedings of the Geologists' Association*, **86**, 549–557.

LAKE, S. D. & KARNER, G. D. 1987. The structure and evolution of the Wessex Basin, southern England: an example of inversion tectonics. *Tectonophysics*, **137**, 347–378.

LINTON, D. L. 1969. The formative years in geomorphological research in south-east England. *Area*, **1**(2), 1–8.

LYELL, C. 1833. *Principles of Geology.* John Murray, London.

MARTIN, E. A. 1920. Glaciation of the South Downs. *Transactions of S.E. Union of Scientific Societies*, **25**, 13–30.

MARTIN, P. J. 1828. *A Geological Memoir on a Part of Western Sussex.* London.

MATHERS, S. J. & ZALASIEWICZ, J. A. 1988. The Red Crag and Norwich Crag formations of southern East Anglia. *Proceedings of the Geologists' Association, London*, **99**, 261–278.

MOFFAT, A. J. & CATT, J. A. 1986. A re-examination of the evidence for a Plio-Pleistocene marine transgression on the Chiltern Hills. III. Deposits. *Earth Surface Processes and Landforms*, **11**, 233–247.

MORTIMORE, R. N. 1986. Stratigraphy of the Upper Cretaceous White Chalk of Sussex. *Proceedings of the Geologists' Association, London*, **97**, 97–139.

—— & POMEROL, B. 1987. Correlation of the Upper Cretaceous White Chalk (Turonian and Campanian) in the Anglo-Paris Basin. *Proceedings of the Geologists' Association, London*, **98**, 97–143.

—— & —— 1991. Upper Cretaceous tectonic disruptions in a placid Chalk sequence in the Anglo-Paris Basin. *Journal of the Geological Society, London*, **148**, 391–404.

OAKLEY, K. P. 1939. Geology and Paleolithic studies. *In*: *A Survey of the Prehistory of the Farnham District (Surrey)*. Surrey Archeological Society.

POMEROL, C. 1989. Stratigraphy of the Palaeogene: hiatuses and transitions. *Proceedings of the Geologists' Association, London*, **100**, 313–324.

PREECE, R. C., SCOURSE, J. D., HOUGHTON, S. D., KNUDSEN, K. L. & PENNY, D. N. 1990. The Pleistocene sea-level and neotectonic history of the eastern Solent, southern England. *Philosphical Transactions of the Royal Society of London*, **B328**, 425–477.

PRESTWICH, J. 1852. On the structure of the strata between the London Clay and the Chalk in the London and Hampshire Tertiary Systems. *Quarterly Journal of the Geological Society, London*, **8**, 235–264.

—— 1890. On the relation of the Westleton Beds, or Pebbly Sands of Suffolk, to those of Norfolk and on their extension inland. *Quarterly Journal of the Geological Society, London*, **46**, 84–154.

RAMSEY, A. 1864. *The Physical Geology and Geography of Great Britain*. Stanford, London.

ROBASZYNSKI, F. & AMÉDRO, F. 1986. The Cretaceous of the Boulonnais (France) and a comparison with the Cretaceous of Kent (United Kingdom). *Proceedings of the Geologists' Association, London*, **97**, 171–208.

ROBINSON, N. D. 1986. Lithostratigraphy of the Chalk Group of the North Downs, southeast England. *Proceedings of the Geologists' Association, London*, **97**, 141–170.

SELLWOOD, B. W., SCOTT, J. & LUNN, G. 1986. Mesozoic basin evolution in Southern England. *Proceedings of the Geologists' Association, London*, **97**, 259–289.

SHERLOCK, R. L. 1929. Origin of the Devil's Dyke, near Brighton. *Proceedings of the Geologists' Association, London*, **40**, 371–372.

SMALL, R. J. 1980. The Tertiary geomorphological evolution of south-east England: an alternative interpretation. *In*: JONES, D. K. C. (ed.) *The Shaping of Southern England*. Institute of British Geographers, Special Publication, **11**, Academic Press, London, 49–70.

—— & FISHER, G. C. 1970. The origin of the secondary escarpment of the South Downs. *Transactions of the Institute of British Geographers*, **49**, 97–107.

SMITH, A. J. 1989. The English Channel – by geological design or catastrophic accident? *Proceedings of the Geologists' Association, London*, **100**, 325–337.

STAMP, L. D. 1921. On the beds at the base of the Ypresian (London Clay) in the Anglo-Franco-Belgian Basin. *Proceedings of the Geologists' Association, London*, **32**, 57–108.

—— 1927. The Thames drainage system and the age of the Strait of Dover. *Geographical Journal*, **70**, 386–390.

STONELEY, R. 1982. The structural development of the Wessex Basin. *Journal of the Geological Society, London*, **139**, 543–554.

SUMMERFIELD, M. A. 1991. *Global Geomorphology*. Longman, Harlow.

THORNES, J. B. & JONES, D. K. C. 1969. Regional and local components in the physiography of the Sussex Weald. *Area*, **1**(2), 13–21.

TOPLEY, W. 1875. *The Geology of the Weald*. Memoir of the Geological Survey, UK.

WHITAKER, W. 1867. On subaerial denudation, and on the cliffs and escarpments of the Chalk and Lower Tertiary Beds. *Geological Magazine*, **4**, 447–454 and 483–493.

WILSON, D. I. 1995. Soils of the central North Downs: their distribution, derivation and geomorphological significance. *Zeitschrift für Geomorphologie* **39**, 433–460.

WOOLDRIDGE, S. W. & GOLDRING, F. 1953. *The Weald*. Collins, London.

—— & HENDERSON, H. C. K. 1955. Some aspects of the physiography of the eastern part of the London Basin. *Transactions of the Institute of British Geographers*, **21**, 19–31.

—— & LINTON, D. L. 1938a. Influence of the Pliocene transgression on the geomorphology of South-East England. *Journal of Geomorphology*, **1**, 40–54.

—— & —— 1938b. Some episodes in the structural evolution of the South-East England considered in relation to the concealed boundary of Meso-Europe. *Proceedings of the Geologists' Association, London*, **49**, 264–291.

—— & —— 1939. *Structure, Surface and Drainage in South-East England*. Institute of British Geographers, Publication **10**.

—— & —— 1955. *Structure, Surface and Drainage in South-East England*. Philip, London.

WORSSAM, B. C. 1973. *A new look at river capture and at the denudation history of the Weald*. Institute of Geological Sciences, Report **73/17**.

WRIGLEY, A. G. 1940. The faunal succession in the London Clay, etc. *Proceedings of the Geologists' Association, London*, **51**, 230–255.

Chattian and Miocene elements in the modern landscape of western Britain and Ireland

PETER WALSH[1], MICHAEL BOULTER[2] & IWONA MORAWIECKA[1]

[1] *Katedra Geomorfologii, Uniwersytet Slaski, ul. Bedzinska 60, 41-200 Sosnowiec, Polska*
[2] *Palaeobiology Research Unit, University of East London, Romford Road, London E15 4LZ, UK*

Abstract: The best preserved denudation surfaces which are benched into the margins of the pre-Permian oldland massifs of western Britain and Ireland are those comparatively close to sea-level. Such surfaces have frequently been regarded as being of late Pliocene or Pleistocene age and of marine origin. Recent discoveries reviewed in this work reinforce a growing view that these low-level planations are much older and may be of terrestrial origin. Evidence from Wales and southwest England suggests that these little-modified planation elements represent landscapes some 15 Ma old. This paper reviews the palaeobotanical, sedimentological and geomorphological evidence for the close association of several bodies of non-marine Chattian and Miocene sediment and saprolite with some of the better known planation features. In these areas, the former vegetation comprised mixed coniferous/deciduous forest, extinct species of modern north temperate genera, well known from contemporaneous deposits in north Europe. Evidently, differential relief in those times was appreciably less than that of the present and it seems probable that much of what is now the western half of the British Isles was then a wide, forested extension of a previous North European Plain, which varied in altitude by no more than a few decametres.

Nearly flat, very nearly horizontal dissected planation surfaces, which truncate quite complex structural foundations of Palaeozoic and Precambrian rocks, form a widespread and distinctive element of the landscape of western Britain and Ireland (WB&I). Those surfaces at elevations of between 50 and 150 m a.s.l. are particularly sharply defined and studies of them have been carried out over the past 150 years (Ramsay 1846; Belbin 1985). Despite the effort which has been expended in trying to establish the age and origin of these landforms, there is still very little agreement as to what they actually represent. Whatever their origins, they can easily be traced by various cartographic or computer-based techniques over areas measuring tens, if not hundreds of square kilometres; most slope seawards at angles of 2° or less and have a very limited altitudinal range (Belbin 1985). The best examples (including some of those mentioned below) are strikingly obvious to the field observer.

Among the better known planated landscape elements to be considered in this paper are: the Menaian Surface of Anglesey and adjacent Welsh mainland (Greenly 1919; Embleton 1964; Battiau-Queney 1980, 1984; Walsh *et al.* 1996); the Bosherston–Castlemartin Plain of South Pembrokeshire (Dixon 1921; George 1974; Battiau-Queney 1980, 1984; Jenkins *et al.* 1995); the Reskajeage Surface of west Cornwall (Everard 1977; Walsh *et al.* 1987; Coque-Delhuille 1987, 1991, 1992); various surfaces in the Bristol Channel area, including that which transects Lundy Island; those preserved in the Gower Peninsula (Goskar & Trueman 1934) and the Durdham Down Surface of the Bristol area; various surfaces bordering the southwest Peninsula of England, of which, perhaps, the Berry Head–Daddyhole Plain–Babbacombe Downs Surface around Torbay (Jukes-Brown 1907; Proctor 1988) is the best known; the surface which has planated the Channel Islands of Alderney, Jersey and Guernsey (Renouf 1993); and the Coastal Peneplain and South of Ireland Surface of Miller (1955), of which Davies & Stephens (1978) wrote, 'In East Cork and Waterford, there is a staircase of planation surfaces which [in their uniformity] is noteworthy by international standards'.

The list is not intended to be comprehensive. Most localities mentioned in the text are shown in Fig. 1.

Previous studies

There is insufficient space here to review all the works in which the age and origin of the 50–150 m-level coastal planation surfaces of WB&I have previously been considered. Therefore, these remarks relate principally to those surfaces bordering the Celtic and Irish Seas in western Britain, i.e. those most relevant to the main thesis of this paper.

The wide expanse of the 50–120 m surface (or surfaces) of west Cornwall can hardly have

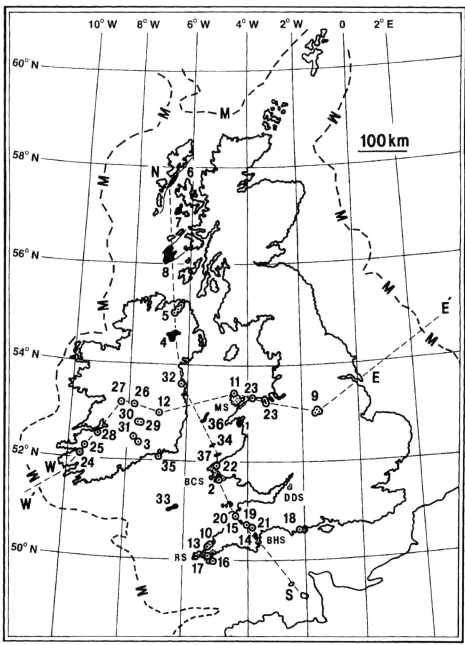

Fig. 1. Map of the British Isles showing the distribution of possible Chattian and Miocene non-marine sedimentary bodies (numbers correspond to the list in Table 1).

escaped the notice of the earliest geologists working in this area. However, it was not until Reid's memoir on the Pliocene rocks of Great Britain (Reid 1890) that the notion that this surface represents a stranded former Atlantic sea-bed, and that the higher ground of west Cornwall represents a former archipelago in this supposed transgression, gained any particular prominence. Later, in co-authorship with Scrivenor, Reid wrote in the Newquay Geological Survey Memoir (Reid & Scrivenor 1906): 'whether the singular contours ... are part of early Tertiary, or even Secondary, date we cannot yet say; but the plateau may have originated long before the oldest of the deposits that now cover it. The surface may well have been merely cleaned up and the features shaped in Pliocene times ... the platform is obviously an ancient sea floor or plane of marine denudation and the hills arising out of it once formed scattered islands, like those of Scilly'.

According to Reid & Scrivenor (1906), the reasoning for the supposed Pliocene age for the platform went thus: the St Erth Outlier, the deposits of which signified a marine transgression to a depth of at least 104 m a.s.l. in the St Ives Bay/Mounts Bay depression, contained an undoubted Pliocene marine fauna; the St Agnes deposits, which lie 21 km NE of St Erth, had not then yielded any fossils, but were so similar in general aspect to the St Erth deposits that they could reasonably be considered to be at least broadly coeval. Therefore, as the St Agnes deposits apparently overlie the upper levels of the 'stranded sea floor', this, too, must be considered to be of Pliocene age, with the corollary that the St Ives/Mounts Bays depression must simply have been a relatively deep water channel in the Pliocene sea. Support for this notion had already been provided by Davies & Kitto (1878), who had recorded what they considered to be beach gravels in the St Agnes sediments, supposedly banked up against an old cliff line 'with caves and sea-stacks' on the east side of the St Agnes Beacon. Apparently, these features were recorded in subsurface mining operations and, as far as we are aware, their precise location has not subsequently been established.

Support for this 'marine/Pliocene' interpretation has persisted. On the basis of petrographic studies, both Milner (1922) and Boswell (1923) accepted the contemporaneity of the St Erth and St Agnes sediments without reservation. Further support was offered by Wilson (1975) and Coque-Delhuille (1987, 1992); the latter made extensive petrographic and sedimentological analyses of the St Agnes deposits and concluded that, despite the absence of a marine macrofauna in these deposits, they nevertheless represented a littoral marine sequence. Also, on the basis of correlation with the stratigraphic sequences in NW France, she considered that they should be regarded as of Pliocene age. With regard to the unlithified gravels which rest at c. 100 m on the highest levels of the Lizard facet of the surface, Flett & Hill (1912) regarded these as a remnant of a former submarine gravel bank, while Gullick (1936) and Everard (1977) considered them to represent a shoreline beach gravel. Thus, the view that the west Cornwall 50–120 m planation surface is of marine origin and Pliocene age is firmly entrenched in the geological literature. It is the view expressed in the British Geological Survey (BGS) regional guide for southwest England (Edmonds et al. 1975) and in the BGS Sheet Memoir for Penzance (Goode & Taylor 1988). The latter suggest that the surface is 'complex' and that it 'may have been submerged on more than one occasion'.

However dominant the 'marine/Pliocene' hypothesis for the origin of the West Cornwall planation may be, it is by no means a universal view. As early as 1832, Hawkins considered the St Agnes beds to be fluviatile. Mitchell (1965) was the first to record that Oligocene pollen was present in a sample collected by H. Dewey in 1932 from the claypits at Beacon Cottage Farm at St Agnes (SW 705501) (BGS sample MR 10401). It is believed that the sample came from the lower levels of the St Agnes sequence. If accurately curated, it indicates that the St Agnes beds cannot be coeval with the St Erth deposits. Later, further examination of the microflora from the BGS sample led Atkinson et al. (1975) to confirm that it was undoubtedly of mid- to late-Oligocene age. Subsequent attempts by Walsh et al. (1987) and Jowsey et al. (1992) to rediscover the original source of the organic clay at Beacon Cottage Farm were unsuccessful, but analysis of organic material from the New Downs Member, in the middle levels of the St Agnes sequence in the New Downs Sandpits (SW 706509) (Walsh et al. 1987), revealed a microflora of undoubted Miocene age. This indicates that the St Agnes sequence (s.s.) cannot be coeval either with that at Beacon Cottage Farm or that at St Erth. Moreover, detailed petrographic and sedimentological analysis by both Walsh et al.(1987) and Jowsey et al. (1992) suggested that the sediments of both outliers were best regarded as aeolian and colluvial terrestrial sequences.

Thus, while there has been little dispute that the St Erth deposits in the St Ives/Mount's Bays depression are marine in character and

Pliocene in age, there is now considerable evidence that we should regard the St Agnes deposits as terrestrial and Miocene and the Beacon Cottage Farm sediments as terrestrial and Oligocene. On borehole evidence, Walsh et al. (1987) demonstrated that the form of the sub-Miocene surface beneath the St Agnes Outlier (s.s.) was virtually identical to that of the adjacent parts of the 50–120 m surface, for which they proposed the term 'Reskajeage Surface'. These two surfaces were thus considered to be physiographic homologues and, inasmuch as they regarded the Miocene sediments of the St Agnes outlier as non-marine, this indicated that the Reskajeage Surface must be of an earlier Miocene, if not pre-Miocene age and, at least in its latest stages of development, it must have been fashioned by terrestrial processes. Subsequent boreholes and trenches into the Beacon Cottage Farm Outlier (Jowsey et al. 1992) indicated that the supposed Oligocene sediments there rest on a surface which is much more irregular than that below the St Agnes Outlier and there is certainly no reason to regard this as yet another facet of the sub-Miocene/Reskajeage Surface.

In respect of the 4 km wide Bosherston–Castlemartin Surface in the southern part of the Pembroke Peninsula of SW Wales, the erosional bevel cut at c. 50 m a.s.l. across its Palaeozoic foundation is directly associated with at least three formations, all presumably of pre-Pleistocene age. Plainly, there is little point in commenting about the age and origin of this surface unless the nature and age of these deposits have been taken into account. These are the so-called Gash Breccias and the Flimston Pipeclays, both of which underlie the surface and are therefore presumably older than it, and an unnamed sheet of siliciclastic gravels which overspread the surface and which are presumably younger. On the basis of general lithological comparisons with the Dolomitic Conglomerate of Keuper age of the West Glamorgan area, 100 km or so to the east, the unfossiliferous Gash Breccias were regarded by Dixon (1921) as cave-roof collapse breccias of Triassic age. However, Thomas (1970) regarded them not as true sediments but large-scale tectonic fracture zones, while Battiau-Queney (1980), supporting this hypothesis, determined that the mineralogy of the matrix of the breccias was quite unlike that of the Dolomitic Conglomerate. Instead, she considered that the clay mineral suite present in the matrix indicated that this originated in a hot, wet climate and surmized that this was probably of mid-Cenozoic age.

The Flimston Clays have universally been regarded as solution subsidence deposits which originated at some unknown level above the Bosherston–Castlemartin Surface. Murchison (1839), Dixon (1921), Thomas (1970) and George (1974) all considered that the Flimston sediments had genetic affinities, both in terms of their sedimentology and age, with the Palaeogene sedimentary fills of the Petrockstow and Bovey Basins of SW England. George (1974) considered that they represented the basal deposits of a large sedimentary basin, the contents of which, except for the solution subsidence masses such as that at Flimston, were removed by the same planation process which created the Bosherston–Castlemartin Plain. Brown (1960) was also in no doubt that the Flimston Pipeclays had been planated at the same time as the host Carboniferous Limestone. George (1974) and Brown (1960) agreed that the planation was undoubtedly marine. However, Battiau-Queney (1980) could find little evidence of marine planation; she reported that the mineralogy of both the Flimston Pipeclays and the unnamed gravel layer which overlies the surface were very similar; like the matrix of the Gash Breccias, these contained illitic-kaolinitic and ferralitic mineral suites which were apparently formed in a terrestrial environment and a hot wet climate. The Palaeogene age generally attributed to the Flimston Pipeclays since first they were described by Murchison (1839), has recently been confirmed. Spoil from what are presumably eighteenth or early nineteenth century workings of the pipeclay, has yielded a sparse, but nevertheless unmistakable, Chattian microflora (Jenkins et al. 1995).

As far as is known, the gravel cover has never yielded any fossils. Dixon (1921) regarded it as 'the alluvium of a Pliocene river'. Thomas (1970) thought it represented a remnant of conglomeratic Namurian material, 'more or less at the point of origin'. Battiau-Queney (1980) considered that the coarse siliceous material was derived principally from the Old Red Sandstone outcrops to the north, and was transported fluvially in wide, variable channels. There seems to be little if any support for a Pleistocene age for this material, although this is theoretically possible.

Some 1000 km^2 of Anglesey and the adjacent Welsh mainland has been spectacularly planated in one or more erosional episodes at between 50–150 m a.s.l. The most frequently used term for the general planation is the 'Menaian Surface or Platform' (Greenly 1919). On the basis of long-distance correlation with either the similar physiographic feature in Cornwall (Greenly 1919) or the so-called Calabrian (i.e. early Pleistocene) Surface of SE England (Brown 1960; Embleton 1964), the surface has been regarded

as being of marine origin (Smith & George 1961) although Greenly (1919) regarded the lower levels of the general planation feature to be 'a base level of subaerial waste'. No pre-Pleistocene deposits of any kind have ever been located on the surface. Until recently, the only rock masses which appeared to have any bearing on its age were Palaeogene dykes (Greenly 1919) – which, being the youngest rocks truncated by it, are plainly older than the Surface – and the large, deep pockets of intensely weathered rock, such as those at Porth Wen (SH 402947) and Porth Swtan (SH 302894). These lie below the Surface but they could theoretically be associated with whatever process created the planation feature. Greenly (1919) considered that the saprolite zones were the product of rotting '... in the course of the long-continued genial climate of the Pliocene'. Battiau-Queney (1980) agreed that these were the products of extensive terrestrial weathering but considered them to be of rather greater antiquity.

Recently, an extensive Miocene palynoflora has been described by Walsh *et al.* (1996) from saprolitic material contained in solution pipes present in Gwna Group limestones at Trwyn y Parc on the northern coast of Anglesey. Hereabouts, no facet of the Menaian Surface has been preserved within several kilometres of the Trwyn y Parc pipes, but both Greenly (1919) and Battiau-Queney (1980) concurred that, in this part of the island, its former position is likely to have been *c.* 50 m a.s.l. On the evidence that the source of the saprolite at Trwyn y Parc was quite different from the host rock containing it, Walsh *et al.* (1996) speculated that the weathering took place well above the modern landscape of northern Anglesey and that the saprolitic fill of the pipes might reasonably be considered to be a foundered mass of organic soils and weathered rock, which locally formed the pre-subsidence Menaian Surface. On this indirect evidence, the Surface could be regarded as pre-end-Miocene in age and terrestrial in origin.

Elsewhere around the coasts of the Irish and Celtic Seas and the Bristol and English Channels, surfaces similar in form and altitude to those already described are often no less clear-cut or extensive. Owing to their evenly truncated form and limited altitudinal range, they have also frequently been regarded as of marine origin and late Cenozoic age. But, with a single possible exception, discussed below, none of these surfaces has ever been found to preserve a cover of marine sediment, nor, apparently, has any marine sediment been preserved directly below them as a result of karstic subsidence or neptunian fissuring.

Denudation chronology studies which sought to explain these 50–150 m a.s.l. surfaces were perhaps at their most popular in the middle decades of the present century. At that time, in practically every scheme in which the authorship offered an opinion regarding the likely age of the individual steps on the planation sequences, all those below the +150 m level were deemed to be of Pleistocene age. As an example of this (one of dozens which we might have selected from that period), Orme's (1964) planation staircase system for the Drum Hills of Co. Waterford recognized clear-cut steps at 215, 183, 158, 148, 120, 108, 90, 78 and 64 m a.s.l. Of these, the 215 m bench was considered to be of terrestrial origin and of late Pliocene/early Pleistocene age, while those below were supposedly marine cut and denoted prolonged stillstands in a progressively falling Pleistocene sea-level. Part of the basis for attributing the sequence to the Pleistocene was the long-distance correlation with the supposed 'Red Crag', early Pleistocene surface in SE England, though Orme admitted that 'in the absence of local correlative deposits, the precise dating of the strandline sequence is virtually impossible'.

Thus, Orme's work, like those of many contemporaries, seems to have been greatly influenced by Wooldridge & Linton's notions concerning the geomorphological evolution of SE England (Wooldridge & Linton 1938, 1955). Indeed, for many geomorphologists working in the mid-century period, such was the dominance of the hypothesis that the early Pleistocene 'Calabrian' Transgression had left an erosional bevel at a height of 180–200 m or so around the whole of the British Isles (as, supposedly, it did in the SE England model of Wooldridge & Linton (1955)), that the possibility that other regions of the British Isles could have had a radically different physiographical development in the late Cenozoic was, for a time, almost completely disregarded. The Calabrian level was ubiquitous, so it was generally supposed; everything else had to fit in with this. On this general theory, any flattish surface less than 180 m a.s.l., whether deemed to be of marine origin or otherwise, must, therefore, be younger than the Calabrian Surface and hence of unquestioned Pleistocene age. As an example of this, we choose Brown (1960), with respect to his views on the physiographical evolution of Wales (though, again, there are dozens of others to whom we might have referred). Sometimes, this logic led to the supposition that the coastal 50–150 m surfaces must have been cut 'at times of high sea level during the interglacial periods' (e.g. Embleton (1964), with respect to how he

envisaged the creation of the Menaian Surface of Anglesey).

The sole exception known to the authors where a 50–150 m planation surface is undoubtedly associated with marine sediment is that at Ballinabranagh in the Barrow valley of Co. Carlow (Irish Grid Reference S 686708). Here, Cole (1912) reported material containing *Fusus*, *Pectunculus* and other fossils, apparently of 'Red Crag' age, from local soils. These are found at an elevation of c. 150 m a.s.l. What Red Crag fossils are doing here, some 53 km from the nearest part of the Irish coast, is not at all obvious; derivation from glacial sediments is, perhaps, the simplest explanation for this occurrence at present.

The notion that the surfaces which lie at less than 150 m a.s.l. were of Pleistocene age and marine in formation led to two particular difficulties for mid-twentieth century geomorphologists. First, if the comparatively narrow and exposed late-Pleistocene raised beach surfaces of western Britain were capable of preserving covers of marine sediments, why didn't the much more extensive supposed early and mid-Pleistocene marine-cut surfaces also have them? Second, how could several kilometre-wide remnants of supposed early to mid-Pleistocene sea-floors possibly have been cut in time spans not exceeding a few hundreds of thousands of years?

With respect to the denudation chronology of the SW Peninsula of England, Balchin (1964) reviewed in some detail the criteria for the recognition of stranded former marine-cut surfaces. He regarded all the major planation benches of the SW Peninsula below 200 m as unquestionably marine-cut, with the probability that at least some of those above are also marine-cut. Nevertheless, he clearly was puzzled that none of those above c. 20 m a.s.l. could be shown to preserve a marine sedimentary cover, even if, in his view, all the other criteria for recognizing stranded ancient sea-floors could be met in all parts of his denudation scheme. To circumvent the difficulty, Balchin explained the general absence of marine covers as simply being due either to the fact that the ancient sea-floors had no marine sediment on them when they were stranded ('as had been found to be the case on the modern sea-floors off West Cornwall'); or, even if they had, such was the erosional vulnerability of the unlithified sediment, that it was unlikely to have survived much beyond the stranding episode.

For those who considered that, in a WB&I context, it was unlikely that such wide surfaces could be cut by marine action in relatively short 'interglacial' periods during the Pleistocene, a 'trimming' process has sometimes been invoked. The implication of this is that, though the time intervals for the cutting were relatively short, not much work had to be done by marine action because the general level of the pre-Pleistocene landscape (perhaps formed from Mesozoic soft rocks) had already been reduced to near-transgression level by earlier erosional episodes. In such interpretations, for which there was never any direct evidence, the supposed marine-cut Pleistocene surfaces were interpreted as representing surfaces which lie only a few metres below sub-Triassic unconformities or the like. Such a 'trimming' effect was implied by Brown (1960), with respect to the evolution of some of the coastal plateaux of Wales; by Embleton (1964), to the creation of the Menaian Surface of Anglesey; by Dixon (1921), to the Bosherston–Castlemartin Surface of Pembrokeshire; by Fryer (1958) and Everard (1977) to the 131 m surface of Cornwall; and by Renouf (1993), to the planations of the major Channel Islands.

It will be seen from the foregoing that, in a WB&I context, the general paucity of truly diagnostic criteria with which to identify any particular erosional or depositional environment in association with the 50–150 m a.s.l. surfaces, has meant that the old controversies about marine/non-marine origins and age determinations can still be debated freely without these debates being constrained by too many hard stratigraphic facts. This is true right up to the present and we accept that it is quite possible that many will see this review as simply maintaining the status quo.

Defining the surfaces

One possible reason for the decline in the popularity of denudation chronology studies in WB&I at the end of the 1960s was the frustration of never having any local stratigraphical markers with which to identify the age of any planation element. However sophisticated the statistical analysis and computer methodology, even if one were able to define very delicate slope changes otherwise undetectable to the naked eye, one still could not place these into any historical sequence.

The various schemes proposed up to the mid-1970s are summarized in the Methuen *Geomorphology of the British Isles* series: Sissons (1976) for Scotland; Jones (1981) for SE England; Straw & Clayton (1978) for the east and Midlands of England; and Davies & Stephens (1978) for Ireland. The most extensive review of the geomorphology of Wales which emphasizes

the ubiquity of the planation sequences of the principality is probably that of Brown (1960), and Coque-Delhuille (1987) for the Cornubian Peninsula.

On the basis of a statistical analysis of Ordnance Survey spot height data and planated spur levels, Hollingworth (1938) considered that it was possible to recognize similar sequences of planation features around many of the oldland massifs of the British Isles, from the Southern Uplands of Scotland to Cornwall, to heights of c. 600 m a.s.l. He considered these to be the products of similar erosional events which affected the various oldlands uniformly. Hence, planation surfaces of comparable elevation thus identified were considered to be coeval. Though he recognized that it was not possible to date the higher members of the sequence, owing to the absence of any stratigraphical control, he opined that the entire sequence up to 600 m a.s.l. must postdate the 'Alpine' tectonic event of southern England. During the middle years of the present century, these views came to have a considerable influence on denudation chronology studies in a British context. By contrast, some more recent views (e.g. Battiau-Queney 1984, 1999) hold that, given our present knowledge of the structure and behaviour of the Earth's crust, it is most unlikely that any particular planation feature could be traced across any single oldland massif, let alone from oldland to oldland, as Hollingworth (1938) had supposed.

In our own studies and in this work, we have had no cause to challenge any of the previously proposed planation sequences, however defined and however extensive the latter might be. We would only comment that, as there is so little stratigraphic control anywhere, denudation chronology studies are best restricted to the local area and, even then, with a very generalized viewpoint. For example, we propose that the geomorphology of west Cornwall may adequately be approached on the assumption that there is a single generalized surface ranging in altitude from c. 75 to 121 m a.s.l. (Walsh *et al.* 1987); in contrast, Everard (1977) had previously concluded that this same area had clearly recognizable imprints of marine benches, with slope rises at 82, 90 and 106 m.

Only if it becomes demonstrable that planation features of similar elevation in several different localities can be dated by fossil or other absolute dating evidence, will it be reasonable to postulate inter-regional correlation of certain erosional events and, perhaps, extend this link to other oldlands where the control is lacking.

Solution-subsidence phenomena and landscape evolution in Britain and Ireland

In the denudation chronology studies of 30–50 years ago, there was very little appreciation that fossiliferous sediments preserved by solution-subsidence mechanisms might become important indicators of landscape evolution. But, in a WB&I context, nothing which has so far been suggested as a useful indicator of landscape evolution has been either predictable or predicted. Indeed, the concept works in reverse: a fortuitous find of fossiliferous material preserved by karstic subsidence enables our appreciation of landscape evolution problems to be placed on a much firmer basis, at least at the local scale. As examples of this, previous to the dating of the Kenslow flora in the Brassington Formation of Derbyshire (Boulter 1971), no-one suspected that what is now part of the crest of the Pennine axis was a major basin of fluvial sedimentation, which presumably lay close to sea-level, only 10 Ma or so ago. Likewise, before the discoveries of Miocene floras in the Barrow valley of Co. Laois (Watts 1985) and Anglesey (Walsh *et al.* 1996), no-one predicted the immediacy of Miocene sediments above the modern landscapes of these areas.

Not only do solution-subsidence fills permit the establishment of a qualitative picture of previous landscapes, but they may, with considerable reservations, be used to erect a quantitative model of the former elevations of surfaces now removed from the local landscape. In theory, a crude estimate can be made of the elevation of former surfaces, now removed, providing that:

1. At least one (ideally several) distinctive (ideally fossiliferous) horizon is present in the fill of the subsidence structure.
2. The thicknesses and stratigraphy of the components of the fill bodies are still recognizable, despite any distortion attributable to solution subsidence.
3. The age of any marker horizons present must be older than the date at which the solution-subsidence began.
4. The downward penetration of the subsidence body must always have outstripped the general erosional lowering of the landscape surrounding the subsidence structure.
5. Either the distinctive material in the solution-subsidence mass can be linked satisfactorily to its counterpart at unsubsided outcrops nearby, or the identity of insoluble residues preserved as part of the fill permits a reconstruction to be made of the host-rock layers

which have been removed from the local landscape.
6. The altitude of the rim of the subsidence structure and the depth of the preserved fill is known.
7. The geological structure of the area surrounding the solution-subsidence outlier is reasonably well understood.

Studies of such features are still rudimentary and opportunities to put the theory into practice are extremely infrequent and almost inevitably contentious. However, it is perhaps worth giving a few examples of how this technique both has been, and might be, applied.

Rhes-y-cae, Flintshire

Data from Maw (1867), Strahan & De Rance (1890), Walsh & Brown (1971) and Battiau-Queney (1980), suggest that solution-subsidence here has reached a depth of $c.$ 80 m for the non-preserved part of the structure and at least 27 m (possibly 54 m; Strahan & De Rance 1890) for that preserved. Calculations are based on a knowledge of the structure of the local Carboniferous host rocks and the matching of distinctive crinoidal cherts which outcrop near the walls of the subsidence mass with the natural outcrop at a distance of 450 m (Walsh & Brown 1971, fig. 4). This enables us to approximate the elevation of the landscape at the time when the subsidence began, although, unfortunately, we have no palaeontological control here to enable us to date this event. The total is thus certainly in excess of 107 m and possibly as much as 134 m. Expressed another way: the total subsidence exceeds the general lowering of the adjacent landscape by a ratio of the order of 5:4 or 7:5.

Bees Nest Pit, Derbyshire

Data from Walsh *et al.* (1972) and Walsh *et al.* (1980) suggest that the lowering of the late Miocene Kenslow Clays in the Bees Nest solution-subsidence structure to their present altitude of $c.$ 315 m a.s.l., is of the order of 165 m. This is based on an identification of fossiliferous material in solution residues which derive from Carboniferous formations no longer present in the immediate area, and which are present in the walls of the subsidence structure, together with an interpretation of the local geological structure (Walsh *et al.* 1972). Notionally, this places the pre-subsidence late Miocene landscape at $c.$ 480 m a.s.l. The stratigraphic sequence below this level in the Bees Nest subsidence is 43 m thick, while the deepest borehole sunk by the owners of the pit to the base of the Brassington Formation is also 43 m. Thus, the total subsidence here is at least 208 m and the ratio of total subsidence to general lowering of the adjacent landscape, $c.$ 5:4. The latter represents a rate of $c.$ 16 mm ka^{-1}.

Natural Pits, Hainaut, Belgium

The palaeokarst setting of the natural pits (otherwise termed 'circular faults') of the Hainaut Coalfield of Belgium has been summarized by Quinif (1989). Of the various known examples, one of the best documented is that at Ghlin, which was penetrated by a borehole to 1585 m b.g.l. (Delmer *et al.* 1982). In the Ghlin structure, the base of the early Cretaceous fill has subsided well over 400 m relative to the equivalent in the unaffected wall-rocks. Most of this took place in the early Cretaceous but, in the same structure, the base of the mid-Cretaceous fill is lowered by $c.$ 150 m, whereas the early Turonian fill, some several decametres. Subsidence was thus spasmodic throughout much of the Cretaceous, but apparently died out after the Turonian.

It should also be pointed out that there are other solution-subsidence structures (in Ireland for example) where it might be construed that there is now enough published information available to attempt to estimate the elevation of the surface at which the fill originated and/or from which subsidence took place. As examples, two subsidence outliers, those at Hollymount, Co. Laois, and Ballygaddy, Co. Offaly, seem especially amenable to analysis.

Hollymount, Co. Laois

The Hollymount Outlier, contained in the Carboniferous Limestone of Co. Laois, has briefly been described by Mitchell (1980, 1981) and its palynoflora by Boulter (1980) and Watts (1985). Most recent opinion favours a late Miocene age for the fill of this solution-subsidence structure (Watts 1985). Very few details of the fill lithology have been published and it is not known whether the fill mass is primarily a saprolite or a sedimentary sequence. However, the record of 'red and white siliceous clays' (Mitchell 1981) suggests a possible derivation as a saprolite from a former cover of insoluble 'Millstone Grit', analogous, perhaps, to those at Rhes-y-cae, Ballygaddy, Flimston and elsewhere. As the

pocket at Hollymount was unbottomed by a hole at 75 m depth, whereas the Barrow valley adjacent to the outlier is only at 60 m a.s.l., the solution subsidence structure must descend to below sea-level.

Sheet 23 of the Geological Survey of Ireland 1:100 000 series geological map (Tietszch-Tyler & Sleeman 1994) shows that the present-day rim of the outlier lies within, but close to the top of, the Ballyadams Formation. Above this lies the 90 m thick Clogrenan Formation, comprising argillaceous limestones which are occasionally cherty. To assume that the Hollymount palaeokarst was generated beneath a substantial cover of basal Namurian mudstones, as in the case of the Bees Nest (Derbyshire), Rhes-y-cae (Flintshire), Flimston (Pembrokeshire) and Ballygaddy (Offaly) subsidence structures (noted elsewhere in this paper), would imply that the part of the subsidence not preserved was of the order of 140 m deep. Our estimate is derived from the known thickness of the Clogrenan Formation plus an assumption that, perhaps, a decametre thickness of the Ballyadams Formation and two decametres of basal Namurian have been involved in the subsidence. Total solution-subsidence was thus notionally $c.$ 210 m and the pre-subsidence (?late Miocene) surface at $c.$ 180–200 m a.s.l. The ratio of total subsidence to general lowering of the adjacent landscape at Hollymount was thus $c.$ 3:2. In the 10 Ma or so of post-late Miocene time, this implies a local rate of landscape lowering of the order of 14 mm ka^{-1}.

Ballygaddy, Co. Offaly

The Ballygaddy Outlier, Co. Offaly, has been described by Beese *et al.* (1983). It has a diameter of the order of 60 m and, as suggested by boreholes and geophysical evidence, a depth of $c.$ 40 m. The rim of the outlier lies at $c.$ 130 m a.s.l. On the evidence of the fossil casts in chert clasts present in the fill mass, Beese *et al.* (1983) regard the fill as a saprolite, derived from decalcification of late-Viséan argillaceous and cherty limestones, the base of which formerly lay some 200 m above the present rim of the pocket (their estimate). Unfortunately, there is as yet no palaeontological control to denote the age of either the saprolite or the subsidence process.

The available data thus suggest that the land surface on which the saprolite was generated was rather higher than that of Hollymount, which lies 58 km to the east, i.e. $c.$ 330 m a.s.l., though its location is more centrally placed within the Irish massif. The ratio of total subsidence to the general lowering of the adjacent landscape was thus of the order of 5:4.

Summary

In summary, we suggest that solution-subsidence phenomena have already offered a considerable objectivity to landscape evolution interpretations, particularly in a WB&I context, and that we should accept the methodology as being one of the principal sources of stratigraphical control. Indeed, we suggest that it will probably become an increasingly important criterion in landscape evolution studies. The possibility that many other subsidence deposits in British and Irish limestone terrains await discovery seems very real and all temporary exposures in these areas should be examined in the hope that some key piece of geomorphologically useful evidence will be revealed. Numerous studies have shown that, even in the limited context of WB&I, subsidences *can* and *do* amount to several hectometres; indeed, there is a growing realization that karstic subsidence is limited in scale only by the thickness of the host carbonate.

The control offered by terrestrial palaeontology

Most of the evidence of biological cover on the landscapes described comes from cored sediments and, over the last twenty years, a large number of new localities have been found to contain fossils of Cenozoic age. There is not much local variation revealed in the floras but a north–south transect over the 10° latitude of the British Isles shows significant changes in vegetation. The cooling boreal realm (after the maximum warm-temperate climate of the Terminal Eocene Event) through the Oligocene and Miocene caused migrations, a large number of family originations and some extinctions of species (and perhaps genera, but not families). The environmental and weather systems also changed substantially, a major fall in sea-level near the end of the Oligocene causing much erosion, and the opening of the North Atlantic to the Arctic Ocean resulting in more frequent windy weather from the west.

In the western British Isles, boreholes have penetrated several Oligocene outliers, yielding evidence of plants and some insect remains (the latter not yet studied), which give further details of these changing landscapes. Oligocene floras

Table 1. *Possible non-marine Chattian and Miocene Outliers in West Britain and Ireland*

Age	Location	Region	Site and ref. number for Fig. 1	Dating Authority
Definite Chattian component	Onshore	Wales	Tremadoc Bay Basin Gwynedd and offshore (1)	Herbert-Smith 1979; Wilkinson & Boulter 1980
			Flimston Outlier (2)	Jenkins et al. 1995
		Ireland	Ballymacadam s.s. Complex, Co. Tipperary (3)	Watts 1962; Wilkinson & Boulter 1980; Boulter 1980
			Lough Neagh Basin, Co. Tyrone (4)	Watts 1962; Wilkinson & Boulter 1980; Boulter 1980
	Offshore	Hebridean Sea	South Harris Basin (6)	McCann 1989
			Canna Basin (7)	Evans et al. 1991
			Blackstones Bank Basin (8)	Evans et al. 1979
				Smythe & Kenolty 1975
Definite Miocene component	Onshore	N. England	Brassington Formation s.s. Complex, Derbys. (9)	Boulter 1971
		SW England	St Agnes Outlier, Cornwall (10)	Walsh et al. 1987
		Wales	Trwyn y Parc s.s. Complex, Anglesey (11)	Walsh et al. 1996
		Ireland	Hollymount Outlier, Co. Laois (12)	Boulter 1980; Watts 1985
Probable Chattian component	Onshore	SW England	Beacon Cottage Farm Outlier, Cornwall (13)	Walsh et al. 1987 Jowsey et al. 1992
			Bovey Basin, Devon	Wilkinson & Boulter 1980; Boulter 1988
			Petrockstow Basin, Devon (15)	Wilkinson & Boulter 1980
Possible Chattian or Miocene component	Onshore	SW England	Crousa Outlier, Cornwall (16)	Walsh et al. 1987
			Polcrebo Outlier, Cornwall (17)	Walsh et al. 1987
			The St Oswald's Bay s.s. Complex, Dorset (18)	House 1995
			Various gravel spreads, e.g. Cadham Farm, (19)	Edwards & Freshney 1982
			Orleigh Court, (20)	Simpson & Rodgers 1937
			Sandy Park, Devon (21)	Coque-Delhuille 1987
			Various silcretes	Bristow 1992, 1996
			Various saprolites	Bristow 1989
		Wales	Treffymon Basin, Pembs. (22)	Allen 1981
			Several pocket deposits in Flintshire, Denbighshire and Conwy (23)	Maw 1867; Strahan & De Rance 1890; Walsh & Brown 1971; Battiau-Queney 1980
			Various gravel spreads	Dixon 1921; Thomas 1970; Battiau-Queney 1980, 1984
			Various saprolites	Battiau-Queney 1980, 1984
Possible Chattian or Miocene component	Onshore	Ireland	Carrigaphuca Outliers, Co. Kerry (24)	Walsh 1965
			Ballyegan Outlier, Co. Kerry (25)	Battiau-Queney & Saucerotte 1985
			Ballygaddy Outlier, Co. Offaly (26)	Beese et al. 1983
			Tynagh s.s. Complex, Co. Galway (27)	Mitchell 1980; Monaghan & Scannell 1991
			Aughinish s.s. Complex, Co. Limerick (28)	Clarke et al. 1981
			Galmoy Outlier, Co. Kilkenny (29)	Coxon & Coxon 1997
			Lisheen Outlier, Co. Tipperary (30)	Coxon & Coxon 1997
			Knockgraffon Outlier, Co. Tipperary (31)	Archer et al. 1996
			Tullyallen s.s. Complex, Co. Meath (32)	Burdon 1987
			Various saprolites	
	Offshore	Celtic and Irish Seas	Stanley Bank, West Lundy & S. Celtic Sea Basins (33)	Bisshop & McCluskey 1948; Mitchell 1980, 1981; Higgs & Beese 1986
			St George's Channel Basins (34)	Boulter & Craig 1979
			Hook Head Basin, off Waterford coast (35)	Tucker & Arter 1987
			?unsampled basins such as the Central Irish Sea Basin	Mitchell 1981
				Tucker & Arter 1987
			Teifi Basin (37)	Warrington & Owens 1977

have been described from offshore Lundy (Boulter & Craig 1979), Devon (Freshney et al. 1982; Boulter 1988; Wilkinson & Boulter 1980), Cornwall (Walsh et al. 1987), Dyfed (Jenkins et al. 1995), Mochras (Wilkinson & Boulter 1980), offshore Canna (Evans et al. 1979) as well as from several sites in Ireland, which have recently been reviewed by Coxon & Coxon (1997) (Table 1). The pollen and spore assemblages are very similar, reflecting dense mixed conifer–broadleaf forest with swamp cypress in low-lying wetland areas: an environment not unlike that of present-day Florida. But, because the terrestrial vegetation was consistent and slow to change, both geographically and evolutionarily, the pollen and spores cannot usually determine the sediment's age to within a stage. Owing to the climatic cooling, changes in this Oligocene vegetation along the north–south transect of the British Isles were far fewer than before the Terminal Eocene Event, but, in this area, there was a varied flora of evergreen Pinaceae forest, ferns and, significantly, more flowering plants than previously. All these localities yield a comprehensive pollen and spore flora reflecting forest or heathland vegetation close by the freshwater site of deposition. Near shore marine deposition can be ruled out due to the lack of dinoflagellate cysts in the assemblages. The palynodebris consists of palynowafers (Boulter 1994) characteristic of low energy local sedimentation.

The lowering of sea-level in the late Oligocene and the consequent hiatus which is found across Europe, was one of the environmental changes which caused marked differences between the Palaeogene and the Neogene floras. Whatever the unknown details of these events may have been, there were many evolutionary changes in the plants across that boundary. However, in the western British Isles this is less obvious than, say, in the Rhine valley, because relatively few fossil localities are known and, owing to the progressive widening of the Atlantic, there was more oceanicity in the flora. This is evidenced by plants such as rhododendron, holly and some palms which occur in both the Cenozoic and modern flora of WB&I. Onshore in the British Isles, pollen and spores have been described from Derbyshire (Boulter 1971), Cornwall (Walsh et al. 1987), Co. Laois (Watts 1985) and Anglesey (Walsh et al. 1996). The different composite angiosperm taxa contrast strongly with the earlier Oligocene forms; some have become extinct in Europe (*Bohlensipollis*, *Gothanipollis*, *Aglaoreadia*) whereas others (e.g. tricolpates and tricolporates) have diversified considerably. The conifers also changed in composition, shown particularly by the introduction of two major kinds of *Tsuga*- and *Pinus*-like pollen and the occurrence of swamp cypress, *Taxodium*-like, in near-marine and freshwater environments.

Visually, the landscape of the British Isles may not have changed very much during this transition. The North Atlantic and cooler temperatures would have caused more storms and, so, less dense forests. In turn, this would have encouraged angiosperms to invade new niches and evolve less-woody herbaceous forms. Grazing and other controls on forests increased through the more numerous Miocene mammals, and other vectors, such as more insects and frosts, also made ecosystems much more varied (Moore et al. 1996).

The Chattian and Miocene sedimentary outliers and saprolite bodies in a WB&I context

Over one hundred isolated rock masses are now known in the study area for which a Chattian or Miocene age is either certain or theoretically possible. Those we know of are listed in Table 1. This number has increased markedly in recent years, especially as a result of new finds in Ireland, where commercial drilling has revealed somewhat surprising results (for example, Sleeman & McConnell (1995) report a find of late Jurassic/Lower Cretaceous clays of 'marginal marine aspect', presumably preserved partly by karstic agencies, in the Carboniferous Limestone at Piltown, Co. Waterford). Currently, new finds of post-Variscan outliers in these oldland areas (not necessarily of Oligo-Miocene age) are running at about two per annum, so, doubtless, the list in Table 1 will quickly be extended.

Links between the 50–150 m a.s.l. surfaces and the Chattian and Miocene deposits

Whereas it has so far been maintained that there is no undisputed association of any of the 50–150 m a.s.l. planation surfaces with marine sediments or any sediment of Plio-Pleistocene age, it is undeniable that there are certain links, however few, between these surfaces and Chattian or Miocene sediment. The critical localities are: (i) the Beacon Cottage Farm and St Agnes Outliers in Cornwall, relative to the Reskajeage Surface; (ii) the Flimston Outlier, Pembrokeshire, relative to the Bosherston–Castlemartin Surface; and

(iii) the pipe fills at Trwyn y Parc, Anglesey, relative to the Menaian Surface.

The Beacon Cottage Farm and St Agnes Outliers

In 1987, details of the geology and geomorphology of these outliers was published, in which it was suggested that, based on the palynological determinations (Walsh *et al.* 1987) the Beacon Cottage Farm sediments were Oligocene and the St Agnes sediments, Miocene. We still subscribe to this view and are not aware of any subsequent contrary opinion, save for that of Coque-Delhuille (1991, 1992), which has already been noted. On the evidence of some 36 boreholes down to Devonian bedrock below the St Agnes Outlier, the form of the sub-Miocene unconformity is revealed as virtually identical to that of the adjacent facet of the Reskajeage Surface. The critical evidence is the very close coincidence of the 'buried cliff line' on the margin of the Outlier bordering the upstanding mass of St Agnes Beacon, and the steep rise of slope on those adjacent parts of the Beacon no longer covered by Miocene sediment. Subsequently, a programme of boreholes and trenches around Beacon Cottage Farm (Jowsey *et al.* 1992, fig. 16) showed that, by contrast, the sub-Oligocene floor of that Outlier was much more irregular than the planar feature below the St Agnes Outlier. By implication, at least part of the planation of west Cornwall, which produced the Reskajeage Surface, took place around the time of the Palaeogene/Neogene (g/n) boundary. This evidence from west Cornwall is central to our general thesis.

The environment in which the St Agnes (*s.s.*) sediments were deposited was discussed at length in Walsh *et al.* (1987). Interpretations were based on palynological, particle size distribution and quartz grain surface texture analyses. These gave no support to previous hypotheses that the sediments were formed in either marine or fluvial environments; on the contrary, quartz grain surface textures and particle size analysis clearly indicated a close similarity of the St Agnes sediments to modern aeolian deposits, in the case of the two sandy members of the sequence (the Doble and Beacon Members), and to modern subtropical colluvium, in respect of the intervening clays and silts of the New Downs Member. The palynoflora preserved in the colluvium denotes a sub-tropical or mediterranean climate at the time of its formation. In the palynological preparations, the sporomorphs were all in excellent condition, there being no trace of reworking; no dinocysts were present. The floral assemblage represents a fully integrated freshwater ecosystem and the palyno-debris comprised unabraded palynowafers. On an individual basis, none of these features is exclusive of the terrestrial environment, but, taken together, it leaves very little room to argue that these sediments were formed anywhere near the ocean. No evidence bearing on the problem of the length of the hiatus between the cutting of the surface and the deposition of the oldest sediments resting on it is available; there is, however, no reason to believe that these two events were not closely consecutive. Where exposed in recent years in the floor of the New Downs Pit, the sub-Miocene surface is smooth over an area of about 1 ha and, for several decimetres below the surface, is heavily stained by iron compounds.

The Flimston Outlier

It seems to be generally agreed that the Flimston Outlier, the outcrop of which is $c.$ $40\,000\,m^2$, rests on, or in, the Carboniferous Limestone of the Bosherston–Castlemartin Plain, which is part of the Pembroke Peninsula of Pembrokeshire. Adjacent to the outlier, the Plain lies at 50 m a.s.l. As with all previous authorities, we regard the Flimston mass as a solution-subsidence outlier, though George (1974) considered that part of its pocket-like form might be controlled by faulting.

Jenkins *et al.* (1995) recorded a Chattian palynoflora from a borehole sample which, as subsequently revealed in trenching operations, came from spoil which dates from 150 years ago, during the exploitation of the Flimston mass for pipeclay. Several trenches were excavated in 1995 to a maximum depth of $c.$ 7 m. These indicated that those parts of the outlier thereby exposed are not a ball clay-type sediment but, rather, a pale grey or mottled saprolite which is presumably derived from the intense chemical weathering of an isolated foundered mass of basal Namurian mudstone, i.e. the Flimston mass is simply the westernmost of several such weathered bodies which are exposed in the trough of the Bullslaughter Bay syncline between here and the eastern cliffs of the Bay (Thomas 1970; Battiau-Queney 1980). Dixon (1921) considered that the rim of the Flimston mass lies in the lower part of the D_1 zone hereabouts, which, together with the overlying D_2 zone, is shown locally to be of the order of 250 m in thickness. This would appear to give a crude estimate of the amount of subsidence of this mass from its

original location, though how much of a pre-Chattian component is included in this figure is not known. It is presumed that the microflora was preserved in an organic palaeosol, rather than a lacustrine clay.

We agree with all previous opinion that the Flimston mass is preserved entirely below the general level of the South Pembrokeshire Surface and also that the saprolite/ball clay mass (whichever it is) was itself planated at the same time as the surrounding limestone outcrop. Thus, the planation was either a late- or post-Chattian event.

Unfortunately, the siliciclastic gravels which cover the Bosherston–Castlemartin Surface remain undated. Most previous opinion (Dixon 1921; George 1974; Thomas 1970; Battiau-Queney 1980) favours a pre-Pleistocene rather than a Pleistocene age; we also subscribe to this view. The petrography of the gravel cover was investigated by Battiau-Queney (1980) who regards this essentially as a pediment which has, somewhat miraculously, survived the effects of snow and ice here during the Pleistocene. She is emphatic that the Bosherston–Castlemartin Surface was not formed by marine transection, though she acknowledges that the quartz fragments which exhibit *émoussés luisant* textures in the gravel spread (as those elsewhere at various levels in Pembrokeshire) could have been formed in a marine environment. If this interpretation is correct, the marine transgression must post-date the cutting and subsequent deformation of the Plain but pre-date a rhexistatic episode when most of the pre-existing early or mid-Cenozoic deep weathering profiles were destroyed and so became the source of the clastic formations which have been trapped below, and which rest on, the Carboniferous Limestone benches in South Pembrokeshire (Y. Battiau-Queney, pers. comm. 1997).

About 35 km to the NNW, the undated gravels of the Treffynnon Outlier, regarded by Allen (1981) as of Cenozoic age, and which rest at *c.* 90 m a.s.l. on a much-weathered Palaeozoic foundation of igneous rocks and greywackes, appear to be possible correlatives of the Bosherston–Castlemartin gravels. Poorly exposed and irregularly distributed siliciclastic gravel covers, none of which have yet been dated, are also reported from a number of localities in the Cornubian Peninsula. Coque-Delhuille (1987) has given the most authoritative account of these. Mostly, they are regarded as terrestrial and formed as pediments or lag concentrates from fluvial bodies. The gravels basal to the Miocene St Agnes Formation on the east side of St Agnes Beacon (Davies & Kitto 1878) could also be an analogue of the Flimston Gravel on the south side of the Celtic Sea (Dixon 1921).

The Trwyn y Parc Solution-Subsidence Complex

The palaeokarst setting of the Trwyn y Parc Solution-Subsidence Complex on the northern coast of Anglesey has been described by Morawiecka *et al.* (1996) and the palynology of a black clay preserved in Pipes 11/12 of that complex by Walsh *et al.* (1996). In view of the steep dip of the host and the lithological contrast of host and fill, it is believed that the total subsidence here is greatly in excess of the 12 m elevation observed in current exposures. On the indirect grounds that 75% of the floral determinations by Skawińska-Wieser are of swamp-forest taxa, an association is proposed with the Menaian Surface (*s.l.*) of Greenly (1919) (that which Battiau-Queney (1980) terms the 'Fundamental Surface of Erosion of Anglesey'). This would seem to be reasonable, given the data set available at present. If correct, this implies that the Menaian Surface was already in existence by the end of the Miocene. During at least part of the Miocene, this surface is considered to have supported a freshwater swamp-forest cover.

The petrography of the pipe infills at Trwyn y Parc was described by Morawiecka *et al.* (1996). These fills, of which intensely weathered igneous rocks appear to form the major constituent, contain in their mineral suites goethite, which is often regarded as a product of severe weathering of various silicate-bearing rocks in continental areas subject to a hot and wet climate, and gibbsite, which is commonly regarded as being produced by desilication of kaolinite in the intense tropical weathering of an iron-poor rock in a fairly well drained upland region (Bosák 1989). Quartz, muscovite, chlorite and imogolite are also present. With regard to the palynoflora, the condition of the sporomorphs is excellent and the flora represents an integrated freshwater ecological system; no marine organisms were recognizable. Thus, neither the mineralogy nor the palynology provides any hint of marine influence during the formation of the fill sediments. The whole complex of pipe fills is regarded by Morawiecka *et al.* (1996) as a mixed sequence of organic palaeosols and saprolites.

Summary

In all three cases, a major facet of landscape planation in the 50–150 m a.s.l. range is associated

with datable terrestrial sediment. All three are remarkably level and extend for several kilometres inland from the present coasts. Clearly, in all three cases, the facets are merely backslope erosional remnants, the original wider surfaces having been much destroyed at their seaward edges by late Cenozoic marine erosion. In two cases, the age of the facet appears to be post-Chattian and in two cases there is a good case for it being pre-end-Miocene. An early Miocene age seems to be indicated and a terrestrial mode of origin, well away from any marine influence, is the best way we have of accounting for the available mineralogical, sedimentological and palaeobotanical evidence. The exact manner in which the surfaces were sculptured is not a particular concern of the present paper. In very broad terms, we believe our model accords quite well with the 'Savanna planation' theory of Cotton (1961).

In recognizing that a number of major landscape elements are of much greater antiquity than was generally supposed up to a few years ago, this view is not incompatible with a number of other recently expressed views on the geomorphological evolution of WB&I during the Neogene (e.g. Naylor 1992; Bowen 1994; Straw 1995; Bristow 1996), nor is it in any sense incompatible with the evidence of the disposition of the other known terrestrial outliers of proven Chattian or Miocene age in the WB&I area (Table 1). Moreover, if a widespread planation surface roughly of g/n boundary age is bevelled into the margins of the various oldland massifs of WB&I, given their gentle seaward slopes, logic dictates that we should expect to find that some of the other datable Chattian and Miocene sedimentary bodies which are now located more centrally within the massifs were formed at levels consonant with these slopes and altitudes. For example, it was reasoned in an earlier section that the sedimentary fill of the late Miocene Hollymount Outlier of Co. Laois, which is nowhere nearer than 52 km from the present coast of Ireland, may have formed when the local landscape lay at about 200 m a.s.l. The various undated saprolitic/palaeosol masses contained in the Carboniferous Limestone outcrop of central Ireland, i.e. those at Ballyegan, Aughinish, Tynagh, Ballygaddy and Tullyallen (Burdon 1987; Coxon & Coxon 1997), also fit quite neatly into a model of a planed landscape, roughly of g/n or early Miocene age, which inclines gently seawards from the interiors of the various modern massifs. In our model, this earlier version of the Central Plain of Ireland was possibly 150 m or so higher than that of the present day.

Even where Chattian sediments are still contained in the freshwater basins of WB&I (e.g. Lough Neagh), the notional Palaeogene/Neogene surface apparently lies well above sea-level, despite the probability of continuing basin subsidence into the Miocene (Cope 1994).

Based on the fundamental control offered by the St Agnes–Flimston–Trwyn y Parc axis, we believe there is a general consistency of form and elevation of the various coastal planation facets from Co. Cork in the west to Bristol in the east and southwards from Anglesey to at least the Channel Islands. Of course, the further away from the central axis, the less reliable the correlation. Above this vast early Neogene 'Celtic Plain', only subdued hill masses, the early forerunners of the Munster, Leinster, Welsh and Cornubian Uplands, projected above the general level. As for the other oldland massifs, there is no evidence that either the Connacht Upland or the Pennine range existed at that time. It is supposed that neotectonism in Plio-Pleistocene times rejuvenated these Uplands and created the Connemara Mountains (Dewey & McKerrow 1963) and the Pennines (Walsh *et al.* 1972) in that period, as has also been postulated for the latest-stage development of the Armorican Massif of Brittany (Battiau-Queney 1993).

An attempt is made in Fig. 2 to illustrate possible relationships between some of the outliers and intervening landforms.

Discussion and conclusions

The main purpose of this paper is to reinforce the concept that there is hardly any evidence that the 50–150 m surfaces benched into the margins of the oldland massifs of WB&I have been created by marine agencies, or that they are of Plio-Pleistocene age. Rather, there is a growing body of evidence that many of these can be shown to relate to terrestrially dominated sediments and processes and that, in terms of their age, there are several cases where a late Palaeogene/early Miocene dating seems inescapable. Of course, the surfaces discussed here may well be polygenetic, the products of several superimposed planations, some, perhaps, dating back to the Mesozoic, each of which successively smoothed the landscape to become the remarkably flat surfaces we see today. However, in the absence of associated Mesozoic or early or mid-Palaeogene sediments, there seems no way of proving this at present and we must assume that, so dominant were the erosional processes which were active in Chattian and Miocene times, any evidence of earlier (perhaps marine) episodes was completely

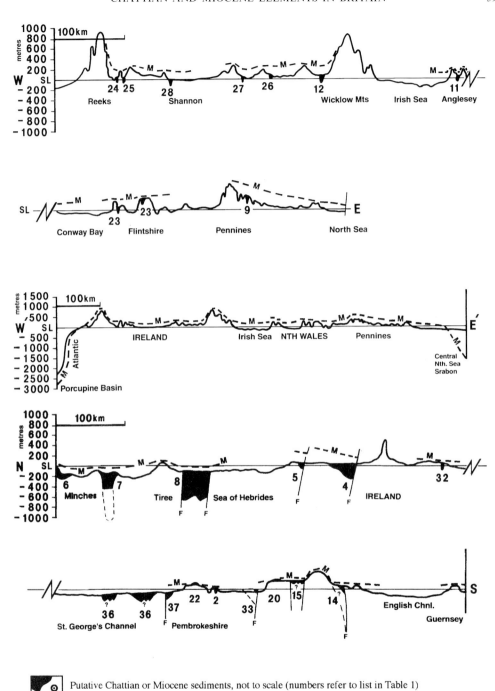

Fig. 2. Sections showing the relationships between sedimentary bodies of possible Chattian or Miocene age and the modern landscape of the British Isles and adjacent continental shelf (numbers correspond to list in Table 1).

obliterated. Nevertheless, there are numerous instances in the stratigraphic record in the west Britain region where planar sub-Mesozoic unconformities are overlain by undoubted marine sequences and it would certainly be interesting to be able to test whether any of the 50–150 m a.s.l. planation surfaces have their roots as stripped sub-Mesozoic unconformities, which formerly lay only a few metres or, at most, decametres above the former. In a SW England context, the sub-Cenomanian surface of east Devon, and the sub-Liassic and sub-Inferior Oolite surfaces of the Mendip region offer realistic possible correlatives.

In a partial reversion to the denudation chronological tenets of 40–50 years ago, we believe that there was an essential simplicity and stability in geomorphological relationships from oldland to oldland across the British Isles, at least around their fringes. But, whereas the geomorphologists of that period saw stability and uniformity from oldland to oldland in terms of a few millions or hundreds of thousands of years before the present, we suggest that the time of relative stability was in a much older period, that roughly between 25 and 15 Ma ago. Since then, while crustal deformation in the study area has generally intensified, the coastal benches appear to have been largely immune from major distortion. Indeed, but for the antecedents of the modern uplands of Leinster, Munster, Wales and Dartmoor, all supposedly rejuvenated in Plio-Pleistocene times, the area of WB&I considered here must then have been a broad forested plain at the g/n boundary interval, every bit as subdued and monotonous as the landscape of the modern North European Plain, or what the present continental shelf would look like if drained and allowed to develop a natural vegetation. In its location on a continental margin, as contrasted to a location in a cratonic continental interior, our model suggests that the oldland fringes of WB&I represent one of the oldest little-changed landscapes in the world. Certainly, we would argue that the St Agnes–Flimston–Trwyn y Parc axis offers a far better starting point for studies of pre-Quaternary landscape evolution in the British Isles than anywhere else, particularly the neotectonically active SE England with its fragmentary and disturbed sedimentary covers of somewhat doubtful age (the Lenham Formation and its Neogene relatives) (Jones 1981).

Of course, very many problems concerning the age and origin of the landscape features of WB&I remain for consideration by future workers. For instance, there is the yet unresolved question of whether soft-rock outliers, such as those at St Agnes, might just have survived Plio-Pleistocene sea-level rises up to 200 m or so, whatever the duration of such a transgression. There is no direct evidence that the late Pliocene St Erth marine transgression of west Cornwall reached higher than 50 m a.s.l., i.e. rather lower than the sea depth envisaged by Reid (1890), or that it left any detectable bevel in the west Cornish landscape (Mitchell et al. 1973; Head 1993).

Further, there remains the question of a possible homeomorphism of such landforms. Future workers must always bear in mind the possibility that identical landforms may be produced by different (sometimes quite different) combinations of geomorphological processes. In this case, in a WB&I context, could a terrestrial planation surface roughly of g/n boundary age be represented by a macro-landform which is identical in all respects to one produced by Plio-Pleistocene marine transection? (In which case, with respect to this particular study area, given that there are so few stratigraphic controls, one has simply been unlucky in not yet finding any associated marine sediments anywhere; hence, one would never be able to discern the true identity of the marine forms.) We know of no research anywhere into this problem and, unless, for example, one can find a major surface which is underlain by a variety of different karstic subsidence deposits, it is difficult to see how this hypothesis might be tested.

We consider that, owing to the sedimentological, karstological and palaeobotanical evidence now at our disposal, significant advances in our understanding of Cenozoic landscape evolution in WB&I have been made in recent years. Nevertheless, those who go looking for complexities in landscape form and evolution will always find them; we would argue that, in a WB&I context, as we still have so little fundamental control through stratigraphy or any other tool, a broad-brush, 'overview' approach is all that will be possible for the foreseeable future.

We are pleased to acknowledge the considerable help provided by the two anonymous referees during the preparation of this paper. P.T.W. is pleased to acknowledge the opportunity to view the UCW Cardiff trenching operations at Flimston in May 1996, at the invitation of the late Dr Graham Jenkins. Keith Elay, University of East London, drew the figures.

References

ALLEN, P. M. 1981. A new occurrence of possible Tertiary deposits in south-western Dyfed. *Geological Magazine*, **118**, 561–564.

ARCHER, J. B., SLEEMAN, A. G. & SMITH, D. C. 1996. *A geological description of Tipperary and adjoining parts of Laois, Kilkenny, Offaly, Clare and Limerick, to accompany the Bedrock Geology 1:100,000 scale Map Series; Sheet 18, Tipperary.* Geological Survey of the Republic of Ireland.

ATKINSON, K., BOULTER, M. C., FRESHNEY, E. C., WALSH, P. T. & WILSON, A. C. 1975. A revision of the geology of the St. Agnes Outlier, Cornwall. *Proceedings of the Ussher Society*, **3**, 286–287.

BALCHIN, W. G. V. 1964. The denudation chronology of southwest England. *In*: HOSKING, K. F. G. & SHRIMPTON, G. J. (eds) *Present Views of Some Aspects of the Geology of Cornwall and Devon*. The Royal Geological Society of Cornwall, Truro. 267–281.

BATTIAU-QUENEY, Y. 1980. Contribution à l'étude géomorphologique du Massif Gallois (G.B.). Thèse lettres, Université de Bretagne Occidentale, Paris.

—— 1984. The pre-glacial evolution of Wales. *Earth Surface Processes and Landforms*, **9**, 229–252.

—— 1993. *La relief de la France – coupes et croquis.* Masson Géographie, Paris.

—— 1999. Crustal anisotropy and differential uplift: their role in long-term landform development. This volume.

—— & SAUCEROTTE, M. 1985. Paléosols pré-glacière de la carrière de Ballyegan (Kerry, Irelande). *Hommes et Terres du Nord*, **3**, 234–237.

BEESE, A. P., BRÜCK, P. M., FEEHAN, J. & MURPHY, T. 1983. A silica deposit of possible Tertiary age in the Carboniferous Limestone near Birr, Co. Offaly, Ireland. *Geological Magazine*, **120**, 331–340.

BELBIN, S. 1985. Long term landform development in NW England: the application of the planation concept. *In*: JOHNSON, R. H. (ed.) *The Geomorphology of North-West England*. Manchester University, 37–58.

BISSHOP, D. W. & MCCLUSKEY, J. A. G. 1948. *Sources of Industrial Silica in Ireland.* Geological Survey of Ireland Emergency Period Pamphlet No. 3, Department of Commerce and Industry, Republic of Ireland, Dublin.

BOSÁK, P. 1989. Clays and sands in palaeokarst. *In*: BOSÁK, P., FORD, D., GLAZEK, J. & HORÁČEK, I. (eds) *Palaeokarst: a systematic and regional review*. Elsevier, Amsterdam, 431–442.

BOSWELL, P. G. H. 1923. The petrography of the Cretaceous and Tertiary outliers of the west of England. *Quarterly Journal of the Geological Society of London*, **79**, 205–230.

BOULTER, M. C. 1971. A palynological study of two of the Neogene plant beds in Derbyshire. *Bulletin of the British Museum of Natural History (Geology)*, **19**, 360–411.

—— 1980. Irish Tertiary fossils in a European context. *Journal of Earth Science of the Royal Dublin Society*, **3**, 1–11.

—— 1988. Tertiary monocotyledons from aquatic environments. *Tertiary Research*, **9**, 133–146.

—— 1994. An approach to a standard terminology for palynodebris. *In*: TRAVERSE, A. (ed.) *Sedimentation of Organic Particles*. Cambridge University, 199–216.

—— & CRAIG, D. L. 1979. A Middle Oligocene pollen and spore assemblage from the Bristol Channel. *Review of Palaeobotany and Palynology*, **28**, 259–272.

BOWEN, D. Q. 1994. Late Cenozoic Wales and southwest England. *Proceedings of the Ussher Society*, **8**, 209–213.

BRISTOW, C. M. 1989. Gravity anomaly between Dartmoor and Bodmin Moor; an alternative explanation. *Proceedings of the Geologists' Association*, **100**, 143.

—— 1992. Silcrete duricrusts west of the Bovey Basin. *Proceedings of the Ussher Society*, **8**, 172–176.

—— 1996. *Cornwall's Geology and Scenery: an introduction.* Cornish Hillside, Truro.

BROWN, E. H. 1960. *The Relief and Drainage of Wales.* University of Wales, Cardiff.

BURDON, D. J. 1987. Some ancient dolines in the karst of Ireland. *In Second Multidisciplinary Conference on Sinkholes and the Environmental Impacts of Karst*, Orlando, Florida, 9–11 February 1987, 65–71.

CLARKE, R. G., GUTMANIS, J. C., FARLEY, A. E. & JORDAN, P. G. 1981. Engineering geology of a major industrial complex at Aughinish, Co. Limerick, Ireland. *Quarterly Journal of Engineering Geology*, **14**, 231–239.

COLE, G. A. J. 1912. The problem of the Liffey valley. *Proceedings of the Royal Irish Academy*, **30B**, 8–19.

COPE, J. C. W. 1994. A latest Cretaceous hotspot and the south-easterly tilt of Britain. *Journal of the Geological Society, London*, **151**, 905–908.

COQUE-DELHUILLE, B. 1987. *Le massif du Sud-ouest Anglais et sa bordure sédimentaire; étude géomorphologique.* Thèse d'Etat de l'Université de Paris. Panthéon-Sorbonne, Paris.

—— 1991. The long-term evolution of the English South-west Massif (U.K.). *Zeitschrift für Geomorphologie*, **N.F. 35**, 65–84.

—— 1992. La plate-forme d'abrasion marine Pliocène du sud-ouest Anglais. *Norois*, **38**, 39–59.

COTTON, C. A. 1961. The theory of savanna planation. *Geography*, **46**, 89–101.

COXON, P. & COXON, C. 1997. A pre-Pliocene or Pliocene land surface in County Galway, Ireland. *In*: WIDDOWSON, M. (ed.) *Palaeosurfaces: Recognition, Reconstruction and Palaeoenvironmental Interpretation*. Geological Society, London, Special Publications, **120**, 37–55.

DAVIES, A. & KITTO, B. 1878. On some beds of sand and clay in the parish of St. Agnes, Cornwall. *Transactions of the Royal Geological Society of Cornwall*, **9**, 196–203.

DAVIES, G. L. H. & STEPHENS, N. 1978. *The Geomorphology of the British Isles: Ireland.* Methuen, London.

DELMER, A., LECLERCQ, V., MARLIÉRE, R. & ROBASZINSKI, F. 1982. Le géothermie du Hainaut et le sondage de Ghlin (Mons – Belgique). *Annales du Societe Géologique du Nord*, **101**, 189–206.

DEWEY, J. F. & MCKERROW, W. F. 1963. An outline of the geomorphology of Murrisk and NW Galway. *Geological Magazine*, **100**, 260–275.

DIXON, E. E. L. 1921. *The geology of the South Wales Coalfield. XIII, The geology of the country around Pembroke and Tenby*. Memoir of the Geological Survey of England and Wales.

EDMONDS, E. A., McKEOWN, M. C. & WILLIAMS, M. 1975. *British Regional Geology – South-west England*. HMSO, London.

EDWARDS, R. A. & FRESHNEY, E. C. 1982. The Tertiary sedimentary rocks. *In*: DURRANCE, E. M. & LAMING, D. J. C. (eds) *The Geology of Devon*. Exeter University Press, 204–237.

EMBLETON, C. 1964. The planation surfaces of Arfon and adjacent parts of Anglesey: a re-examination of their age and origin. *Transactions of the Institute of British Geographers*, **35**, 17–26.

EVANS, D., HALLSWORTH, C., JOLLEY, D. W. & MORTON, A. C. 1991. Late Oligocene terrestrial sediments from a small basin in the Little Minch. *Scottish Journal of Geology*, **27**, 33–40.

——, WILKINSON, G. C. & CRAIG, D. L. 1979. The Tertiary sediments of the Canna Basin, Sea of the Hebrides. *Scottish Journal of Geology*, **15**, 329–332.

EVERARD, C. E. 1977. *Valley direction and geomorphological evolution in West Cornwall*. Occasional Paper No. 10, Department of Geography, Queen Mary College, London.

FLETT, J. S. & HILL, J. B. 1912. *Geology of the Lizard and Meneage*. Memoir of the Geological Survey of the UK (Sheet 359), HMSO, London.

FRESHNEY, E. C., EDWARDS, R. A., ISAAC, K. P., WITTE, G., WILKINSON, G. C., BOULTER, M. C. & BAIN, J. A. 1982. A Tertiary basin at Dutsun, near Launceston, Cornwall, England. *Proceedings of the Geologists' Association*, **93**, 395–402.

FRYER, G. 1958. Evolution of the landform of Kerrier. *Transactions of the Royal Geological Society of Cornwall*, **19**, 122–153.

GEORGE, T. N. 1974. The Cenozoic evolution of Wales. *In*: OWEN, T. R. (ed.) *The Upper Palaeozoic and Post-Palaeozoic Rocks of Wales*. University of Wales Press, Cardiff, 341–371.

GOSKAR, K. C. & TRUEMAN, A. E. 1934. The coastal plateaux of South Wales. *Geological Magazine*, **71**, 468–477.

GOODE, A. J. J. & TAYLOR, R. T. 1988. *Geology of the country around Penzance*. Memoir of the British Geological Survey, HMSO, London.

GREENLY, E. 1919. *The geology of Anglesey*: 2 vols. Memoir of the Geological Survey of England and Wales.

GULLICK, C. F. W. R. 1936. A physiographical survey of west Cornwall. *Transactions of the Royal Geological Society of Cornwall*, **16**, 380–399.

HAWKINS, J. 1832. On a very singular deposit of alluvial matter on St. Agnes Beacon, and on the granitical rock which occurs in the same situation. *Transactions of the Royal Geological Society of Cornwall*, **4**, 135–144.

HEAD, M. J. 1993. *Dinoflagellates, sporomorphs and other palynomorphs from the Upper Pliocene St. Erth Beds of Cornwall, south-western England*. Memoir of the Palaeontological Society, No. 31.

HERBERT-SMITH, M. 1979. *The age of the Tertiary deposits of the Llanbedr (Mochras Farm) Borehole as determined from palynological studies*. Reports of the Institute of Geological Sciences, **78/24**, 13–29.

HIGGS, K. & BEESE, A. P. 1986. A Jurassic microflora from the Colbond Clay of Cloyne, Co. Cork. *Scientific Proceedings of the Royal Dublin Society*, **7**, 99–110.

HOLLINGWORTH, S. E. 1938. The recognition and correlation of high level erosion surfaces in Britain; a statistical study. *Quarterly Journal of the Geological Society of London*, **94**, 55–84.

HOUSE, M. R. 1995. Dorset dolines: Part 3, Eocene pockets and gravel pipes in the Chalk of St. Oswald's Bay. *Proceedings of the Dorset Natural History and Archaeological Society*, **117**, 109–116.

JENKINS, D. G., BOULTER, M. C. & RAMSAY, A. T. S. 1995. The Flimston Clay, Pembrokeshire, Wales: a probable late Oligocene lacustrine deposit. *Journal of Micropalaeontology*, **15**, 66.

JONES, D. K. C. 1981. *The Geomorphology of the British Isles: South-east and Southern Britain*. Methuen, London.

JOWSEY, N. L., PARKIN, D. L., SLIPPER, I. J., SMITH, A. P. C. & WALSH, P. T. 1992. The geology and geomorphology of the Beacon Cottage Farm Outlier, St. Agnes, Cornwall. *Geological Magazine*, **129**, 101–121.

JUKES-BROWN, A. J. 1907. The age and origin of the plateaux around Torquay. *Quarterly Journal of the Geological Society*, **63**, 106–123.

McCANN, N. 1989. Lignite in Ireland. *British Geologist*, **15**, 23–24.

MAW, G. 1867. On the distribution beyond the Tertiary districts of white clays and sands subjacent to the boulder clay. *Geological Magazine*, **4**, 241–251 and 299–307.

MILLER, A. A. 1955. The origin of the South of Ireland Peneplain. *Irish Geography*, **3**, 79–86.

MILNER, H. B. 1922. The nature and origin of the Pliocene deposits of the county of Cornwall and their bearing on the Pliocene geography of the south west of England. *Quarterly Journal of the Geological Society of London*, **78**, 348–377.

MITCHELL, G. F. 1965. The St. Erth Beds – an alternative explanation. *Proceedings of the Geologists' Association*, **76**, 345–366.

—— 1980. The search for Tertiary Ireland. *Journal of Earth Sciences of the Royal Dublin Society*, **3**, 13–33.

—— 1981. Other Tertiary events. *In*: HOLLAND, C. H. (ed.) *A Geology of Ireland*. Scottish Academic Press, Edinburgh, 231–234.

——, CATT, J. A., McMILLAN, N. F., MARGAREL, J. P. & WHATLEY, R. C. 1973. The late Pliocene marine formation at St. Erth, Cornwall. *Philosophical Transactions of the Royal Society of London*, **B266**, 1–37.

MONAGHAN, N. T. & SCANNELL, M. J. 1991. Fossil cypress wood from Tynagh Mine, Loughrea, Co. Galway. *Irish Naturalists' Journal*, **23**, 377–378.

MOORE, P. D., CHALONER, W. G. & STOTT, P. 1996. *Global Environment Change*. Blackwell, London.

MORAWIECKA, I. M., SLIPPER, I. J. & WALSH, P. T. 1996. A palaeokarst of probable Kainozoic age preserved in Cambrian marble at Cemaes Bay, Anglesey, North Wales. *Zeitschrift für Geomorphologie*, **N.F. 40**, 47–70.

MURCHISON, R. I. 1839. *The Silurian System* (2 vols). John Murray, London.

NAYLOR, D. 1992. The post-Variscan history of Ireland. *In*: PARNELL, J. (ed.) *Basins of the Atlantic Seaboard: petroleum geology, sedimentology and basin evolution*. Geological Society, London, Special Publication, **62**, 255–275.

ORME, A. R. 1964. Planation surfaces in the Drum Hills, Co. Waterford and their wider implications. *Irish Geography*, **5**, 48–72.

PROCTOR, C. J. 1988. Sea level related caves on Berry Head, South Devon. *Cave Science*, **15**, 39–49.

QUINIF, Y. 1989. Palaeokarsts of Belgium. *In*: BOSÁK, P., FORD, D., GLAZEK, J. & HORÁČEK, I. (eds) *Palaeokarst: a systematic and regional review*. Elsevier, Amsterdam, 38–50.

RAMSAY, A. C. 1846. *The denudation of South Wales and adjacent English counties*. Memoir of the Geological Survey of England and Wales.

REID, C. 1890. *Pliocene deposits of Great Britain*. Memoir of the Geological Survey of the UK, HMSO, London.

—— & SCRIVENOR, J. B. 1906. *The geology of the country near Newquay*. Memoir of the Geological Survey of the UK, HMSO, London.

RENOUF, J. T. 1993. Solid Geology and tectonic background. *In*: KEEN, D. H. (ed.) *The Quaternary of Jersey*. Quaternary Research Association, London, 1–11.

SIMPSON B. & RODGERS, I. 1937. The flint gravel deposits of Orleigh Court, Buckland Brewer, North Devon. *Geological Magazine*, **74**, 309–316

SISSONS, J. B. 1976. *The Geomorphology of the British Isles: Scotland*. Methuen, London.

SLEEMAN, A.G. & MCCONNELL, B. 1995. *A geological description of East Cork, Waterford and adjoining parts of Tipperary and Limerick to accompany the Bedrock Geology 1:100,000 Scale Map Series, Sheet 22 East Cork and Waterford*. Geological Survey of the Republic of Ireland.

SMITH, B. & GEORGE, T. N. 1961. *British Regional Geology: North Wales*. HMSO, London.

SMYTHE, D. K. & KENOLTY, N. 1975. Tertiary sediments in the Sea of the Hebrides. *Journal of the Geological Society, London*, **131**, 227–233.

STRAHAN, A. & DE RANCE, C. E. 1890. *The geology of the neighbourhood of Flint, Mold and Ruthin*. Memoir of the Geological Survey of England and Wales.

STRAW, A. 1995. Aspects of the geomorphology of Exmoor. *In*: BINDING, H. (ed.) *The Changing Face of Exmoor*. Exmoor Books, Tiverton, 13–25.

—— & CLAYTON, K. 1978. *The Geomorphology of the British Isles: Eastern and Central England*. Methuen, London.

THOMAS, T. M. 1970. Field meeting of the South Wales group on the Stacks Rocks to Bullslaughter Bay section of the South Pembrokeshire coast. Report of the Director. *Proceedings of the Geologists' Association*, **81**, 241–248.

TIETZSCH-TYLER, D. & SLEEMAN, A. G. 1994. *Geology of Carlow-Wexford. A geological description to accompany the Bedrock Geology 1:100,000 Scale Map Series. Sheet 19*. Geological Survey of the Republic of Ireland.

TUCKER, R. M. & ARTER, G. 1987. The tectonic evolution of the North Celtic Sea and Cardigan Bay Basins with special reference to tectonic inversion. *Tectonophysics*, **137**, 291–307.

WALSH, P. T. 1965. Possible Tertiary outliers from the Gweestin Valley, Co. Kerry. *Irish Naturalists' Journal*, **15**, 100–104.

—— & BROWN, E. H. 1971. Solution subsidence outliers containing probable Tertiary sediment in North-east Wales. *Geological Journal*, **7**, 299–320.

——, ATKINSON, K., BOULTER, M. C. & SHAKESBY, R. A. 1987. The Oligocene and Miocene outliers of West Cornwall and their bearing on the geomorphological evolution of Oldland Britain. *Philosophical Transactions of the Royal Society of London*, **A323**, 211–245.

——, BOULTER, M. C., IJTABA, M. & URBANI, D. M. 1972. The preservation of the Neogene Brassington Formation of the southern Pennines and its bearing on the evolution of Upland Britain. *Journal of the Geological Society, London*, **128**, 519–559.

——, COLLINS, P., IJTABA, M., NEWTON, J. P., SCOTT, N. H. & TURNER, P. R. 1980. Palaeocurrent directions and their bearing on the origin of the Brassington Formation (Mio-Pliocene) of the southern Pennines, Derbyshire, England. *Mercian Geologist*, **8**, 47–62.

——, MORAWIECKA, I. M. & SKAWIŃSKA-WIESER, K. 1996. A Miocene palynoflora preserved by karstic subsidence in Anglesey and the origin of the Menaian Surface. *Geological Magazine*, **133**, 713–719.

WARRINGTON, G. & OWENS, B. 1977. Micropalaeontological biostratigraphy of offshore samples from southern Britain. Report of the Institute of Geological Sciences, 77/7.

WATTS, W. A. 1962. Early Tertiary pollen deposits in Ireland. *Nature*, **1931**, 600.

—— 1985. Quaternary vegetation cycles. *In*: EDWARDS, K. J. & WARREN, W. P. (eds) *The Quaternary History of Ireland*. Academic, London, 153–185.

WILKINSON, G. C. & BOULTER, M. C. 1980. Oligocene pollen and spores from the western part of the British Isles. *Palaeontographica*, **B 175**, 27–83.

WILSON, A. C. 1975. A late-Pliocene marine transgression at St. Erth, Cornwall, and its possible geomorphic significance. *Proceedings of the Ussher Society*, **3**, 289–292.

WOOLDRIDGE, S. W. & LINTON, D. L. 1938, 1955. *Structure, Surface and Drainage in South-East England*. Institute of British Geographers, Publication 10, Philip, London.

Crustal anisotropy and differential uplift: their role in long-term landform development

Y. BATTIAU-QUENEY

Department of Geography, University of Science and Technology of Lille, F-59655 Villeneuve d'Ascq Cedex, France

Abstract: The British Isles belong to a passive continental margin where recently acquired geophysical data have confirmed the great heterogeneity of the crust. In such an area, models based on time-related landform evolution leading to planation fail to explain the long-term development, at least since the North Atlantic began to open. The concept of a morphotectonic system taking into account a continuous input of energy (due to plate tectonics) and a state of stress controlled by pre-existing crustal discontinuities is well adapted to the British area and other similar areas. The spatial units of a morphotectonic system are defined by the crustal framework and do not generally coincide with drainage basins. Regional-scale landforms are not time-related but depend on crustal properties, and denudation rates might strongly differ on either side of main crustal hinges, whatever the proximity of the sea.

It is true to say that long-term landform evolution has been much neglected since the 1970s and that simultaneously the study of processes has become pre-eminent. One reason is probably that most classical landscape models which were proposed at this time were incompatible with the plate tectonics theory that was being introduced into the Earth sciences including the concept of the cycle of erosion and the idea that long-term evolution is time-dependent and leads to planation. In contrast, models or concepts which were previously not widely accepted appeared much more appropriate to new tectonic theories, for example, Hack's (1960) dynamic equilibrium and Wayland's (1933) etched plain (see reviews by Thomas 1994).

In all of these studies of long-term land and landscape development, the role of tectonics has been diversely treated. In Davis' (1899) model, a rapid initial uplift is the main source of energy which is inferred to trigger a cycle of erosion. For Penck (1952), tectonic quiescence is seen to occur with time and King's (1967) model, at least in the first versions, is not fundamentally different even if isostatic rebound is inferred each time a denudation threshold has been reached. In all these models, the morphogenic system is considered to approximate a closed system with declining potential energy and landform evolution is seen as time-dependent and leading to a low-relief surface developed in relation to base-level. In contrast to these models, Gilbert (1877) and Hack (1960, 1975) considered that a 'dynamic equilibrium' exists between rates of denudation and elevation, an idea which was also defended by Ahnert (1970) and subsequently developed by Thomas (1994) as a concept of 'dynamic etchplanation'. Hack's model came too early to take into account the concept of plate tectonics, but it was prescient when it claimed that the geomorphologic system is an open system, more or less independent of time, receiving a continuous input of energy.

Since the advent of these ideas any geomorphologists who have considered long-term landform development have therefore had to integrate plate tectonics and recently acquired geophysical and geological data into models consistent with the superficial geology and geomorphology. This has been particularly the case for passive continental margins (e.g. Battiau-Queney 1978, 1981, 1983, 1991; Birot 1982; Godard 1982; Ollier 1985; Ollier & Marker 1985; Summerfield 1985, 1988; Bishop 1988; Peulvast 1988; Lecoeur 1991) and the 'British margin' deserves a special place in these researches, as it is one of the most intensively studied regions in the world. Despite its geological complexity, it throws light on the links between deep and superficial properties which might help to explain the evolution of other margins, the deep structure of which is less well known. As with most rifted continental margins of the North Atlantic, the British margin lacks a 'great escarpment' standing inland and facing the sea. Instead it exemplifies a type of rifted margin where a complex pre-rifting structure exercises a strong control on the geomorphological development of the subsequent passive margin (Fig. 1). This paper examines the nature of these controls and uses them to comment on

From: SMITH, B. J., WHALLEY, W. B. & WARKE, P. A. (eds) 1999. *Uplift, Erosion and Stability: Perspectives on Long-term Landscape Development.* Geological Society, London, Special Publications, **162**, 65–74. 1-86239-030-4/99/ $15.00 © The Geological Society of London 1999.

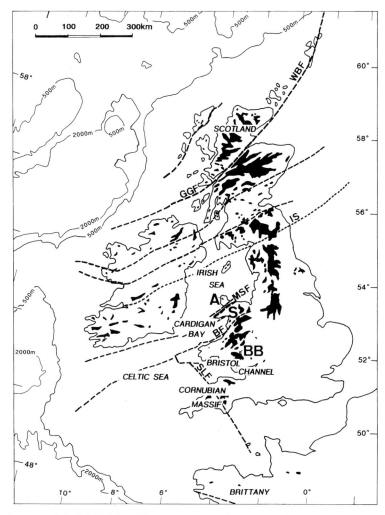

Fig. 1. Location map of the British Isles with some structural features. Black areas indicate areas with elevation greater than 300 m. BF, Bala Fault; GGF, Great Glen Fault; IS, Iapetus suture; MSF, Menai Straits Fault system; SLF, Sticklepath–Lustleigh Fault; WBF, Walls Boundary Fault. **A**, Anglesey–Arfon block; **BB**, Black Mountain–Fforest Fawr–Brecon Beacons scarp; **S**, Snowdonia block.

the validity of landscape development models which do not incorporate the long-term tectonic history of an area.

Anisotropy of the continental crust: a consequence of plate tectonics

Since the 1980s, there has been a rapid growth in our knowledge of the British continental crust, especially thanks to BIRPS (British Institutions Reflection Profiling Syndicate) deep seismic-reflection profiling (Matthews and BIRPS Group 1990; Blundell 1990). Two main aspects have to be considered for the present purpose.

Sharp lateral discontinuities in the crust

The continental crust beneath the British Isles has sharp lateral discontinuities between different structural units (Bamford *et al.* 1978; Blundell 1981; Bott *et al.* 1985; Chadwick *et al.* 1989). The present situation results from a long and complex geological history which started in the Precambrian. Several times, the British area was involved in mobile belts associated with

active margins, subduction or continental collision (cf. Ziegler 1990; Dunning 1992).

Owing to its buoyancy, the continental crust permanently records the successive geological events which have affected it. Reflection signatures of past events are recognized on deep profiles such as major strike-slip faults, palaeo-ocean sutures and thrusts inherited from continental collision. These include the following.

1. The Iapetus suture which is manifested as a north-dipping reflector which intersects the whole crust and the upper mantle (Beamish & Smythe 1986; Klemperer & Matthews 1987; Freeman et al. 1988; Klemperer 1989; Klemperer et al. 1990; Soper et al. 1992).
2. The Great Glen fault, a major strike-slip fault, identified as a nearly vertical reflector which dips from the surface into the lower crust and the upper mantle (Brewer et al. 1983; Lemiszki & Brown 1988).
3. Many faults which crop out at the surface and which are identified on deep profiles by reflectors which dip at relatively low angles to mid-crustal levels, where they merge into a reflective lower crust, e.g. the Bala Fault and the Menai Straits Fault System seen on the SWAT (South West Approach Transverse) and WINCH (Western Isles North Channel) profiles in the Celtic Sea/Cardigan Bay area (BIRPS & ECORS 1986; Le Gall 1991).

More generally, the basement appears as a complex mosaic of different terranes (including the Lewisian foreland, Caledonian orogen and Midlands microcraton), each with specific properties of density, mechanical resistance and heat flow (Bott et al. 1985; Chadwick et al. 1989; Brown & Whelan 1995). Among these, the Cornubian terrane appeared during the Variscan orogeny and remained a prominent massif unaffected by subsidence throughout the Mesozoic (Day et al. 1989) compared to surrounding basins (Celtic Sea basins, Bristol Channel basin, Plymouth basin and others), the development of which was controlled by Variscan or pre-Variscan thrusts and faults.

In such a complex, typically anisotropic crust, stresses are strongly controlled by pre-existing lines of weakness (Butler et al. 1997) and each crustal block is presumed to have its own morphotectonic equilibrium (as defined below). The resultant structural pattern and crustal rheology differ strongly from a basement made of units which could jointly and uniformly react to external stresses.

A lithosphere with a typical vertical structure

The British continental lithosphere usually presents a brittle upper crust, a ductile lower crust and an upper mantle which is brittle at the top but progressively more ductile with depth (Bott 1982; Blundell 1991) (Fig. 2). In the brittle crust, a common feature observed on BIRPS profiles is the reactivation of pre-existing faults (normal faults converted into thrusts and thrusts converted into normal faults) when compression followed extension and vice versa (Chadwick 1986). More generally, inherited structures (Precambrian, Caledonian or Variscan) have often controlled the geometry of recent structures. For example, the Sticklepath–Lustleigh Fault Zone, which is characterized by a series of small pull-apart basins, is a Tertiary sinistral

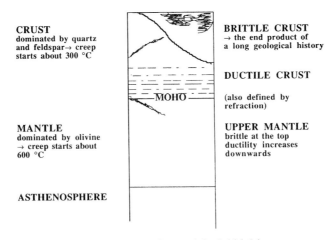

Fig. 2. Typical structure of the lithosphere beneath and around the British Isles.

reactivation of a Variscan dextral strike-slip fault (Holloway & Chadwick 1986). Other examples of ancient fault reactivation have been described in the Cardigan Bay Basin (Turner 1997).

It has already been noted that most faults dip at relatively low angles from the surface into a reflective lower crust. Only a few large-scale strike-slip faults, such as the Walls Boundary Fault, in the Shetland Islands, dip into the lower crust and the upper mantle. This has been interpreted as the northward extension of the Great Glen Fault (McGeary 1989). However in some areas, exemplified by the London–Brabant Platform, no sub-horizontal reflections have been observed in the lower crust. More generally, a 'transparent' lower crust seems to reflect the presence of stable blocks which have a cratonic behaviour.

The role of the reflective ductile lower crust seems to be very important in the tectonic deformation of the whole crust. Sandwiched between the brittle upper crust and topmost mantle, it is a weak zone 'where concentrations of strain might occur and where detachment of upper crustal crystalline nappes can take place' (Brewer & Smythe 1986). At the base of the ductile crust, the Moho might also be actively involved in the post-rift structural evolution of the continental lithosphere (Blundell 1990; Bois 1993).

In contrast to the lower crust, which deforms along a complex system of ductile shears and which is nearly independent of the superficial geology, the brittle crust preserves the signature of previous tectonic events. When stresses are imposed on the continental crust, the brittle and ductile crusts behave differently and in the former, pre-existing fractures, faults and other lines of weakness might be reactivated. Ductile creep, mantle intrusion and partial melting into eclogite have been envisaged in the lower crust (Bois 1993). This could explain why the whole crust is in isostatic equilibrium, despite heterogeneity within the upper crust and a nearly flat Moho. A mechanism of dynamic isostatic compensation due to various processes in the lower crust (Meissner *et al.* 1987; Bois 1993) could play a major role in the development of passive margins.

Morphotectonic systems receiving a continuous input of endogenic energy

A continuous input of energy

As a consequence of plate tectonics, the state of stress along the British margin is due to dynamic processes, including mid-Atlantic ridge push forces, seafloor spreading and lithostatic pressure differences at the transition between continental and oceanic crust, which act continually on the shifting Eurasian plate. There is a continuous input of endogenic energy. The quantity of energy has varied with time according to the rate of the Atlantic Ocean spreading, plate motion and other factors, e.g. the Alpine collision, the Rhine graben rifting and the influence of a mantle plume (Cope 1994; Kent 1995), but pressure has continued to be exerted. Because of this, in our present state of knowledge, the idea of short periods of rapid uplift followed by long phases of stillness cannot be sustained. At any instant of time, the crustal structure results from a combination of past events and normally becomes increasingly heterogeneous as major tectonic events accumulate. The result is a complex mosaic of blocks which are variously shaped and generally quite small. Stresses are closely controlled by pre-existing discontinuities. It is impossible to predict future evolution, but a decline of crustal heterogeneity with time is not to be expected because the current record suggests that all preceding events are not eradicated.

Crustal discontinuities and the origins of major escarpments

The concept of the cycle of erosion, linked with a fluvial system controlled by base-level, implies that landforms evolve towards a low and uniform relief through a predictable sequence of events. It cannot explain why some areas suffer a permanent uplift more or less balanced by erosion (e.g. the Cornubian massif throughout the Mesozoic), contrary to others which were consistently subsiding (e.g. the South Celtic Sea basin). Moreover, fluvial processes and superficial geological features are unable to explain some major escarpments.

One of the best examples is the 500–700 m high escarpment which runs from the Black Mountain to the Fforest Fawr and the Brecon Beacons in South Wales. The scarp front does not follow the strike of Palaeozoic strata and cannot be explained by differential weathering, as the same strata crop out variably at the foot of the scarp, on top of it or on its back slopes. Nor is it a simple fault scarp, since nothing other than transverse faults have ever been recognized in this area. It is important to recognize that the Upper Devonian and Lower Carboniferous strata which cap several summits of this escarpment belong to a marginal facies and must have been deposited very close to the southern edge

of St George's Land; the present relief is completely inverted with respect to the Palaeozoic landscape and a substantial recession of the scarp as implied in a cyclic scheme is unrealistic. In this case, the only explanation for the reversal of the relative topographic position of Mynydd Eppynt and Brecon Beacons is an uplift of the Black Mountain–Brecon Beacons block relative to the Mynydd Eppynt block (Battiau-Queney 1984). A lack of deep seismic profiles in this area means that any sharp crustal discontinuity which might control the supposed hinge has yet to be identified. Nevertheless, an active hinge between two crustal blocks appears to be a hypothesis consistent with all currently available data.

Morphotectonics systems

In view of the difficulty in explaining some major topographic features in terms of surface geology and a cyclic denudation model, a new concept of the morphotectonic system is proposed (Fig. 3). This takes into account crustal anisotropy and the state of stress due to plate tectonics, erosion and isostatic rebound. In a morphotectonic system, regional-scale landforms are not fundamentally time-related but are controlled mainly by crustal properties. The spatial limits of morphotectonic systems are defined by a crustal framework formed from units which have relatively homogeneous rheologic properties. In the British Isles area, thanks to a long and complex geological history, crustal units are generally small. Usually, they do not coincide with drainage basins and one of them, Anglesey–Arfon, is less than $1000 \, km^2$ in extent.

At the surface, adjacent crustal blocks are separated by either gentle slopes or sharp escarpments according to local circumstance. In most cases, it is not easy to define the morphological boundary by marked fault lines. In fact, on the deep seismic-reflection profiles which have been acquired all over the world, some of the major structures which have a good reflection signature in the brittle and ductile crust are more frequently associated with complex tectonic features at the surface rather than a simple fault plane. For example, the Brevard Fault Zone, 'a broad zone of faults and fault slices' on the western edge of the Appalachian Piedmont (Hack 1980, 1982), appears on deep COCORP profiles as a major thrust diverted from a sub-horizontal décollement

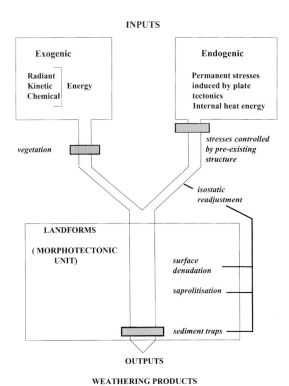

Fig. 3. Scheme of the morphotectonic system.

(Battiau-Queney 1989). The main exceptions come from a few large-scale strike-slip faults which dip steeply into the lower crust, e.g. the Great Glen fault. In each crustal unit, a dynamic morphotectonic equilibrium tends to occur between rock resistance, erosional processes and uplift rate.

The concept of the morphotectonic system in North Wales

The northeast trending 500–800 m high mountain front of Snowdonia, in North Wales, is a prime example of a large topographic feature produced along a major crustal hinge between one stable block (Anglesey–Arfon) and a mobile block (Snowdonia) (Battiau-Queney 1984, 1989). It has long been appreciated that the low elevation of Anglesey–Arfon, almost entirely less than 120 m except for a few hills between 155 and 220 m, as compared with Snowdonia where several summits exceed 1000 m, is not due to the presence of less resistant rocks there. The relative positions of Anglesey and Snowdonia were inverted between the Palaeozoic and late Mesozoic times: Snowdonia was an area of subsidence until at least the early Carboniferous. Uplift started in the Cretaceous but it was at first balanced by denudation, so that the relief remained low. Abundant clasts from Snowdonia rocks found in the Palaeogene sediments of Cardigan Bay mark the onset of rapid uplift and mountain building in this block (Harrison 1971). By contrast, Anglesey remained as a low platform over the same period. The island had already been eroded down to its present topographic level in the Triassic and it has clearly experienced very slow denudation over the last 200 million years (Battiau-Queney 1978).

Such a low platform has long been seen as a marine-cut Pliocene (Greenly 1919) or Pleistocene (Embleton 1964) surface. However, the preservation of Miocene saprolites and organic palaeosols in karstic pipes, at Trwyn y Parc, Cemaes Bay, at sea-level (Walsh *et al.* 1996), does not support these interpretations, even if the Miocene sediments in the pipes have suffered a subsidence of several decametres. The low altitude of the 'Menaian Surface' seems best related to a long period of structural stability and subaerial evolution with pedogenetic processes acting almost continually at the same level.

The existence of a major crustal discontinuity between the two blocks along the northwestern front of the Snowdonian mountain was presumed first from surface analysis (cf. Rast 1969) and later confirmed by geophysical data from the WINCH 4 seismic profile crossing the South Irish Sea, St Georges and Cardigan Bay basins (Brewer *et al.* 1983). On this, the Menai Straits Fault System appears as a complex of reflectors dipping WNW at 25° to a depth of 15–18 km. One of these marks the southern boundary of the St Georges Channel Basin. The Moho reflectors differ on both sides. It must be a major terrane boundary (Gibbons 1987) which dates back to the late Precambrian and which marks the northwestern edge of the early Palaeozoic Welsh basin. There can be little doubt that it was reactivated several times, in the Carboniferous, the Permo-Triassic and, lastly, in Cenozoic times with Palaeogene dykes which are offset sinistrally by as much as 1.5 km across the fault zone (Bevins *et al.* 1996).

In this part of North Wales, the relief contrast (500–800 m) seems out of proportion to the apparent displacement of the faults. The WINCH 4 profile was taken offshore because such surveys are much cheaper than those onshore. Unhappily, it does not give a direct image of the crust beneath the front scarp. Nevertheless, it can be identified that two different types of crust characterize the Anglesey–Arfon, on one side, and the Snowdonia–Harlech Dome on the other. In addition, a complex fault zone (the Menai Straits Fault Zone), which has a very long history of movement, is present in the brittle upper crust at the contact of both blocks, but, by itself, the Menai Straits Fault Zone seems incapable of explaining the 500 to 800 m high mountain front.

The explanation could be that the fault zone in the brittle crust is the superficial, discrete track of a major crustal discontinuity which has mainly developed as a sharp (though relatively flexible) deformation of the basement, accompanied by strong uplift of Snowdonia as a block. The nature of the crust beneath the volcanic Ordovician rocks of the Snowdonia block requires further investigation, but the Lleyn Peninsula earthquake of 19 July 1984 gave some relevant information concerning the deep structural geology and the regional stress field (Turbitt *et al.* 1985). Fracturing took place by a combination of thrust, strike-slip and oblique faulting with hypocentres at a depth of more than 20 km. This suggests that the brittle crust, slightly to the NW of the front, is thicker than is usual in that area. The Menaian block appears to be a small cratonic terrane which has suffered neither significant subsidence nor uplift since at least Upper Palaeozoic times.

Several macroscale landforms suggest a late prolongation of warping along the major structural hinge between both blocks (Battiau-Queney 1984). Escarpments of this type tend to

escape significant back retreat because they are closely linked with deep crustal properties and are constantly rejuvenated.

Some geomorphological consequences

Models of landscape development which emphasize time-dependent landforms and base-level control consider that low platforms, at or near sea-level, are typically recent erosion surfaces produced by denudation processes and back-wearing of higher relief. The model which is proposed here is radically different, whereby low altitude does not imply a Pliocene or Pleistocene age. In the British Isles, as in other high and mid-latitude marginal regions, it is not uncommon to find palaeoforms and saprolites preserved on low coastal platforms: for example, in Wales (Battiau-Queney 1984), eastern Scotland (Hall 1985, 1986), western Scotland (Lecoeur 1989), western Ireland (Coxon & Coxon 1997), southern Sweden (Lidmar-Bergström *et al.* 1997), southern Brittany and Vendée (Barbaroux & Sellier 1989; Chevalier & Borne 1989). The mineralogical properties and depth of several tens of metres of some of these saprolites cannot relate to the present climatic environment, or even to Interglacial periods, but a pre-Pliocene age which is presumed (in South Wales and Scotland) or proved (in North Wales, Brittany and Vendée) is difficult to reconcile with a late Cenozoic age for the surface which bears these saprolites.

One solution to this apparent contradiction could be to imagine a recent exhumation, but in most cases there is no trace of overburden sediments, despite the presence, in some places, of karstic basins which could easily have trapped them. All these difficulties disappear if it is accepted that these low platforms have developed as true etchplains, in Palaeogene times or even before, and survived up to the present time on stable blocks. They have remained close to sea-level since the opening of the Atlantic Ocean and suffered a very slow rate of lowering by a mechanism of dynamic etchplanation (Thomas 1994).

Northwestern Wales is an example of this because it juxtaposes two crustal units which have developed in different morphotectonic systems and suffered contrasting denudation rates, in spite of the fact that both lie very close to sea-level. The landscape contrast between Anglesey and Snowdonia is not time-dependent and not related to two successive cycles of erosion. The higher altitude of Snowdonia clearly results from a morphotectoni equilibrium which is appreciably different to that northwest of a major structural hinge and crustal discontinuity. Its further evolution, at least in the near future, will depend not on fluvial processes, but on the state of stress in the deep crust and mantle.

The morphotectonic system as applied to the British Isles area could be tested against other complex passive continental margins, especially where well documented geological data suggest reactivation of ancient structures, as in southeastern Australia (Foster & Gleadow 1992) or northeastern Brazil (Gallagher & Brown 1997). It should be interesting to apply apatite fission track analysis (AFTA) in the British heterogeneous crust, because it could help to corroborate (or deny) a differentiated uplift history across what are supposed to be major crustal discontinuities, as has been done in northeast Bavaria (Bischoff *et al.* 1993). It will be more difficult to estimate the amount of denudation from the thermal history of outcropping rocks because the geothermal gradient cannot be independently ascertained (Green 1986, 1989; Holliday 1993; Jaspen 1997). However, in southeastern Australia, for example, the use of a constant palaeogeothermal gradient ($25°C\,km^{-1}$ to $30°C\,km^{-1}$), from the beginning of cooling, leads to an apparent conflict between AFTA data and geomorphological studies (Bishop & Goldrick 1998).

Conclusions

Some general conclusions may be proposed.

1. Morphotectonic systems are organized in homogeneous crustal units which do not generally coincide with drainage basins.
2. The British passive margin receives a continuous input of energy directly related to the opening of the North Atlantic. Long-term landform evolution is not a function of time, but depends on energy input, the state of stress and isostatic adjustment inside the morphotectonic system.
3. In a fragmented crustal province like that of the British Isles, a widespread planation surface is very unlikely to appear as long as present dynamic processes act on the shifting Eurasian plate. Estimates of mean denudation rates are worthless if they are not related to a specific crustal unit.
4. Inherited crustal discontinuities strongly influence the present landscape. The concept of the cycle of denudation has to be replaced by that of a cumulative evolution where any previous event contributes to the present system.

References

AHNERT, F. 1970. Functional relationships between denudation relief and uplift in large, mid-latitude drainage basins. *American Journal of Science*, **268**, 243–263.

BAMFORD, D., NUNN, K., PRODEHL, C. & JACOB, B. 1978. LISPB-IV. Crustal structure of northern Britain. *Geophysical Journal of the Royal Astronomical Society*, **54**, 43–60.

BARBAROUX, L. & SELLIER, D. 1989. Les minéraux argileux des altérites et des sédiments pliocènes et quaternaires du Pays Nantais (France). Essai de bilan géomorphodynamique et paléoclimatique. *Géologie de la France*, **1–2**, 285–311.

BATTIAU-QUENEY, Y. 1978. *Contribution à l'étude géomorphologique du Massif gallois*. Thèse d'Etat, Université de Bretagne Occidentale, Honoré Champion, Paris (published in 1980).

—— 1981. Le Pays-de-Galles: un massif ancien de la marge atlantique. *Hommes et Terres du Nord*, 1–12.

—— 1983. Les mouvements verticaux associés aus marges passives: l'exemple du Nord-Est de l'Atlantique. *Physio Géo*, **7**, 21–36.

—— 1984. The pre-glacial evolution of Wales. *Earth Surface Processes and Landforms*, **9**, 229–252.

—— 1989. Constraints from deep crustal structure on long-term landform development of the British Isles and Eastern United States. *Geomorphology*, **2**, 53–70.

—— 1991. Les marges passives. *Bulletin de l'Association de Géographes Français*, **2**, 91–100.

BEAMISH, D. & SMYTHE, D. K. 1986. Geophysical images of the deep crust: the Iapetus suture. *Journal of the Geological Society, London*, **143**, 489–497.

BEVINS, R.E., HORAK, J. M., EVANS, A. D. & MORGAN, R. 1996. Palaeogene dyke swarm, NW Wales: evidence for Cenozoic sinistral fault movement. *Journal of the Geological Society, London*, **2**, 177–180.

BIROT, P. 1982. Quelques réflexions sur l'origine des bourrelets montagneux des marges passives. *Hommes et Terres du Nord*, **3**, 1–8.

BIRPS & ECORS. 1986. Deep seismic reflection profiling between England, France and Ireland. *Journal of the Geological Society, London*, **143**, 45–52.

BISCHOFF, R., SEMMEL, A. & WAGNER, G. A. 1993. Fission-track analysis and geomorphology in the surroundings of the drill site of the German continental deep drilling project (KTB)/Northeast Bavaria. *Zeitschrift für Geomorphologie, N.F., Suppl. Bd.* **92**, 127–143.

BISHOP, P. 1988. The Eastern Highlands of Australia: the evolution of an intra-plate highland belt. *Progress in Physical Geography*, **12**, 159–182.

—— & GOLDRICK, G. 1998. Eastern Australia. *In*: SUMMERFIELD, M. A. (ed.) *Global Tectonics and Geomorphology*. Wiley, Chichester, Chapter 11.

BLUNDELL, D. J. 1981. The nature of the continental crust beneath Britain. *In*: ILLING, L. V. & HOBSON, G. D. (eds) *Petroleum Geology of the Continental Shelf of North-West Europe*. Heyden, London, 58–64.

—— 1990. Seismic images of continental lithosphere. *Journal of the Geological Society, London*, **147**, 895–913.

—— 1991. Some observations on basin evolution and dynamics. *Journal of the Geological Society, London*, **148**, 789–800.

BOIS, C. 1993. Orogenic belts and sedimentary basins. Thoughts on crustal evolution suggested by deep seismic reflection images. *Bulletin de la Société Géologique de France*, **164**, 327–342.

BOTT, M. H. P. 1982. *The Interior of the Earth, its Structure, Constitution and Evolution*, 2nd edition. Elsevier, London.

——, LONG, R. E., GREEN, A. S. P., LEWIS, A. H. J., SINHA, M. C. & STEVENSON, D. L. 1985. Crustal structure south of the Iapetus suture beneath northern England. *Nature*, **314**, 724–727.

BREWER, J. A. & SMYTHE, D. K. 1986. Deep structure of the foreland to the Caledonian orogen, NW Scotland: results of the BIRPS WINCH profile. *Tectonics*, **5**, 171–194.

——, MATTHEWS, D. H., WARNER, M. R., HALL, J., SMYTHE, D. K. & WHITTINGTON, R. J. 1983. BIRPS deep seismic reflection studies of the British Caledonides. *Nature*, **305**, 206–210.

BROWN, C. & WHELAN, J. P. 1995. Terrane boundaries in Ireland inferred from the Irish Magnetotelluric profile and other geophysical data. *Journal of the Geological Society, London*, **154**, 523–534.

BUTLER, R. W. H., HOLDSWORTH, R. E. & LLOYD, G. E. 1997. The role of basement reactivation in continental deformation. *Journal of the Geological Society, London*, **154**, 69–71.

CHADWICK, R. A. 1986. Extension tectonics in the Wessex Basin, southern England. *Journal of the Geological Society, London*, **143**, 465–488.

——, PHARAOH, T. C. & SMITH, N. J. P. 1989. Lower crustal heterogeneity beneath Britain from deep seismic reflection data. *Journal of the Geological Society, London*, **146**, 617–630.

CHEVALIER, M. & BORNE, V. 1989. Remise en question de l'attribution 'Sables rouges pliocènes' aux formations détritiques du domaine du lac de Grand-Lieu (loire-Atlantique). *Géologie de la France*, **12**, 277–284.

COPE, J. C. W. 1994. A latest Cretaceous hotspot and the southeasterly tilt of Britain. *Journal of the Geological Society, London*, **151**, 905–908.

COXON, P. & COXON, C. 1997. A pre-Pliocene or Pliocene land surface in County Galway, Ireland. *In*: WIDDOWSON, M. (ed.) *Palaeosurfaces: Recognition, Reconstruction and Palaeoenvironmental Interpretation*. Geological Society, London, Special Publications, **120**, 37–55.

DAVIS, W. M. 1899. The geographical cycle. *Geographical Journal*, **14**, 481–504.

DAY, G. A., EDWARDS, J. W. F. & HILLIS, R. R. 1989. Influences of Variscan structures off southwest Britain on subsequent phases of extension. *In*: TANKARD, A. J. & BALKWILL, H. R. (eds) *Extensional Tectonics and Stratigraphy of the North Atlantic Margins*. APG, Memoir **46**, 111–129.

DUNNING, F. W. 1992. Structure. In: DUFF, P. MC L. D. & SMITH, A. J. (eds) Geology of England and Wales. Geological Society, London, 523–561.

EMBLETON, C. 1964. The planation surfaces of Arfon and adjacent parts of Anglesey: a re-examination of their age and origin. Transactions of the Institute of British Geographers, **35**, 17–26.

FOSTER, D. A. & GLEADOW, A. J. W. 1992. Reactivated tectonic boundaries and implications for the reconstruction of southeastern Australia and northern Victoria Land, Antarctica. Geology, **20**, 267–270.

FREEMAN, B., KLEMPERER, S. L. & HOBBS, R. W. 1988. The deep structure of northern England and the Iapetus Suture Zone from BIRPS deep seismic reflection profiles. Journal of the Geological Society, London, **145**, 727–740.

GALLAGHER, K. & BROWN, R. 1997. The onshore record of passive margin evolution. Journal of the Geological Society, London, **154**, 451–457.

GIBBONS, W. 1987. Menai Straits fault system: an early Caledonian terrane boundary in North Wales. Geology, **15**, 744–747.

GILBERT, G. K. 1877. Report on the Geology of the Henry Mountains. US Geographical and Geological Survey of the Rocky Mountain Region, Washington, DC.

GODARD, A. (ed.) 1982. Les bourrelets marginaux des hautes latitudes. Bulletin de l'Association de Géographes Français, **489**, 239–259.

GREEN, P. F. 1986. On the thermo-tectonic evolution of Northern England: evidence from fission track analysis. Geological Magazine, **123**, 493–506.

—— 1989. Thermal and tectonic history of the East Midlands shelf (onshore UK) and surrounding regions assessed by apatite fission track analysis. Journal of the Geological Society, London, **146**, 755–774.

GREENLY, E. 1919. The Geology of Anglesey, 2 vols. Memoir of the Geological Survey of Great Britain.

HACK, J. T. 1960. Interpretation of erosional topography in humid temperate regions. American Journal of Science, **258A**, 80–97.

—— 1975. Dynamic equilibrium and landscape evolution. In: MELHORN, W. N. & FLEMAL, R. C. (eds) Theories of Landform Development. Allen & Unwin, London, 87–102.

—— 1980. Rock control and tectonism, their importance in shaping the Appalachian Highlands. United States Geological Survey Professional Paper, **1126B**.

—— 1982. Physiographic divisions and differential uplift in the Piedmont and Blue Ridge. United States Geological Survey Professional Paper, **1265**.

HALL, A. M. 1985. Cenozoic weathering covers in Buchan, Scotland and their significance. Nature, **315**, 392–395.

—— 1986. Deep weathering patterns in north-east Scotland and their geomorphological significance. Zeitschrift für Geomorphologie, N.F., **30**, 407–422.

HARRISON, R. K. 1971. The petrology of the upper Triassic rocks in the Llanbedr (Mochras Farm) borehole. In: WOODLAND, A. W. (ed.) The Lanbedr Borehole. Institute of Geological Science, Report **71/18**, 37–72.

HOLLIDAY, D. W. 1993. Mesozoic cover over northern England: interpretation of apatite fission track data. Journal of the Geological Society, London, **150**, 657–660.

HOLLOWAY, S. & CHADWICK, R. A. 1986. The Sticklepath–Lustleigh fault zone: Tertiary sinistral reactivation of a Variscan dextral strike-slip fault. Journal of the Geological Society, London, **143**, 447–452.

JASPEN, P. 1997. Regional Neogene exhumation of Britain and the western North Sea. Journal of the Geological Society, London, **154**, 239–247.

KENT, R. W. 1995. Magnesian basalts from the Hebrides, Scotland: chemical composition and relationship to the Iceland plume. Journal of the Geological Society, London, **152**, 979–983.

KING, L. C. 1967. Morphology of the Earth, 2nd edition. Oliver & Boyd, Edinburgh.

KLEMPERER, S. L. 1989. Seismic reflection evidence for the location of the Iapetus suture west of Ireland. Journal of the Geological Society, London, **146**, 409–412.

—— & MATTHEWS, D. H. 1987. Iapetus suture located beneath the North Sea by BIRPS deep seismic reflection profiling. Geology, **15**, 195–198.

——, HOBBS, R. W. & FREEMAN, B. 1990. Dating the source of lower crustal reflectivity using BIRPS deep seismic profiles across the Iapetus suture. Tectonophysics, **173**, 445–454.

LECOEUR, C. 1989. La question des altérites profondes dans la région des Hébrides internes (Ecosse occidentale). Zeitschrift für Geomorphologie N.F., Suppl. Bd, **72**, 109–124.

—— 1991. L'évolution morphologique d'une marge fragmentée: la bordure écossaise. Bulletin Association Géographes Français, **2**, 109–116.

LE GALL, B. 1991. Crustal evolutionary model for the Variscides of Ireland and Wales from SWAT seismic data. Journal of the Geological Society, London, **148**, 759–774.

LEMISZKI, P. J. & BROWN, L. D. 1988. Variable crustal structure of strike-slip fault zones as observed on deep seismic reflection profiles. Geological Society of America Bulletin, **100**, 665–676.

LIDMAR-BERGSTRÖM, K., OLSSON, S. & OLVMO, M. 1997. Palaeosurfaces and associated saprolites in southern Sweden. In: WIDDOWSON, M. (ed.) Palaeosurfaces: Recognition, Reconstruction and Palaeoenvironmental Interpretation. Geological Society, London, Special Publication, **120**, 95–124.

MCGEARY, S. 1989. Reflection seismic evidence for a Moho offset beneath the Walls Boundary strike-slip fault. Journal of the Geological Society, London, **146**, 261–269.

MATTHEWS, D. H. & the BIRPS Group. 1990. Progress in BIRPS deep seismic reflection profiling around the British Isles. Tectonophysics, **173**, 387–396.

MEISSNER, R., WEVER, TH. & FLÜH, E. R. 1987. The Moho in Europe – Implications for crustal development. Annales Geophysicae, **5B**, 357–364.

OLLIER, C. D. 1985. Morphotectonics of continental margins with Great Escarpments. In: MORISAWA, M. & HACK, J. (eds) Tectonic Geomorphology. Allen & Unwin, Boston, 1–25.

—— & MARKER, M. E. 1985. The Great Escarpment of Southern Africa. *Zeitschrift für Geomorphologie, Suppl. Bd*, **54**, 37–56.

PENCK, W. 1952. *Morphological Analysis of Landforms*. Macmillan, London.

PEULVAST, J. P. 1988. Mouvements verticaux et genèse du bourrelet Est-groenlandais dans la région de Scoresby Sund. *Physio Géo*, **18**, 87–105.

RAST, N. 1969. The relationship between Ordovician structure and volcanicity in Wales. *In*: WOOD, A. (ed.) *The Pre-Cambrian and Lower Palaeozoic rocks of Wales*. University of Wales, Cardiff, 305–335.

SOPER, N. J., ENGLAND, R. W., SNYDER, D. B. & RYAN, P. D. 1992. The Iapetus suture. *Journal of the Geological Society, London*, **149**, 697–700.

SUMMERFIELD, M. A. 1985. Plate tectonics and landscape development on the African continent. *In*: MORISAWA, M. & HACK, J. T. (eds) *Tectonic Geomorphology*. Allen & Unwin, Boston, 27–51.

—— 1988. Global tectonics and landform development. *Progress in Physical Geography*, **12**, 389–404.

THOMAS, M. 1994. *Geomorphology in the Tropics. A study of weathering and denudation in low latitudes*. Wiley, Chichester.

TURBITT, T., BARKER, E. J., BROWITT, W. A. *et al.* 1985. The North Wales earthquake of 19 July 1984. *Journal of the Geological Society, London*, **142**, 567–571.

TURNER, J. P. 1997. Strike-slip fault reactivation in the Cardigan Bay basin. *Journal of the Geological Society, London*, **154**, 5–8.

WALSH, P., MORAWIECKA, I. & SKAWINSKA-WIESER, K. 1996. A Miocene palynoflora preserved by karstic subsidence in Anglesey and the origin of the Menaian Surface. *Geological Magazine*, **133**, 713–719.

WAYLAND, E. J. 1933. Peneplains and some other erosional platforms. *Annual Report and Bulletin, Protectorate of Uganda Geological Survey*. Department of Mines, **1**, 77–79.

ZIEGLER, P. A. 1990. *Geological Atlas of Western and Central Europe*, 2nd edition. Shell Internationale Petroleum Maatschappij. B.V. 2t.

Pre-Devonian landscape of the Baltic Oil-Shale Basin, NW of the Russian Platform

VÄINO PUURA, REIN VAHER & IGOR TUULING

Institute of Geology, Tallinn Technical University, 7 Estonia Avenue, 10143 Tallinn, Estonia

Abstract: The erosional relief of Ordovician and Silurian deposits in Estonia was developed during the continental period in late Silurian and early–middle Devonian times. The uplift of the area and marine regression were induced by compressional tectonics in the continental interior related to the closure of the Iapetus and Tornqvist Oceans.

In the northern part of the Baltic sedimentary basin (Estonia), on the gentle southerly dipping slope between the Fennoscandian Shield (Finland) and Baltic Syneclise (Latvia), a pre-Devonian, slightly rugged erosional relief with few cuesta was developed. The pre-Devonian erosional landforms – hills, depressions and escarpments reaching 150 m in height – were probably buried under the Devonian deposits and then partly re-exposed by pre-Quaternary erosion. These landforms are described in detail using data from several thousands of cores drilled in the course of oil-shale and phosphorite exploration and mining.

Erosional relief on Ordovician and Silurian deposits in the northwest of the East European Craton was developed during the continental period of exposure in the late Silurian and early–middle Devonian. Uplift of the territory and marine regression were induced by compressional tectonics in the continental interior due to the closure of the external Iapetus Ocean to the northwest (Phillips *et al.* 1976), and the Tornqvist Ocean to the southwest (Ziegler 1990; Schwan 1997).

At the beginning of pre-Devonian erosion, the land surface was of generally low relief due to Ordovician and Silurian near-shore marine sedimentation. Structurally induced local landforms were almost non-existent before the late Caledonian phase of compression. Before the late Silurian–early Devonian phase of deformation, the whole East Baltic area was a shallow, intracontinental seabed or near-shore continental flat with few structural complications in the early Palaeozoic sedimentary cover. The compressional tectonics along the craton margins caused uplift of the region and deformation of the crystalline basement and sedimentary cover in the western corner of the Russian Platform.

East of the present Baltic Sea, at least two areas of different types of structure were formed. In the central part of the Baltic sedimentary basin, i.e. in Latvia, tectonic deformation was most extensive, and fault throws reached 700 m (Suveizdis 1979). After early Devonian erosion, the fault- and dome-related residual escarpments and hills had altitudes of up to 150 m (Polivko & Ulst 1969). In the northern part of the sedimentary basin (Estonia), within the gentle southerly dipping slope between the Fennoscandian Shield (Finland) and Baltic Syneclise (Latvia), a flatter hilly relief developed. The influence of faults with throws of up to 50 m on the formation of the pre-Devonian topography was negligible. These oldest erosional landforms are now partly buried under Devonian deposits and partly re-exposed during late-Phanerozoic erosion. Following the previous preliminary interpretations published in Estonian and Russian (Puura 1974; Puura *et al.* 1987; Tuuling 1988), this paper discusses sedimentary bedrock structure–topography relationships and reconstruction of the pre-Devonian landscape.

Geological and geomorphological setting

The study area is situated in the northwestern part of the Russian Platform of the East European Craton south of the Fennoscandian Shield (Fig. 1). Oil-shale deposits of the Baltic Oil-Shale Basin occur in northeastern Estonia and adjacent areas of northwestern Russia extending from 25°E 59°N to 29°E 59°N (Fig. 2). It is the best known portion of the basin because of extensive exploration for oil shale and phosphorite (over 15000 boreholes in 5000 km^2). The oil-shale basin is located in the sub-surface of a slightly hilly plain south of the coastal cliffs of the Gulf of Finland. The highest point of the plain is 166 m above sea level on the Pandivere Heights (PH; Fig. 2). The Devonian is distributed around the periphery of the bedrock hills south of the Pandivere Heights and between the Ahtme Elevation (AE) and Izhora Heights (IH).

From: SMITH, B. J., WHALLEY, W. B. & WARKE, P. A. (eds) 1999. *Uplift, Erosion and Stability: Perspectives on Long-term Landscape Development.* Geological Society, London, Special Publications, **162**, 75–83. 1-86239-030-4/99/ $15.00 © The Geological Society of London 1999.

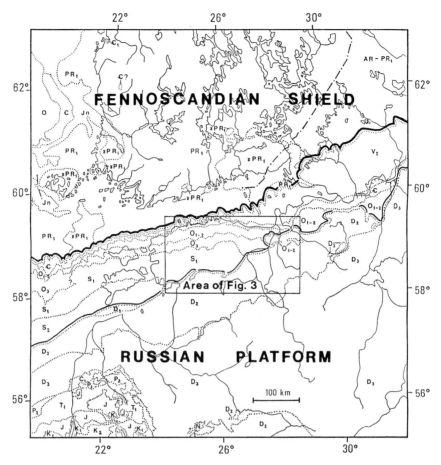

Fig. 1. Distribution of pre-Quaternary rocks. N, Neogene: sand, silt; K_2, Upper Cretaceous: chalk, marl, sandstone, sand, siltstone, silt, clay; K_1, Lower Cretaceous: sand, silt, clay; J, Jurassic: claystone, marl, limestone, sandstone; T_1, Lower Triassic: claystone, marl, limestone, sandstone, siltstone; P_2, Upper Permian: limestone, dolomite, marl; C_1, Lower Carboniferous: sandstone, claystone, marl, dolomite; D_3, Upper Devonian: dolomite, limestone, marl, claystone, sandstone, siltstone, gypsum, anhydrite; D_2, Middle Devonian: sandstone, siltstone, claystone, marl, dolomite, gypsum; D_1, Lower Devonian: sandstone, claystone, marl, dolomite; S_2, Upper Silurian: limestone, dolomite, marl, siltstone; S_1, Lower Silurian: limestone, dolomite, marl, claystone; O_3, Upper Ordovician: limestone, dolomite, marl, sandstone; O_{1-2}, Lower and Middle Ordovician: limestone, dolomite, marl, sandstone; $\mathrm{\small{E}}$, Cambrian: sandstone, siltstone, claystone; V_2, Upper Vendian: sandstone, siltstone, claystone; Jn, Mesoproterozoic, Jotnian: sandstone; γPR_1, Palaeoproterozoic: rapakivi granite, porphyry granite; AR-PR_1, Precambrian crystalline basement.

The strongly disturbed and metamorphosed Precambrian basement is unconformably overlain by up to 400 m of little-deformed and gently tilted sedimentary strata. These comprise Vendian and Cambrian sandstones, siltstones and claystones, and Ordovician and Silurian carbonate rocks (limestone, marl, dolomite). A 1–12 m thick phosphorite layer belongs to the Upper Cambrian and lowermost Ordovician (Tremadocian) (Mens et al. 1993). A 1–3 m thick commercial oil-shale seam of the Estonian and Leningrad deposits lies 9–12 m below the 1.8–2.2 m thick oil-shale layer of the Tapa deposit. Both deposits belong to the upper part of Llandeilo, Middle Ordovician (Puura 1986; Bauert & Puura 1990). The Devonian is represented mostly by clastic rocks (sandstones and siltstones) discordantly overlying Middle Ordovician to Lower Silurian rocks. The solid bedrock is covered by 1–200 m of unconsolidated Quaternary deposits, mostly of glacial origin (Raukas & Teedumäe 1997).

The surface of the basement and the sedimentary bedrock strata have a very gentle

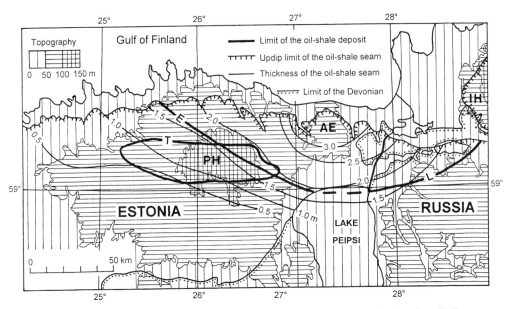

Fig. 2. Location of oil-shale deposits of the Baltic Oil-Shale Basin: E, Estonian; L, Lenigradian; T, Tapa Elevations: AE, Ahtme Elevation; PH, Pandivere Heights; IH, Izhora Heights.

(0.15°–0.19°) southward regional dip (Fig. 3e) (Puura & Mardla 1972). Local variations of dip are well seen on section A–A' (Fig. 3d). Small, dome-like plains-type folds, 1–6 km in diameter and 30–130 m high (Puura 1986) occur occasionally (Fig. 3c). The net result of the folding is a relative local elevation without a corresponding depression, so that there are only anticlines and domes rising above the regional dip. The folds become more pronounced at depth, the lowermost strata wedge out on the flanks of the folds, and there is a thinning of the strata above the crests of the folds.

The narrow, 1–4 km wide, linear zones of disturbance (Puura 1986) divide the area into a number of different-sized blocks. A typical zone of disturbance is represented as a flexure above a basement fault, combined with an anticline on the upthrown side, and with a syncline on the downthrown side. Such folds in the zones of disturbance are very gentle and not high, usually 5–10 m and never greater than 50 m. The uplifts are often called monoclines, because the opposite flank is nothing but a flexure. Anticlinal belts are ascribed mostly to fault movements in the basement because their shape is strongly asymmetrical (Puura & Vaher 1997).

The most prominent topographic features of the bedrock, which can also be observed in the present-day landscape, are the Pandivere Heights (Fig. 3a) where the bedrock surface reaches 132 m (Tavast & Raukas 1982). The bedrock surface of the Ahtme Elevation rises 79 m above sea level and eastward, in northeastern Russia, the bedrock of the Izhora Heights is more than 130 m above sea level. To the east of the Baltic Oil-Shale Basin, in the Narva–Luga Lowland between Ahtme Elevation and Izhora Heights, the pre-Quaternary bedrock surface generally lies between 20–40 m above sea-level (Tuuling 1988).

In the Narva–Luga Lowland, pre-Devonian erosion has cut the Ordovician rocks at the deepest level which is at present some 20 m below sea-level. Devonian deposits cover past erosional depressions and related minor positive landforms of Ordovician rocks. This is the best place to demonstrate the age and origin of erosional landforms.

Pre-Devonian landforms

We have analysed the pre-Devonian landforms of the Baltic Oil-Shale Basin on the basis of three non-reconstructed, present-day contour maps and structural cross-sections because the post-Devonian deformation is negligible. We have taken into consideration that, during and after the Devonian sedimentation, bedrock strata were additionally tilted SSE at an angle of 15' (one-quarter of a degree!).

In the eastern part of the oil-shale basin, where the Devonian rocks are distributed, the pre-

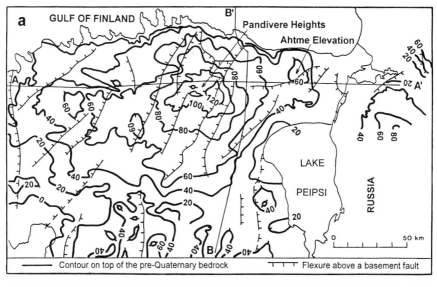

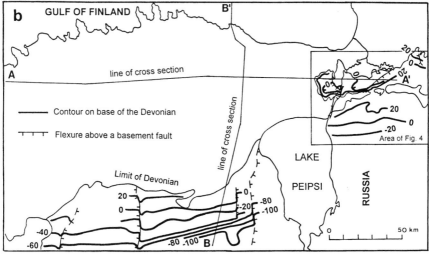

Fig. 3. Topography (**a**) and structure (**b, c**) of the pre-Quaternary bedrock, maps and sections (**d, e**). Contour interval 20 m. Q, Quaternary; D_2, Middle Devonian; S_1, Lower Silurian; O_3, Upper Ordovician; O_2, Middle Ordovician; O_1, Lower Ordovician, ε, Cambrian; V_2, Upper Vendian.

Devonian landscape has been studied in detail (Tuuling 1988). The Narva–Luga Depression can be seen on the base of the Devonian (Fig. 4a), but is missing at the base of the oil-shale seam (Fig. 4b). Consequently, the depression is seen as an erosional landform which was truncated into the Ordovician rocks and subsequently buried under Devonian deposits. This is the best known of the pre-Devonian landforms. Another prominent ancient landform is an escarpment bordering the Narva–Luga Depression to the south (Fig. 4a). In local literature this is referred to as the 'Wesenberg Escarpment' or 'Wesenberg Klint'. This escarpment is also missing at the base of the oil-shale seam (Fig. 4b). It is an erosion scarp covered by Devonian deposits. The E–W striking escarpment is located between the Ahtme Elevation and Izhora Heights, and has its highest section at the central part of the Narva–Luga Depression. In the centre of the depression, the base of the Devonian lies about 20 m below present-day sea-level. The steep, cliff-like slope is 30–45 m high, but the northern slope of the depression is

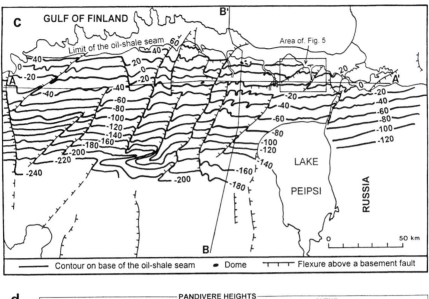

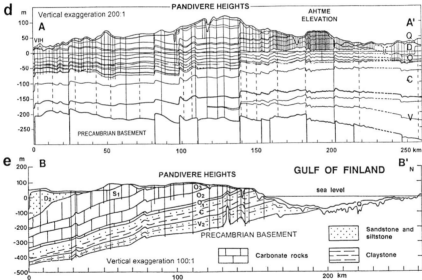

Fig. 3. (*continued*)

very gentle. At the eastern end of the escarpment, the depression opens to the south (Tuuling 1988; Fig. 4a) probably indicating the outflow of the old drainage system. The southern edge of the depression can be distinguished on the geological maps as a sub-surface, pre-Quaternary outcrop area of Ordovician rocks within the main Devonian field. South of the escarpment small hills and hollows are present in the pre-Devonian topography.

Before Devonian sedimentation, most probably in the late Silurian, the Ordovician and Silurian carbonate rocks were deformed by a system of disturbances (Fig. 3). However, they have no substantial influence on the position of the residual heights, or erosional depressions. On the contrary, the disturbed zones cut diagonally across elevations and depressions hardly ever controlling ancient topographic features. In northeastern Estonia only one, the Viivikonna Anticlinal Zone (Puura *et al.* 1987; Fig. 5), of eighteen well studied zones of disturbance, is expressed in the bedrock topography (Raukas *et al.* 1988).

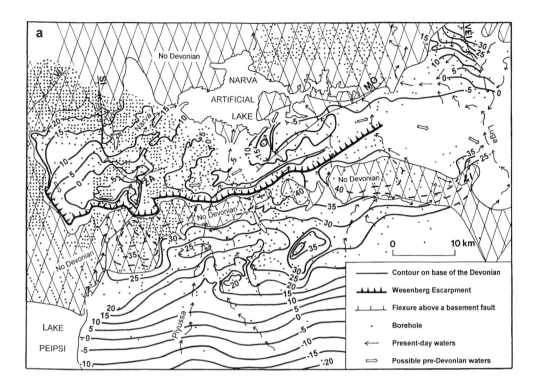
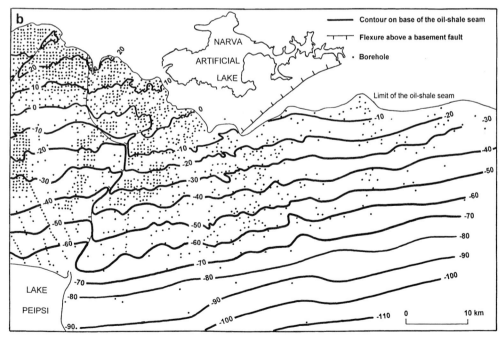

Fig. 4. Bedrock structure in the eastern part of the Baltic oil-shale basin. (a) Base of the Devonian (Tuuling 1988). Contour interval 5 m. (b) Base of the oil-shale seam, Llandeilo. Contour interval 10 m.

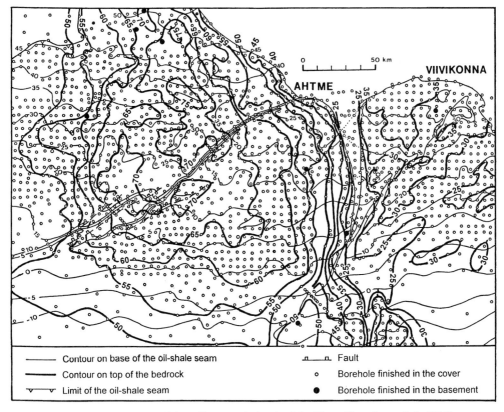

Fig. 5. Structure and topography of the sedimentary bedrock of the Ahtme Elevation and the Viivikonna Anticlinal Zone, northeastern Estonia (Raukas & Teedumäe 1997). Contour interval 5 m.

The best example of the typical disagreement between a medium-sized bedrock surface feature and structural features of the sedimentary bedrock is the Ahtme Elevation in northeastern Estonia (Fig. 5). Against a background of the homoclinal dip of the oil-shale seam, the Ahtme Zone of disturbance is distinctly observable with its total vertical displacement of up to 18 m. This zone diagonally intersects the elevation but is not expressed on the bedrock surface. There are no tectonic disturbances following the slopes of the elevation. The sedimentary bedrock structure–topography relationships thus show that the Ahtme Elevation is not due to horst activity, but is of erosional origin. The bedrock surface on top of the hill is at present about 90 m above the level of the base of the Narva–Luga Depression (Fig. 3d). Thus, we conclude that the Ahtme Elevation was part of the pre-Devonian landscape.

The largest feature, the Pandivere Heights, is intersected by six zones of disturbance trending mainly NE–SW, with steeper-dipping flanks mostly to the NW. They divide the area into several blocks with a throw relative to each other of 10–20 m which are not related to bedrock topographic features (Fig. 3a). Thus, Pandivere Heights is also an erosional remnant. In the centre of the Pandivere Heights, the bedrock surface is c. 130 m above sea level (Fig. 3), hence the pre-Devonian Pandivere Heights rose at least 150 m above the Narva–Luga Depression.

During mid-Devonian sedimentation, the landforms were at least of the same altitudes as they are at present and were probably even higher. The age of the depression between the Pandivere Heights and Ahtme Elevation cannot be determined. Probably this depression was considerably deepened during Quaternary glaciations.

Discussion and conclusions

In the study area, the sub-surface or exposed pre-Devonian erosional landforms are composed mostly of Middle to Upper Ordovician carbonate rocks and Silurian rocks have survived

only in the southern part of the area. The Ordovician is represented by limestones and calcareous mudstones that are occasionally dolomitized. Originally, at the beginning of the late Silurian–early Devonian erosion, nearshore Upper Ordovician and, possibly, Lower Silurian carbonate rocks formed a near-coastal plain (Männil 1966; Kaljo 1970). During the late Silurian compression, the area was uplifted, slightly tilted (not more than 0.1°) in a SSE direction, and gently faulted. Surficial karst processes began after the marine regression and traces of solution of carbonates have been found in south Estonia where erosion was rather weak. Karst forms have been documented at a depth of up to 50 m in the oil-shale basin in northeastern Estonia (Gazizov 1971; Heinsalu & Andra 1975). Lithological alternation of hard limestones and soft calcareous mudstones in the carbonate sequence of the gently tilted regional structure favoured the development of escarpments and residual hills and heights.

The uplift-generated erosion in the lithologically alternating carbonate outcrop area, and on a gentle SSE slope, produced a system of elevations and depressions. Conforming with the strike of Ordovician strata, an E–W orientated cuesta developed south of the Narva–Luga Depression. The formation of this type of relief was poorly influenced by tectonic disturbance of the carbonate rocks; indeed, erosion even smoothed the slight tectonically induced ruggedness.

The scarcity of direct structural control appears to be a specific feature of the Baltic Oil-Shale Basin. Compared to the pre-Devonian topography developed in the central part of the Baltic Syneclise (Latvia), substantial differences occur here. These differences may be dependent on the different geological history. In Latvia, the Silurian and Devonian are separated by a much shorter hiatus, during which tectonically induced landforms were substantially higher than in Estonia due to upthrows of as much as 700 m. In Latvia the upthrown structures were eroded downwards, but still remain as forms of up to 150 m high (Polivko & Ulst 1969). Besides anticlinal folds, reverse faults also remain as topographic escarpments and are clear examples of direct structural control over the subsurface pre-Devonian landforms. In southern Estonia areas of the pre-Devonian surface are gently dissected (Puura 1974).

Taking the affinities of Devonian strata into consideration, it is supposed that the heights in northern Estonia were also buried under Devonian sediments. During the late-Phanerozoic, Devonian sediments were eroded and the heights re-exposed. The hardness of Ordovician and Silurian carbonate rocks, compared to the soft sandstones, siltstones and calcareous mudstones of the Devonian, was the reason for the re-exposure of the pre-Devonian landforms. Karst developed in Ordovician and Silurian carbonate rocks during the pre-Devonian continental period and these old karst forms and infillings (clayey and argillaceous sediments in caves) have survived under the overlying Devonian deposits, and in the vicinity of the present Devonian outcrop area. In contrast, in western Estonia and elsewhere, there are large areas of Ordovician and Silurian outcrops, where, due to the latest erosion, no evidence of pre-Devonian karst has been preserved.

To summarize, at least three different topographic types of the pre-Devonian erosional surface can be distinguished in the Baltic region. They are well expressed in surface beneath Devonian sediments but areally in the sub-Quaternary surface as re-exposed, more or less fully preserved remnants:

1. Hills and cuestas in homoclinal areas of the northern marginal parts of the Baltic Palaeozoic sedimentary basin (northern Estonia);
2. gentle hilly plains in the same situation (Estonia);
3. structurally controlled hills and escarpments in the internal part of the Baltic basin (Latvia).

The authors wish to thank the reviewers and editors for helpful comments and recommendations which considerably improved this manuscript. This study was supported by Estonian Science Foundation Grant 2191.

References

BAUERT, H. & PUURA, V. 1990. Geology of the Baltic Oil Shale Basin. *In*: KALJO, D. & NESTOR, H. (eds) *Field Meeting, Estonia, 1990. An Excursion Guidebook*. Tallinn, 40–45.

GAZIZOV, M. S. 1971. *Karst and its Influence on Mining*. Nauka, Moscow (in Russian).

HEINSALU, Ü. & ANDRA, H. 1975. *Jointing in the Estonian oil shale basin and geophysical research methods for its study*. Valgus, Tallinn (in Russian, summary in English).

KALJO, D. (ed.) 1970. *The Silurian of Estonia*. Valgus, Tallinn (in Russian, summary in English).

MÄNNIL, R. 1966. *The History of the Baltic Basin in Ordovician*. Valgus, Tallinn (in Russian, summary in English).

MENS, K., VIIRA, V., PAALITS, I. & PUURA, I. 1993. Upper Cambrian biostratigraphy of Estonia. *Proceedings of the Estonian Academy of Sciences Geology*, **42**, 4, 148–159.

PHILLIPS, W. E. A., STILLMANN, C. J. & MURPHY, T. 1976. A Caledonian plate tectonic model. *Journal of the Geological Society of London*, **132**, 579–609.

POLIVKO, I. & ULST, R. 1969. Geological situation in the Baltic area at the end of the Silurian and the beginning of the Devonian. *In: Voprosy regional'noi geologii Pribaltiki i Belorussii*. Zinatne, Riga, 171–182 (in Russian, summary in English).

PUURA, V. 1974. *The structure of the southern slope of the Baltic Shield*. Thesis, Institute of Geology of the Academy of Sciences of the Estonian SSR, Tallinn (in Russian).

—— (ed.) 1986. *Geology of the Kukersite-Bearing Beds of the Baltic Oil Shale Basin*. Valgus, Tallinn (in Russian, summary in English).

—— & MARDLA, A. 1972. The structural dissection of the sedimentary cover of Estonia. *Proceedings of the Academy of Sciences of the Estonian SSR Chemistry Geology*, **21**, 1, 71–77.

—— & VAHER, R. 1997. Linear tectonic disturbances of the basement and the cover, Estonian Homocline. *In: Extended Abstract Book. 59th EAGE Conference and Technical Exhibition*, Geneva, Switzerland, 26–30 May 1997. Petroleum Division, European Association of Geoscientists & Engineers, 503.

——, TAVAST, E. & VAHER, R. 1987. Bedrock structure and topography at Kurtna. *In: Kurtna järvestiku looduslik seisund ja selle areng. 3.–5. November 1986. a.toimunud ametkondadevahelise nõupidamise kogumik*. Valgus, Tallinn, 15–24.

RAUKAS, A. & TEEDUMÄE, A. (eds) 1997. *Geology and Mineral Resources of Estonia*. Estonian Academy, Tallinn.

——, TAVAST, E. & VAHER, R. 1988. Some structural and geomorphological aspects of recent crustal movements in Estonia. *In: BOULANGER, Yu. D., HOLTEDAHL, S. & VYSKOCIL, P. (eds) Recent Crustal Movements. Journal of Geodynamics*, **10**, 295–300.

SCHWAN, W. 1997. Europe. *In: MOORES, E. M. & FAIRBRIDGE, R. W. (eds) Encyclopedia of European and Asian Regional Geology*. Chapman & Hall, London, 201–227.

SUVEIZDIS, P. 1979. *Baltic Tectonics*. Mokslas, Vilnius (in Russian, summary in English).

TAVAST, E. & RAUKAS, A. 1982. *The Bedrock Relief of Estonia*. Valgus, Tallinn (in Russian, summary in English).

TUULING, I. 1988. Pre-Devonian and pre-Quaternary relief in the eastern part of the Baltic phosphorite-oil shale basin (Luuga-Narva lowland). *Proceedings of the Academy of Sciences of the Estonian SSR Geology*, **37**, 4, 145–152 (in Russian, summary in English).

ZIEGLER, P. A. 1990. *Geological Atlas of Western and Central Europe*. Shell International Petroleum, Maatschappij B., The Hague.

Uplift histories revealed by landforms of the Scandinavian domes

KARNA LIDMAR-BERGSTRÖM

Department of Physical Geography, Stockholm University, S-10691 Stockholm, Sweden

Abstract: It has long been recognized that the Scandes (Scandinavian Mountains) consist of one dome in southern Norway and one in northern Scandinavia. In this paper a low dome is also identified in southern Sweden. The latter is associated with remains of sedimentary covers of different ages, which makes it possible to date different topographies with associated saprolites and to understand the denudational and uplift history of the dome.
  The eastern flanks of the two domes in the Scandes show distinctly different topographies. The flank of the southern dome is characterized by an undulating hilly relief similar to exhumed Cretaceous surfaces in south Sweden, while the flank of the northern dome is characterized by a stepped sequence of plains with residual hills similar to Tertiary surfaces in the south. The difference in relief probably indicates different uplift histories for the two domes.

A Tertiary uplift of Scandinavia was discussed by geomorphologists at the beginning of this century (Reusch 1901; Ahlmann 1919) and lately a renewed interest has evolved from geologists working offshore (Doré & Jensen 1996; Riis 1996; Stuevold & Eldholm 1996) and fission track studies (Rohrman *et al.* 1995). In this paper the origin and distribution of some major landforms will be used to elucidate the uplift history of Scandinavia.

Cratonic areas exhibit a great variety of landforms, which can provide many clues to Phanerozoic denudation and tectonics. The long-term evolution of landforms was one of the main themes of geomorphic research until the 1950s in which denudation chronology was focused on correlations between morphological steps and changing sea-levels (Baulig 1935). However, the landforms of shield areas show many other features which may be used to interpret denudational history. Fairbridge & Finkl (1980), for example, demonstrated the importance of stacked veneers in Australia which, over long periods of time, protected the basement surface from denudation. Remnants of old saprolites and cover rocks also show that bedrock surfaces in many parts of Fennoscandia are of etch character (cf. Thomas 1994; Migon 1997) and that large areas comprise very old surfaces exhumed from Palaeozoic or Mesozoic covers (Rudberg 1954; Mattsson 1962; Gjessing 1967; Lidmar-Bergström 1982, 1995, 1996; Lidmar-Bergström *et al.* 1997). In this paper the Tertiary uplift of Scandinavia will be discussed in terms of characteristic landform assemblages within the Precambrian rocks on the eastern flanks of the Scandes (the Scandinavian Mountains) and their different histories used to correlate development across the area.

The forms of the bedrock surface in Fennoscandia are comparatively well known and have long been a focus for detailed analysis. The first classification of Swedish bedrock forms was made by De Geer (1910, 1913, 1926), and Rudberg (1960) broadened the analysis to all of Fennoscandia. However in Norway, early emphasis was on differences between the high plains of south Norway, labelled the Palaeic surface by Reusch (1901), and deeply incised valleys.

The sub-Cambrian peneplain

The bedrock of Scandinavia consists of a Precambrian basement in the east and Caledonian basement in the west. It is now generally accepted that all of the Precambrian shield had a cover of Palaeozoic rocks (Männil 1966; Hjort & Sundquist 1979; Jaeger 1984; Zeck *et al.* 1988) and remnants of an Upper Vendian–Lower Palaeozoic cover can be seen in south Sweden, the Baltic, the Gulf of Bothnia, and along the front of the Caledonian nappes. In parts of southern and eastern Sweden the present land surface is an exhumed, extremely flat sub-Cambrian peneplain (Högbom 1910; Rudberg 1954; Lidmar-Bergström 1996) which in parts of south Sweden is well preserved up to 300 m (Lidmar-Bergström 1996). Small areas of the sub-Cambrian surface also occur at about 1300 m at Hardangervidda, southern Norway (Schipull 1974) and it is evident that diastrophic events have affected this surface making it an important reference surface for morphotectonic interpretations (Lidmar-Bergström 1996).

The three domes of Scandinavia

Holtedahl (1953) described Scandinavia as a tilted tectonic block and pointed to similarities

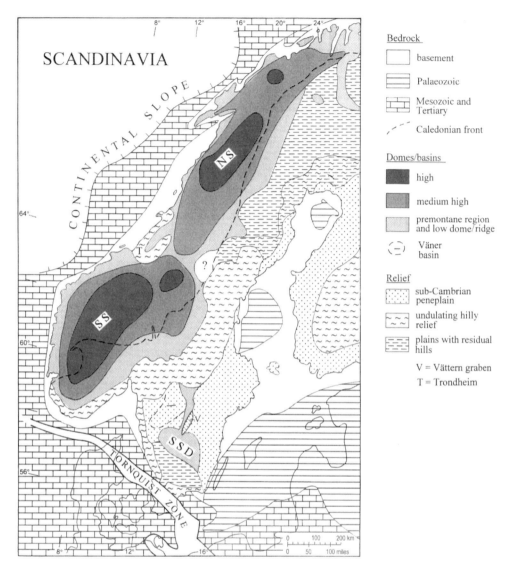

Fig. 1. Scandinavia consists of two high domes in the west, the Northern Scandes (NS) and the Southern Scandes (SS) and a low dome in the south, the South Swedish Dome (SSD). The different relief types on the flanks and their importance for interpretation of uplift history are discussed in the text.

with Baffin Island, which he saw as a counterpart on the other side of Greenland. However, Gjessing (1967) constructed a contour map of a Scandinavian envelope surface, which showed that the Scandes are separated into two high areas which more closely resemble crustal upwarps or domes than tilted tectonic blocks. A third dome was recognized in southern Sweden (Lidmar-Bergström 1988, 1991, 1993). These three Scandinavian domes, namely the South Swedish Dome (SSD), the Southern Scandes (SS) with a sub-area in the northeast, and the Northern Scandes (NS) with a second dome maximum northeast of Lofoten, are shown in Fig. 1. The SSD reaches 377 m, the SS 2452 m (sub-area 1796 m), and the NS 2111 m (the second maximum 1833 m). From the SSD a central high extends northwards and reaches 400 m in its northern part. Three crustal basins can also be identified, two in the Gulf of Bothnia and the Väner Basin to the north of the SSD (Lidmar-Bergström 1996).

The South Swedish Dome – a key area

The low elevation of the SSD and the covers of different ages lying directly on the basement (Cambrian in the north and east, Jurassic and Cretaceous in the south and west) make it a key area for understanding relief evolution in Fennoscandia. On its northern and eastern flanks, the sub-Cambrian peneplain can be traced up onto the summits. In the southwest, the basement is exhumed from below Jurassic and mainly Cretaceous covers. Here the sub-Cambrian, flat bedrock surface was exposed under warm humid climatic conditions during the Mesozoic. Deep weathering penetrated along fracture zones and, depending on the duration of the exposure, the sub-Cambrian flat surface was transformed into landscapes with joint-aligned valleys or undulating hilly relief (Lidmar-Bergström 1995). Remnants of mature kaolinitic saprolites are preserved in these areas and can be up to 60 m thick (Lidmar-Bergström 1989; Lidmar-Bergström et al. 1997). The now-exhumed sub-Mesozoic etchsurfaces extend from below the covers up to 100–150 m. Here they are cut off from above by plains with residual hills, which were developed during the Tertiary from exposed parts of the Mesozoic etchsurface (Lidmar-Bergström 1995, 1996). Deep weathering does not seem to have been significant during this period and the end result of Tertiary denudation was plains with residual hills. Thus, there are three types of relief with distinctly different ages (Fig. 2): the extremely flat sub-Cambrian peneplain, sub-Mesozoic irregular etchsurfaces, and Tertiary plains with residual hills.

It is assumed that uplift causes erosion of cover rocks and as the sub-Cambrian peneplain is well preserved up to 300 m in parts of the SSD, the last rise of the dome must have occurred after the Mesozoic deep weathering event at some time during the Tertiary (Lidmar-Bergström 1991, 1993). As a result, both the sub-Cambrian and sub-Mesozoic surfaces are inclined, while the Tertiary plains with residual hills are almost horizontal. The post-uplift development of the plains with residual hills must have occurred before the onset of the cold climate conditions in the Pliocene.

Sub-Mesozoic etchsurfaces

Jurassic and Cretaceous covers are encountered directly on the basement in and along southwestern Sweden and along the coast of Norway (Fig. 1), with a few inliers close to the coast. The present landsurface on the lower parts of the southwestern flank of the SSD has been identified as an exhumed sub-Jurassic and sub-Cretaceous etchsurface characterized by an

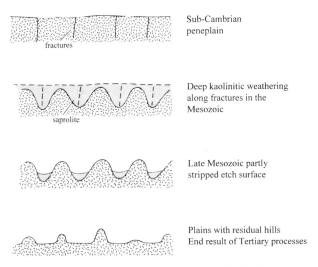

Fig. 2. There are three basic relief types within the Fennoscandian shield. (1) The exhumed and extremely flat sub-Cambrian peneplain, which (together with sub-Vendian and sub-Ordovician facets) has been the starting surface for all relief within the shield. (2) The exhumed sub-Mesozoic etchsurfaces, characterized by an undulating hilly relief and remnants of a kaolinitic saprolite and Mesozoic cover rocks. (3) Plains with residual hills, the end result of Tertiary surface denudation.

undulating hilly relief. This type of relief can be followed northwards into central Sweden and southern Norway (Rudberg 1960) where it is tentatively interpreted as exhumed sub-Mesozoic relief (Reusch 1903; Lidmar-Bergström 1995, 1996). A major erosional event has been indicated by apatite fission track (AFT) analysis in the southern part of the eastern flank of the SS during the Triassic and Jurassic and in its northern part during the Cretaceous (Rohrman 1995). This supports the interpretation of basement exposure during the Mesozoic. Some areas along the Norwegian coast, particularly around Trondheim, have been assigned to the same relief type by Rudberg (1960), and recent findings of Jurassic rocks (Bøe & Bjerkli 1989) in this area strengthen the view that the relief is sub-Mesozoic here (Fig. 1).

The Scandes and their eastern flanks

The upper parts of the two domes in the Scandes are characterized by palaeoplains and palaeovalleys (Gjessing 1967) at heights between about 800 and 2000 m. Both domes are much dissected by valleys, but the southern dome retains a larger intact part in its centre (Ahlmann 1919). The premontane region is characterized by a topography of pronounced valleys, which are more deeply incised on the western sides of the domes, where they show similarities with so-called 'Great Escarpments' (cf. Ollier 1982), particularly along the southern dome.

On the eastern flank of the northern dome the valleys can be included in a system of multicycle relief from the mountains in the west to the coast in the east (Rudberg 1954). East of the premontane region the steps are wide and consist of plains with residual hills (Rudberg 1954, 1960) identified as the Muddus plains by Wråk (1908) and Lidmar-Bergström (1995, 1996). These plains and residual hills are thought to have formed under comparatively dry climates in the Tertiary (Rudberg 1954) and probably in the Early Tertiary (Lidmar-Bergström 1996). This dating relies on the occurrence of redeposited marine Eocene diatoms in Quaternary strata (Cleve-Euler 1941; Hirvas & Tynni 1976). Denudation occurred under stable tectonic conditions and development probably continued into the Late Tertiary (Lidmar-Bergström 1996).

Conversely, the eastern flank of the southern dome is characterized by undulating hilly relief, tentatively interpreted as an exhumed sub-Cretaceous etchsurface. Here it is impossible to map stepped surfaces (Lidmar-Bergström 1996). These differences in morphology between the eastern flanks of NS and SS might reflect a differential Tertiary doming of Scandinavia.

Discussion

The Northern Scandes

An early attempt to ascertain the relief chronology of northern Scandinavia was made by Wråk (1908), who noted an old divide close to the Norwegian coast in the northwest and the inclination of summit surfaces to the southeast. He considered the main summit surface, the Tuipal Surface, to be over 10 Ma old and therefore at least of Miocene age and proposed an area of 'lost' land west of Norway. Conversely, Rudberg (1954) identified almost horizontal summit surfaces further to the south and concluded that uplift must have been achieved by peripherally placed faults and flexures.

As the NS must have been uplifted before the development of the Muddus Plains and its residual hills, a main uplift phase was tentatively dated by Lidmar-Bergström (1996) to the Late Cretaceous/Palaeocene and by Riis (1996), who concluded that a Neogene rise had also affected coastal areas from Lofoten northwards. This may have significance in explaining the differences between Wråk and Rudberg concerning the tilt of the highest surfaces within the NS. The upper surfaces observed by Rudberg are close to the centre of the main dome, while Wråk's research area extended southeastwards from the coastal area in northernmost Norway, which may have been more seriously affected by the Neogene uplift.

The Southern Scandes

The SS are characterized by high plains at different elevations. The outer margins of these plains are dissected by deeply incised valleys. Early geomorphologists suggested that an old surface had been uplifted in the Late Tertiary, prompting incision (Reusch 1901; Ahlmann 1919). This view was also held by Peulvast (1985) and supported by studies of the sedimentary record in Denmark (Spjeldnaes 1975) and offshore Norway (Rundberg & Smalley 1989) and the interpretation of AFT data (Rohrman et al. 1995).

A late uplift could explain why plains with residual hills did not develop on the eastern flank during the Tertiary (Lidmar-Bergström 1996). Instead, the undulating hilly relief was probably protected by a Cretaceous cover. Any

late uplift would have caused erosion of the cover rocks and old saprolites as well as valley incision as the climate at the time was cool and wet.

It is not totally clear from AFT modelling whether the Mesozoic erosional event exposed the sub-Cambrian surface (Cederbom 1997). However, it is difficult to explain the present relief, which has the character of a hilly etch-surface, if the basement was not exposed until the Late Tertiary. Parts of the sub-Cambrian peneplain within the SSD, which were exposed late, still retain the character of an extremely flat surface.

Different uplift histories?

Wråk (1908) correlated the summit surfaces of the NS with those of the SS and thus regarded the Scandes as a single entity with a common tectonic and erosional development, a view shared by Holtedahl (1953). However, it now seems that there is evidence which points to a complex uplift history. The Northern Scandes are, for example, more dissected than the Southern Scandes and the plains and residual hills east of the Northern Scandes contrast with the undulating hilly relief east of the Southern Scandes. Riis (1996) also suggested a correlation between the Tertiary Muddus Plains in the north with the tentatively identified main Palaeic surface on the SS. The correlation by Riis (1996) was based solely on elevation data without any confirmatory morphological analysis but is worthy of further exploration.

The Tertiary uplift of Scandinavia was also discussed during the 1980s by Peulvast (1985, 1986) and, more recently, interest in the uplift history has been fuelled by studies of offshore geology, although it is clear that interpretations differ (Stuevold & Eldholm 1996). Most authors treat Scandinavia as a single unit, although often two uplift phases are suggested, one in the Cretaceous/Palaeogene and one in the Neogene. However, views range from Rundberg (1990), who identified six uplift phases, to Jensen & Schmidt (1992) and Doré & Jensen (1996) who proposed that all uplift occurred in the Neogene.

Approximate calculations of the amounts of Neogene and Palaeogene uplift for different parts of Fennoscandia (Riis 1996) identified Late Cretaceous and Palaeogene uplift as the difference between a postulated Mesozoic peneplain (the summit surface) and a Palaeogene surface. The latter has been mapped in northern Sweden (Rudberg 1954; Lidmar-Bergström 1996), but no corresponding mapping exists for southern Norway. Until palaeosurfaces within the Scandes are more precisely identified and mapped, any such calculations must remain tentative.

Similar problems occur in southern Sweden regarding the interpretation by Riis (1996) of the South Swedish Dome and its palaeosurfaces: he calculated an uplift for this area of over 1000 m in the Plio-Pleistocene with erosion of cover rocks and exhumation of a sub-Cretaceous basement surface over areas totally lacking any sub-Cretaceous features. This is in contrast to other studies that have suggested reshaping of the landscape during the Tertiary with the formation of plains with residual hills (Lidmar-Bergström 1982, 1988, 1995, 1996) (Fig. 1). There has only been very limited uplift along the western rim of these plains after their formation (Lidmar-Bergström 1991, 1993, 1996) in the order of 100–150 m in the south, increasing in amount to the north. The uplift histories of the SS and the SSD also appear to be different with differential uplift of the sub-Cambrian peneplain of about 1000 m between the two areas (cf. above).

Conclusions and implications

It is possible to identify characteristic relief types of different age and origin over basement rocks but hitherto these have only been elaborated for restricted areas. Relief can give indications of when basement areas were exposed and covered and thus can be used to roughly date uplift events. In this respect it is an important basis for comparison with results from AFT modelling.

The different relief types on the eastern flanks of the two Scandinavian domes support the idea of different uplift histories for the north and south of the area with a main phase of uplift in the Early Palaeogene in the north and a major uplift in the Neogene in the south. Future correlations between surfaces and valley generations for the two domes will hopefully provide a better base for calculations of amounts of uplift.

This use of landform interpretation to provide relative chronologies for landscape development is an important traditional task for geomorphologists and has previously been combined with analysis of the stratigraphic record. In future it is likely that apatite fission track analysis will become a third and increasingly significant element in the elucidation of the tectonic and denudational history of Scandinavia and similar basement shield areas.

I wish to thank my colleagues in Stockholm, Johan Kleman and Arjen Stroenven, for critical reading of the manuscript and the referees for their valuable comments and suggestions. The study was supported by a grant awarded by the Swedish Natural Science Research Council.

References

AHLMANN, H. W. 1919. Geomorphological studies in Norway. *Geografiska Annaler*, **1–2**, 1–210.

BAULIG, H. 1935. *The changing sea level*. Institute of British Geographers, Publication **3**, 1–46.

BØE, R. & BJERKLI, K. 1989. Mesozoic sedimentary rocks in Edøyfjorden and Beistad-fjorden, central Norway: implications for the structural history of the Møre-Trøndelag fault zone. *Marine Geology*, **87**, 287–299.

CEDERBOM, C. 1997. *Fission track thermochronology applied to Phanerozoic thermotectonic events in central and southern Sweden*. Thesis for Licentiate degree, Earth Sciences Centre, Göteborg.

CLEVE-EULER, A. 1941. Alttertiäre Diatomeen und Silicoflagellaten im inneren Schwedens. *Palaeontographica*, **92A**, 165–208.

DE GEER, S. 1910. *Map of landforms in the surroundings of the great Swedish lakes. Scale 1:500,000. With explanation*. Sveriges Geologiska Undersökning. Serie Ba, **7**.

—— 1913. *Beskrivning till översiktskarta över södra Sveriges landformer*. Sveriges Geologiska Undersökning, Serie Ba, **9**.

—— 1926. Norra Sveriges landformsregioner. *Geografiska Annaler*, **8**, 125–134.

DORÉ, A. G. & JENSEN, L. N. 1996. The impact of late Cenozoic uplift and erosion on hydrocarbon exploration: offshore Norway and some other uplifted basins. *Global and Planetary Change*, **12**, 415–436.

FAIRBRIDGE, R. W. & FINKL, C. W. Jr. 1980. Cratonic erosional unconformities and peneplains. *Journal of Geology*, **88**, 69–86.

GJESSING, J. 1967. Norway's paleic surface. *Norsk Geografisk Tidsskrift*, **21**, 69–132.

HIRVAS, H. & TYNNI, R. 1976. Tertiary clay deposits at Savukoski, Finnish Lappland, and observations of Tertiary microfossils. Preliminary report (in Finnish with English summary). *Geologi*, **28**, 33–40.

HJORT, C. & SUNDQUIST, B. 1979. *Geologi och miljö*. Natur och Kultur, Stockholm.

HÖGBOM, A. G. 1910. Precambrian geology of Sweden. *Bulletin of the Geological Institution of Upsala*, **10**, 1–80.

HOLTEDAHL, O. 1953. On the oblique uplift of some northern lands. *Norsk Geologisk Tidsskrift*, **14**, 132–139.

JAEGER, H., 1984. Einige Aspekte der geologischen Entwicklung Südskandinaviens im Altpaläozoikum. *Zeitschrift für Angewandte Geologie*, **30**, 17–33.

JENSEN, L. N. & SCHMIDT, B. J. 1992. Late Tertiary uplift and erosion in the Skagerrack area magnitude and consequences. *Norsk Geologisk Tidsskrift*, **72**, 275–279.

LIDMAR-BERGSTRÖM, K. 1982. *Pre-Quaternary Geomorphological Evolution in Southern Fennoscandia*. Meddelanden från Lunds Universitets Geografiska Institution, Avhandlingar, **91**/Sveriges Geologiska Undersökning Serie C, **785**.

—— 1988. Denudation surfaces of a shield area in south Sweden. *Geografiska Annaler*, **70A**(4), 337–350.

—— 1989. Exhumed Cretaceous landforms in south Sweden. *Zeitschrift für Geomorphologie NF. Suppl.* **72**, 21–40.

—— 1991. Phanerozoic tectonics in southern Sweden. *Zeitschrift für Geomorphologie NF*, Suppl. **82**, 1–16.

—— 1993. Denudation surfaces and tectonics in the southernmost part of the Baltic Shield. *Precambrian Research*, **64**, 337–345.

—— 1995. Relief and saprolites through time on the Baltic Shield. *Geomorphology*, **12**, 45–61.

—— 1996. Long term morphotectonic evolution in Sweden. *Geomorphology*, **16**, 33–59.

——, OLSSON, S. & OLVMO, M. 1997. Palaeosurfaces and associated saprolites in southern Sweden. *In*: WIDDOWSON, M. (ed.) *Palaeosurfaces: Recognition, Reconstruction and Palaeoenvironmental Interpretation*. Geological Society, London, Special Publications, **120**, 95–124.

MÄNNIL, R. 1966. *Evolution of the Baltic Basin during the Ordovician* (in Russian with English summary). Eesti NSV Teaduste Akadeemia Geoloogia Instituti, Tallin.

MATTSSON, Å. 1962. *Morphologische Studien in Südschweden und auf Bornholm über die nichtglaziale Formenwelt der Felsenskulptur*. Meddelanden från Lunds Universitets Geografiska Institution. Avhandlingar, **39**.

MIGON, P. 1997. Tertiary etchsurfaces in the Sudetes Mountains, SW Poland: a contribution to the pre-Quaternary morphology of Central Europe. *In*: WIDDOWSON, M. (ed.) *Palaeosurfaces: Recognition, Reconstruction and Palaeonvironmental Interpretation*. Geological Society, London, Special Publications, **120**, 187–202.

OLLIER, C. D. 1982. The Great Escarpment of eastern Australia: tectonic and geomorphic significance. *Journal of the Geological Society of Australia*, **29**, 13–23.

PEULVAST, J. P. 1985. Postorogenic morphotectonic evolution of the Scandinavian Caledonides during the Mesozoic and Cenozoic. *In*: GEE, D. G. & STURT, B. A. (eds) *The Caledonide Orogen – Scandinavia and Related Areas*. Wiley, Chichester, 979–995.

—— 1986. Structural geomorphological development in the Lofoten-Vesterålen area, Norway. *Norsk Geografisk Tidsskrift*, **40**, 135–161.

REUSCH, H. 1901. Nogle bidrag till forstaaelsen af hvorledes Norges dale og fjelde er blevne til. *Norges Geologiske Undersøgelse*, **32** [for 1900], 124–263.

—— 1903. Betrachtungen über das Relief von Norwegen. *Geographische Zeitschrift*, **9**, 425–435.

RIIS, F. 1996. Quantification of Cenozoic vertical movements of Scandinavia by correlation of morphological surfaces with offshore data. *Global and Planetary Change*, **12**, 331–357.

ROHRMAN, M. 1995. *Thermal evolution of the Fennoscandian region from fission track thermochronology. An integrated approach.* Academisch proefschrift, Vrije Universiteit, Amsterdam, Netherlands Research School of Sedimentary Geology (NSG) Publication **950901**.

ROHRMAN, M., VAN DER BECK, P., ANDRIESSEN, P. & CLOETINGH, S. 1995. Meso-Cenozoic morphotectonic evolution of southern Norway: Neogene domal uplift inferred from apatite fission track thermochronology. *Tectonics*, **14**, 704–718.

RUDBERG, S. 1954. *Västerbottens berggrundsmorfologi.* Geographica, Uppsala, **25**.

—— 1960. Geology and geomorphology. *In*: SØMME, A. (ed.) *A Geography of Norden*, 27–40.

RUNDBERG, Y. 1990. *Tertiary sedimentary history and basin evolution of Norwegian North Sea between $60°–62°N$ – An integrated approach.* Dr Ing. Thesis, Geology, Norwegian Institute of Technology/University of Trondheim.

—— & SMALLEY, P. C. 1989. High-resolution dating of Cenozoic sediments from northern North Sea using $^{87}Sr/^{86}Sr$ stratigraphy. *AAPG Bulletin*, **73**, 298–308.

SCHIPULL, K. 1974. Geomorphologische Studien in zentralen Südnorwegen mit Beitragen über Regelungs- und Steuerungssysteme in der Geomorphologie. *Hamburger Geographischer Studien*, **31**.

SPJELDNAES, N. 1975. Palaeogeography and facies distribution in the Tertiary of Denmark and surrounding areas. *Norges Geologiske Undersögelse*, **316**, 289–311.

STUEVOLD, L. M. & ELDHOLM, O. 1996. Cenozoic uplift of Fennoscandia inferred from a study of the mid-Norwegian margin. *Global and Planetary Change*, **12**, 359–386.

THOMAS, M. 1994. *Geomorphology in the Tropics.* Wiley, Chichester.

WRÅK, W. 1908. Bidrag till Skandinaviens reliefkronologi. *Ymer*, **28**, 141–191.

ZECK, H. P., ANDRIESSEN, P. A. M., HANSEN, K., JENSEN, P. K. & RASMUSSEN, B. L. 1988. Paleozoic paleo-cover of the southern part of the Fennoscandian Shield – fission track constraints. *Tectonophysics*, **149**, 61–66.

Inherited landscapes of the Sudetic Foreland (SW Poland) and implications for reconstructing uplift and erosional histories of upland terrains in Central Europe

PIOTR MIGOŃ

Department of Geography, University of Wrocław, pl. Uniwersytecki 1, 50-137 Wrocław, Poland

Abstract: This paper provides evidence of the very varied denudation topography that had developed in the Sudetic Foreland by the end of the Palaeogene and which has survived up to the present day because of partial burial by Neogene sediments. The Sudetic Foreland comprised different landscape units and the range of relative relief within it may have been up to 600 m; 150–200 m seems to be a figure typical for the mid-Tertiary. This stands in apparent contrast to the widely held opinion that the landscape subjected to the Neogene uplift represented a planation surface. It is argued that much of the present-day topography of Central European uplands and mountains, including the adjacent Sudetes, may be inherited from the Early Cenozoic. In many cases there is generally no need to invoke complicated uplift histories, for which scarce independent geological evidence exists, to explain the origin of landsurfaces.

Large regions of elevated terrain in Central Europe are interpreted as having been subjected to considerable epeirogenic and block uplift in the Late Tertiary, with subsequent denudation resulting in the present-day pattern of uplands, mountain massifs and basins. However, little detailed information is available concerning the actual history of both vertical movements and subsequent erosion. This is basically due to the scarcity of correlative deposits in the neighbouring areas and the predominance of crystalline rocks in the basement that do not provide us with stratigraphic clues to identify fault zones. Therefore, a number of contrasting models of geomorphic evolution of the area have been proposed. They range from the proposal that substantial surface uplift has occurred in the latest Cenozoic, as would be indicated by the high-level position of planation surfaces (cf. Kopecky 1972; Meyer *et al.* 1983; Vilimek 1994), to the opinion that actually very little, if any, surface uplift has been involved and the present-day morphology may be explained almost entirely in terms of its long-term erosional history (Šebesta 1993).

The aforementioned scarcity of sedimentary rock cover and persistent difficulties in identification of fault zones and determination of their ages, have directed attention to landforms as possible indicators to help unravel uplift and subsidence histories. The essence of this 'geomorphological' approach is the assumption that prior to the period of uplift some characteristic landscapes existed and their remnants, if positively identified, can now be used as marker horizons suitable for correlation.

Thus, it seems that one of the key issues in the geomorphological approach to uplift and erosional histories in intraplate settings is the proper reconstruction of pre-uplift topography. Two types of landscapes deserve particular attention if a closer look at ancient topographies is to be made: these are exhumed and partially buried surfaces. Both may be fairly precisely dated and comparison can be made between them and their subsequently modified counterparts in terms of relative heights, characteristic landforms, regional slope etc. In this way, the degree of inheritance in epeirogenically uplifted landscapes may be recognized and the amount and timing of uplift and erosion more realistically assessed. For example, identification of types of exhumed relief has recently proved to be of considerable assistance in reconstructing long-term morphotectonic evolution in Scandinavia (Lidmar-Bergström 1996).

This paper will concentrate on the area located in the northwestern corner of the Bohemian Massif, Central Europe, an area that comprises the mountain massif of the Sudetes and its NE foreland known as the Sudetic Foreland (Fig. 1). The latter will be examined in more detail because partial burial of its ancient denudational landscape in the Neogene and Quaternary has resulted in survival of many facets of the pre-Neogene morphology. Hence, it gives us a clearer insight into the nature of pre-uplift topography. Implications for reconstructions of the pattern of the Neogene uplift in the adjacent area of the Sudetes and other European uplands will then be evaluated.

From: SMITH, B. J., WHALLEY, W. B. & WARKE, P. A. (eds) 1999. *Uplift, Erosion and Stability: Perspectives on Long-term Landscape Development*. Geological Society, London, Special Publications, **162**, 93–107. 1-86239-030-4/99/ $15.00 © The Geological Society of London 1999.

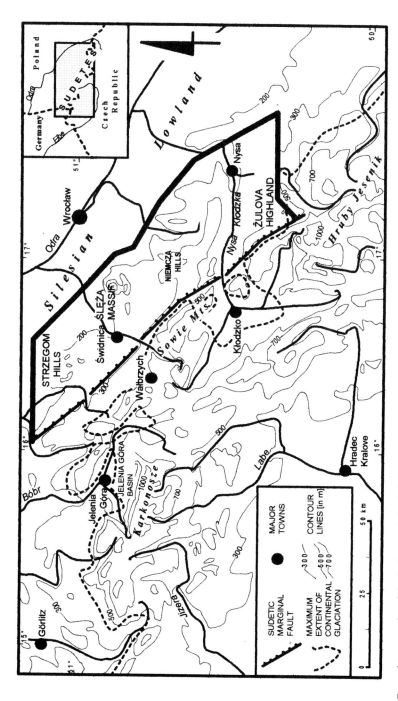

Fig. 1. General topography of the Sudetes and the Sudetic Foreland. Thick solid line shows approximate boundaries of the Sudetic Foreland.

Study area

Location and general topography

The Sudetic Foreland is a hilly landscape developed on the crystalline basement of the Bohemian Massif and forms the NW-SE trending piedmont zone in front of the Sudetes, c. 200 km long and 20-50 km wide (Fig. 1). The altitude of the Sudetic Foreland varies from 180 to 350 m, with some individual hills rising to as much as 718 m. Towards the N, NE and E the crystalline rocks are covered by Neogene and Pleistocene deposits. The latter are mainly of glacial origin, with the present-day landscape reflecting the accumulation and denudation of unconsolidated Pleistocene sediments with no inherited component. Although large parts of the foreland also bear a cover of Cenozoic deposits, it is usually less than 50 m thick and it may be estimated that bedrock outcrops constitute c. 20% of the Sudetic Foreland area. The highest parts of the adjacent elevated unit of the Sudetes are over 1400 m (the highest point is 1603 m), but most individual mountains attain elevations of 700-1100 m and these are separated or surrounded by foothills and intramontane basins with elevations of 300-400 m (Fig. 1).

Geological framework

The crystalline basement rock of the Sudetic Foreland is considerably varied and comprises many rock suites of different ages, origins and relationships to each other (Fig. 2). The largest part of the area is underlain by metamorphic rocks of Proterozoic and Early Palaeozoic age (Cwojdziński & Żelaźniewicz 1995). Among these, gneiss and schists predominate, but migmatites, amphibolites, quartzites, greenschists and marbles are also present. In Devonian times two ophiolite complexes (Ślęża Massif and Grochowa Massif) originated and these consist mainly of gabbro and serpentinite. The Carboniferous period witnessed large granitoid intrusions of complex structure, one in the NW (Strzegom-Sobótka Massif) and another one in the SE (Žulova Massif); a few smaller granitoid bodies are also known from that time. The basement geology of the Sudetes is very similar to that of the foreland, with extensive mesometamorphic complexes dominated by gneiss and mica schists as the most important components (Don & Żelaźniewicz 1990). The ultimate consolidation of this lower structural unit took place during the Hercynian orogeny.

The legacy of subsequent geological development is different in each part of the study area. In the Sudetic Foreland the lower structural unit is separated from the upper Late Cenozoic unit by a major hiatus encompassing the whole Permian, Mesozoic and almost the entire Palaeogene. By contrast, sedimentary rocks from the Carboniferous, Permian, Lower Triassic and Upper Cretaceous are widespread in the present-day Sudetes. It is envisaged that the Sudetic

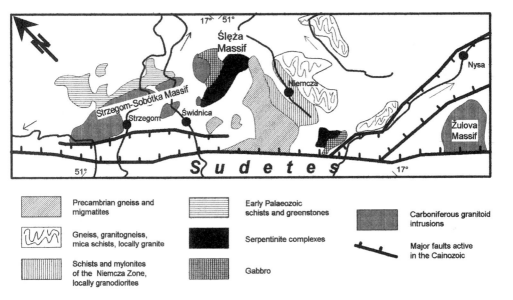

Fig. 2. Outline of the geological structure of the Sudetic Foreland. Blank areas indicate Tertiary and Quaternary sediments.

Foreland was a long-lived elevated area, temporarily even an island, that acted as a source of detrital material for Permo-Mesozoic sedimentary cover (Teisseyre 1960).

Morphotectonics and Cenozoic history

The most prominent expression of morphotectonics in the study area is the boundary between the Sudetes and the Sudetic Foreland, formed by the mountain front which is c. 160 km long and 100–600 m high. Its origin is ascribed to major Neogene–Quaternary uplift of the Sudetes relative to the Foreland (Oberc & Dyjor 1969; Krzyszkowski et al. 1995) and its onset is dated to the latest Oligocene (Grocholski 1977). This topographic feature is related to a fault line of regional importance, called the Sudetic Marginal Fault. Faulting was associated with the origin of deep grabens in front of the mountain area, filled mostly by Miocene clastic sediments (Oberc & Dyjor 1969; Dyjor 1993). Since the time of separation, the Sudetes and the Sudetic Foreland have followed two different pathways of landscape evolution, with predominance of denudation in the former and deposition in the latter.

Neogene sedimentary cover in the Sudetic Foreland consists mainly of clays and silts and locally quartz sands, deposited in fluvial, lacustrine and swampy environments. Frequent intercalations of lignites of various Miocene ages (Sadowska 1977) allow for fairly precise dating of these sediments. To the east, marine incursions of Middle Miocene age left clays, mudstones and marls. In the Pliocene, coarse-grained sedimentation replaced the fine-grained material and this shift is usually interpreted as a reflection of growing intensity of uplift and erosion in the source area (Dyjor 1993). During the Pleistocene, the Sudetic Foreland was twice reached by inland ice-sheets extending from Scandinavia, presumably during stages 12 and 6 (Badura et al. 1992; Krzyszkowski, in press). The ice-sheets left various till and outwash deposits, with thicknesses of up to 40 m.

At much higher altitudes within the Sudetes, a complicated pattern of mountain ranges and basins and an almost total absence of Neogene sediments, has long fuelled speculations about the dominant part played by relatively recent surface uplift. Unquestioned evidence for vertical movements has come from the Middle Sudetes, southwest of Kłodzko, where isolated patches of Turonian (Upper Cretaceous) shallow marine sediments can now be found at different altitudes, from 850–900 m down to 350–400 m (Dumicz 1964). However, in many other parts of the Sudetes the morphological approach was the only one available and was adopted by geomorphologists and geologists alike (Klimaszewski 1958; Kopecky 1972; Demek 1975; Dyjor 1975; Oberc 1975). Different models have been introduced, yet they all focus on identification of remnants of an alleged primary planation surface that would have once dominated the whole landscape. These have been identified at different altitudes, hence differential uplift and subsidence have been invoked and the present-day pattern of elevations and depressions has been seen to reflect the intensity of endogenic factors. Calculations of the amount of relative uplift have involved the simple procedure of subtracting the respective figures showing the altitude of summit surfaces. Accordingly, late Cenozoic uplift of 1200–1500 m has been suggested for the most elevated parts of the Sudetes. Further consequences of this approach have included claims for the existence of deep-seated active faults or crust updomings (cf. Kopecky 1972). Stepped topographies in some massifs and the co-existence of flat surfaces at various altitudes have often been interpreted as being the result of episodic uplift separated by periods during which planation was resumed (Klimaszewski 1958; Jahn 1980). In particular, the concept of three planation surfaces separated by three phases of uplift has gained favour in Polish geomorphology (Klimaszewski 1958; Walczak 1972; Jahn 1980). These discrete uplift phases have been tentatively linked with tectonic events in the Eastern Alps or the Carpathians (Dyjor 1993). Finally, vigorous erosion of bedrock and scarp retreat subsequent to uplift has been thought to be responsible for the fragmentary survival of the Palaeogene planation surface or its replacement by hilly topography.

Pre-Neogene landscape of the Sudetic Foreland

Deep weathering and pre-Neogene landscape evolution

It can be argued that the most reliable key to deciphering the origin of landscape, whether ancient or recent, is through recognition of processes that have shaped it. This approach has proved possible in the Sudetic Foreland because burial of the pre-Neogene landscape by Cenozoic sediments has resulted in preservation not only of ancient landforms, but also of the record of some contemporaneous geomorphic processes.

It has been shown in many places that bedrock, if found beneath the Neogene sediments,

is deeply weathered and this fact points to the considerable part played by deep weathering in landscape evolution. In the Strzegom–Sobótka granitoid massif, granite is altered to a kaolinitic weathering mantle to depths of 40–70 m (Kural 1979; Gawroński 1982). This clayey residual mantle is usually underlain by a zone of slight decomposition (arenaceous weathering) with a kaolinite content up to 15–20%, the thickness of which may be another 20 m. Kaolin occurrences up to 40 m thick are also common in the Żulova Massif. The depth of deep weathering on gneiss has been reported as being up to 40 m and similar values have been encountered on mica schists (Ciuk & Piwocki 1979). Deep weathering also affected serpentinite complexes, where eluvial 'red earths' enriched in nickel-bearing minerals and iron ores are up to 80 m thick (Niśkiewicz 1967).

Although deep weathering is so widespread, the thickness of the weathering mantle is by no means uniform, varying considerably over very short distances. North of Świdnica, deeply weathered surfaces with a residual mantle up to 60 m thick are surrounded by bedrock elevations with no signs of advanced weathering. Around Strzegom, densely spaced boreholes have shown that the depth of the weathering front is very uneven with a range of height difference of 50 m (Kural 1979). The thickness of the weathering mantle on serpentinites in Szklary, south of Niemcza, is even more variable, with changes from 10 m to 80 m in depth occurring over a distance of 200–300 m (Niśkiewicz 1982).

Thus, from the genetic point of view, the denudational landscape of the Sudetic Foreland is best described as an etchsurface, i.e. as having been formed principally by processes of selective deep weathering and evacuation of the resultant saprolite. However, it is intrinsically complex and, according to the terminology introduced by Thomas (1989), comprises mantled, partly stripped, stripped and buried components. In addition, the topographic setting of the Tertiary deposits indicates that mantled, partly stripped and stripped etchsurfaces of the present day do not differ significantly from etchsurfaces that existed at the end of the Palaeogene.

Characteristic landform assemblages of etched origin

A number of etchsurfaces within the foreland will now be examined in order to show their characteristic landforms, the nature of rock control and the range of relative heights present. The most important of these features are presented in Table 1.

Żulova Highland. In the inselberg landscape of the Żulova Highland there are about 30 individual hills that rise to 525 m. Their relative heights are between 30 and 125 m and decrease to the north. Most of the hills are dome-shaped, locally superimposed on a conical base and are frequently crowned by tor-like features, with a variety of smaller weathering forms including rillenkarren, tafoni, alveoles and weathering pits (Czudek et al. 1964). Although large natural exposures are rather scarce, joint patterns exposed in numerous quarries cut into hillslopes suggest that the occurrence of widely spaced, upward-convex fractures has afforded sufficient resistance to withstand the attack of sub-surface weathering and superficial erosional processes after exposure. In addition to inselbergs, numerous low rock elevations occur of between 10 and 30 m in height. They used to be described as roche-moutonnés (Czudek et al. 1964), but were subsequently revisited by Ivan (1983) and are now viewed as minor features of the weathering front. In between the elevations, small-scale basins carved out of bedrock occur and their irregular distribution again suggests that they were formed as depressions at the weathering front guided by fracture patterns and that they owe little to fluvial erosion or glacial excavation (Ivan 1983).

Strzegom–Sobótka granitoid massif. The Strzegom–Sobótka massif is a composite granitoid body in the western part of the Sudetic Foreland (Fig. 1) and shows a number of distinct landscapes. The hilly area around Strzegom is another example of an inselberg-like landscape, where the highest point is 353 m and where individual hills are between 30 and 110 m high. Towards the east these are replaced first by an association of elevations and basins, typically developed north of Świdnica (Fig. 3), and then by a deeply weathered plain. The former consists of broad ridges comprising the two-mica granite and granodiorite, rising to 200–260 m and separated by low-lying areas with altitudes of 180–200 m. These depressions, however, have been infilled with Neogene and Quaternary deposits, characteristically 20–60 m in thickness (Gawroński 1982). Similar pre-Neogene depressions and rises occur to the north with the same range of relative relief. Thus, the total height difference that existed in this area by the end of the Palaeogene may be estimated to be up to 200 m.

The change in morphology from west to east is positively correlated with internal differentiation

Table 1. *Summary of geomorphological characteristics of pre-Neogene etchsurfaces in the Sudetic Foreland and the Sudetes*

Area	Lithology	Landforms	Associated weathering products	Relative heights
Sudetic Foreland				
Žulová Highland	Granite	Inselbergs, multiconvex relief, marginal depressions, joint-controlled basins	Kaolinitic weathering Sandy and grus weathering	50–200 m
Strzegom Hills and surroundings	Composite granite intrusion	Widely spaced inselbergs, broad watershed ridges and basins, deeply weathered plains	Thick (>70 m) kaolinitic weathering	50–200 m
Ślęża–Radunia massif	Gabbro, serpentinite, amphibolite, granite	High, closely spaced inselbergs and elongated ridges	Deeply weathered serpentinites	200–600 m
Niemcza Hills	Gneiss, migmatites, amphibolites, granodiorite, serpentinites	Ridge-and-valley topography, gorges	Lateritic-like deep (>80 m) weathering of serpentinite Sandy weathering of granitoids	50–180 m
Sudetes				
Jelenia Góra Basin	Porphyritic granite, aplogranite, gneiss	Multiconvex relief, marginal depressions, border scarps	Grus weathering	100–250 m
Wałbrzych area	Rhyolites, tuff, sandstones, mudstones, conglomerates	Upland surfaces separated by scarps, rhyolite hills and ridges	No deep weathering found	200–400 m
Kłodzko area	Granitoids, greywackes, hornfelses	Hilly landscape, basins, marginal escarpments facing granite area	Grus weathering	100–300 m

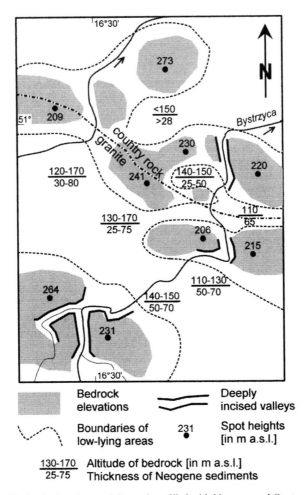

Fig. 3. Spatial pattern of bedrock elevations and depressions filled with Neogene and Quaternary sediments in the area north of Świdnica. Geological data after various sheets of detailed geological map of the Sudetes, 1:25 000.

of the granitoid body into several smaller intrusions of different composition and texture (Migoń 1996, 1997). The inselberg-like landscape has developed on biotite–hornblende, potassium-rich granite while the central part of the massif comprises the two-mica granite, and towards the east, biotite granodiorite prevails (Puziewicz 1990). Therefore, morphology may be regarded as reflecting the unequal susceptibility of different granite variants to weathering, both in terms of depth of saprolite and spatial pattern.

Ślęża–Radunia Massif. Ślęża–Radunia Massif comprises a group of closely spaced gabbro and serpentinite hills that rise over 500 m above the surrounding plain (Fig. 4) is, however, in part a depositional feature. The thickness of the Neogene–Quaternary cover around Mt Ślęża is up to 80 m (Różycki 1968), therefore the total height difference due to long-term bedrock lowering may be estimated as being 550–600 m. Moreover, the footslopes of the hills are deeply weathered with the residual weathering mantle being 10–50 m thick (Fedak & Niśkiewicz 1979; Gawroński 1982).

In this case, however, correlation between bedrock structure and topography is imperfect, occurring in only a few places. For example, the southern hill/footslope junction of the Radunia hill is elongated W–E and runs parallel to the serpentinite/gneiss boundary all along the contact. Therefore, it may reflect the greater resistance to weathering of serpentinite relative to the gneiss, which in turn would be due to the more massive, unoriented texture and fine-grained structure of the former. The maximum height of the serpentinite slope is 350 m with an

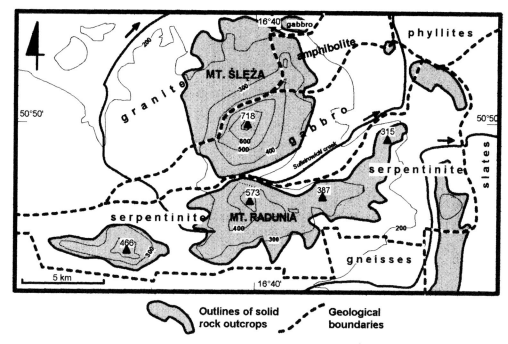

Fig. 4. Residual topography and its relationship to geological structure in the Ślęża Massif.

average of 150–200 m. By contrast, the highest hill within the group (Mt Ślęża – 718 m) does not appear to be influenced by lithology or structure in any significant way. The hilltop comprises coarse-grained gabbro, yet seemingly the same gabbro underlies the broad depression to the southeast, where the altitude is less than 250 m. Thus, long-term etching and stripping have proved capable of producing local relief of between 500 and 600 m although in this particular case bedrock control is rather limited.

Niemcza Hills. The Niemcza Hills provide an example of ridge-and-valley topography (Fig. 5), where the pre-Neogene age is confirmed by the occurrence of Miocene clays, quartz sands and lignites within the major valleys. The valleys, however, seem too broad to be the result of normal fluvial erosion, especially near their headwater areas, and are more likely to be landforms caused by weathering and stripping. Some are thought to be tectonic grabens but although minor faulting has been locally recognized, no border faults have been identified. The overall pattern of valleys and ridges is parallel and elongated trending N–S. To the north, both valleys and ridges are cut by the late Neogene Strzelin Fault, along which the pre-Neogene topography has been downthrown by *c.* 50–100 m (Dyjor 1993).

The ridges are 7–16 km long and 2.5–6 km wide and they attain altitudes of 300–407 m with different relative heights of between 50 and 180 m. The westernmost ridge is surmounted by a few elongated hills 40–50 m high while the others have rather flattened watersheds. One of the ridges is cut through by a few deeply incised (up to 70 m) valleys and in one of these the occurrence of Miocene clays has been reported. Geologically, the ridges comprise a variety of lithologies, including different types of gneiss, migmatites, granodiorites, mica schists and amphibolites. Locally, deeply weathered serpentinite bedrock occurs in watershed settings and its thickness may be as much as 80 m (Niśkiewicz 1967). The reason for its survival in this elevated position seems to be due to the presence of a silica-enriched crust on top of the saprolite.

The present-day floors of the intervening valleys are at altitudes of 200–260 m but the actual valley floors lie considerably deeper as the thickness of the Neogene–Quaternary infill is 10–50 m (Fig. 5). Beneath the infill, bedrock kaolinization of up to 40 m in depth is known to occur. The valleys are developed chiefly in granodiorites and schists. In the former case, the occurrence of the valley is probably due to greater resistance of the country rock subjected to contact metamorphosis.

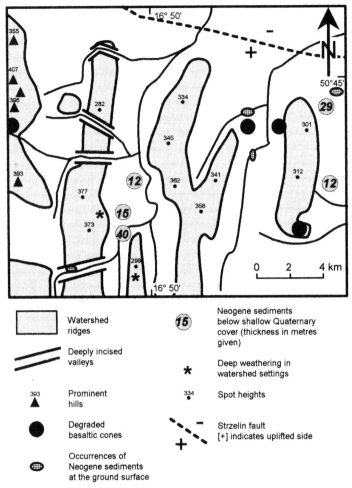

Fig. 5. Valley-and-ridge topography of the Niemcza Hills. Note the occurrence of Neogene deposits between the ridges.

In summary, local relief up to at least 150 m with occasional inselbergs and gorges typifies the pre-Neogene ridge-and-valley topography in the eastern part of the Sudetic Foreland.

Etchsurfaces in the Sudetic Foreland – a general view

The examples described in the previous sections show the variety of pre-Neogene landscapes, the variable extent of litho-structural control and, in most cases, the formation of a quite significant relief within these landscapes. The latter stands in apparent contrast to the picture portrayed in earlier papers, where the pre-Neogene landscape is seen as a rather subdued topography, an end-product of protracted planation (cf. Klimaszewski 1958; Walczak 1972).

It can be argued that at the onset of morphotectonic differentiation within the northeastern part of the Bohemian Massif the landscape of what is now the Sudetic Foreland was in a state of 'dynamic etchplanation' and particular landform assemblages were transient results of a continuous interplay between bedrock properties, geomorphic processes and time. Structural complexity, the selective nature of dominant geomorphic processes and the long time span available for geomorphic evolution of the area provided a setting for considerable relief differentiation. The altitudinal difference within the Sudetic Foreland could have been as much as 600 m, although the figure 150–200 m is more likely to be typical (Table 1).

Implications for adjacent elevated areas of the Sudetes

The evidence from the Sudetic Foreland, presented above, and the realization of the common history of geomorphic evolution prompts a new look at landscape evolution in the Sudetes. It may be hypothesized that its topography prior to morphotectonic differentiation of the region was complicated, reflecting an extremely complex geological structure, with no regional planation surface dominating the area. There are at least a few places within the Sudetes where this assumption can be confirmed and the degree of litho-structural control assessed.

Intramontane basin of Jelenia Góra

The Jelenia Góra Basin is located in the western part of the Sudetes (Fig. 1) and is underlain by Middle Carboniferous granite that was intruded into an older gneissic-schistose massif of Proterozoic–Early Palaeozoic age (Mierzejewski & Oberc-Dziedzic 1990). The basin itself is of complex origin with both differential erosion and tectonic movements involved in its development (Jahn 1980). A cross-section through the basin floor, its northwestern margin and the adjacent gneissic upland gives an insight into the nature of structural control upon the topography (Fig. 6).

The basin floor has developed entirely in granite and comprises a hilly landscape with inselberg-like hills, ruwares and secondary basins. The total height difference within the basin exceeds 300 m, with most of the individual hills being between 50 and 100 m high. Different relationships between landforms and bedrock properties may be identified and these include the association of some valleys with master joints, higher relief on potassium-rich and fine-grained variants of granite, and the development of hills upon dome-like granite compartments (Migoń 1996, 1997). The occurrence of grus weathering mantles around many of the hills strengthens the idea that the landscape of the basin floor is an almost completely exposed weathering front, and thus a stripped etch-surface.

The upland surface adjacent to the NW of the Jelenia Góra Basin shows many similarities to the landscape developed on metamorphic rocks in the Niemcza Hills, described earlier. It comprises broad watershed ridges separated by wide valleys where locally, small basins of triangular outline occur and in the floors of which remnants of kaolinitic weathering mantle have been found (Budkiewicz 1971). Although slopes within the gneissic area are only occasionally steeper than 10°, the total height difference approaches 200 m.

The morphological boundary between these two contrasting landscapes is marked by an escarpment 20 km long and 100–150 m high, the course of which closely follows the boundary of a granite body. In most cases the base of the escarpment is associated with the granite/country rock contact, whilst the escarpment face itself comprises gneiss or granito-gneiss and is steeper if the latter outcrops. No embayments or outliers are present along the scarp, hence in this case scarp retreat is considered negligible. Consequently, there is little to support the view that the escarpment may separate two planation surfaces of different ages, with the lower being younger and developing at the expense of the upper. Rather, the occurrence of the topographic scarp reflects more intense and/or deeper weathering and denudation of the granite area relative to the gneissic area. This view may be corroborated by patchy occurrences of kaolin covers on the gneissic surface indicating a higher degree of inheritance.

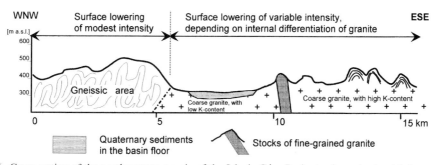

Fig. 6. Cross-section of the northwestern margin of the Jelenia Góra Basin, to show structural influences upon relief development in both macro- (granite/gneiss boundary) and mesoscale (granite hills).

Middle Sudetes around Wałbrzych

The landscape of the Middle Sudetes around Wałbrzych has developed on Carboniferous and Early Permian sedimentary rocks intruded by rhyolites and basalts of volcanic and subvolcanic origin and of roughly the same age as the sediments. Sedimentary series are up to 4000 m thick and consist of conglomerates, sandstones and mudstones of different composition and strength, laid down in a terrestrial environment (Dziedzic & Teisseyre 1990).

The present-day morphology is characterized by the occurrence of basins, ridges, uplands and domes and the range of height difference comes up to 400 m (Fig. 7). Rhyolite domes with their extremely steep slopes, locally up to 40°, are the most prominent landmarks overlooking adjacent basins and valleys by as much as 250–400 m. Minor lithological influences can be recognized within sedimentary rocks; depressions developed on sandstones with mudstones are deeper than those underlain by conglomerates with sandstones, while distinct scarps are formed along outcrops of quartzose conglomerates. Escarpments have also formed where fine-grained units are replaced in the stratigraphical column by coarse-grained sediments and along elongated basalt bodies; the latter, however, are much steeper. These lithologically controlled scarps separate uplands and basins and their spatial pattern is thus likely to reflect differences in denudation rates, which are spatially variable because of lithological heterogeneity.

Kłodzko region

A fine example of lithological control upon landscape development is provided by the Kłodzko-Złoty Stok Granitoid Massif and its surroundings. This Carboniferous granitoid intrusion is bordered by Early Carboniferous greywackes on the northwestern side and Proterozoic gneiss, mica schists and amphibolites on the southeastern side. Both boundaries follow very irregular courses and frequently change their directions (Fig. 8). All along the boundaries the country rock is altered by contact metamorphosis into hornfelses and andalusite schists. Within the intrusion, several types of igneous rocks occur including monzonites, granodiorites and syenodiorites. Generally, these granitoids represent the most basic variants among the Sudetic granitoid massifs.

The lithological contacts between the granite mass and the country rock are remarkably well expressed in the present-day landscape. Metamorphosed greywackes on the northwestern side form a mountain ridge with steep slopes 250–300 m high facing the granitoid area, where the piedmont slope coincides with the change in lithology. The eastern morphological boundary shows less correlation with the bedrock lithology, although at a larger scale the association of higher ground with country rock is more obvious (Fig. 8). The relief of the granitoid area itself is influenced by lithological variations within the bedrock. The central part of the massif is formed by an arcuate elevation trending SW–NE, along which numerous outliers of thermally metamorphosed country rock occur. Their relative height above adjacent valleys and basins cut in the granite is 50–150 m.

The examples reviewed above show that lithologically and structurally controlled denudation can account for at least 100–200 m, and in some specific places for as much as 400 m, of

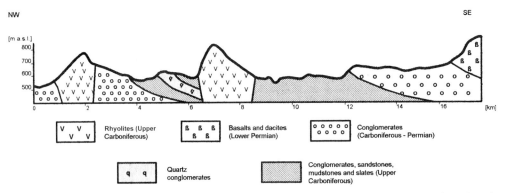

Fig. 7. Lithologically controlled relief on Upper Palaeozoic sedimentary and volcanic rocks around Wałbrzych. Rolling landsurfaces at different altitudes and tops of volcanic hills used to be regarded as products of different cycles of planation in response to episodic uplift of the Sudetes. It is more likely that they simply reflect differences in long-term denudation rates caused by unequal resistance of particular lithological units.

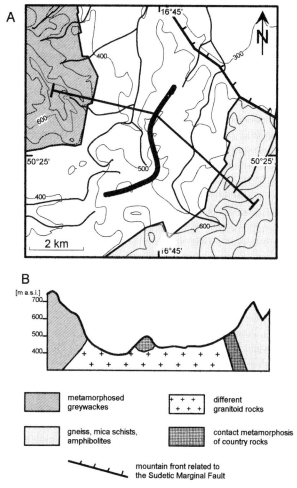

Fig. 8. Bedrock-controlled topography of the Eastern Sudetes near Kłodzko, as seen on the map (**A**) and cross-section (**B**). Thick solid line indicates the central elevation within the granitoid area, related to the occurrence of outliers of contact-metamorphosed country rock.

local relief in the Sudetes. The occurrence of such a relief should now provide a framework in relation to which suggested uplift histories may be reassessed. It seems unnecessary to introduce multiple uplift episodes and phases of renewed, but not completed, planation to explain the occurrence of surfaces of low relief at different elevations and to assign different ages to them because they may well have formed simultaneously. The concept of intensive erosion and scarp retreat in bedrock as a response to uplift can be replaced by the concept of inheritance and legacy of the long-term geomorphic development in heterogeneous substrata. Moreover, stripping of saprolite seems to be more likely than bedrock erosion and is in better accordance with the geological record of the Neogene in the Sudetic Foreland. Composition of the Neogene sediments, with their abundance of allogenic clay, kaolinite and residual quartz sand, points clearly to the destruction of the weathering mantle.

Uplift of the Sudetes relative to the Foreland is not questioned and many morphological contrasts within the former would be difficult to explain without invoking differential vertical movements of adjacent areas. However, it seems that the inherited component in the present-day landscape, of largely pre-Neogene age, is of the same importance as that related to the Neogene–Quaternary tectonics. Although the gross pattern of elevated and subsided areas

is likely to reflect differential crustal movements in the late Cenozoic, most of the secondary facets in the landscape are related to long-term etching and stripping.

Regional context

Neogene uplift is thought to have affected many ancient landsurfaces in central and western Europe, and Palaeogene and Miocene planation surfaces dissected as a response to it have frequently been identified (cf. Demek 1984; Désiré-Marchand & Klein 1987; Andres 1989). In some areas, as many as eight surfaces arranged in a stepped manner have been recognized and linked with a punctuated nature of crustal instability (Junge 1987). The occurrence of a landscape in the state of advanced planation by the end of the Palaeogene is still regarded as a convenient starting point to analyse morphotectonic evolution in the late Cenozoic (e.g. Kral 1968; Demek 1975; Demoulin 1988; Andres 1989).

However, evidence is accumulating, indicating that prior to the Neogene, the landscape in many parts of Europe had already been considerably differentiated. Hall (1986) provided details of Scottish topography in the Tertiary and found a variety of denudational landforms, with inselbergs up to 200 m high and basins of different shape and depth. Upland granite areas in SW England, including Dartmoor, have occupied elevated settings since at least the Eocene, as they acted as source areas for surrounding Palaeogene sedimentary basins (Edwards & Freshney 1982). Greater resistance of granite to denudation in relation to the country rock has eventually resulted in the origin of a topographic scarp 100–200 m high separating two low relief surfaces. The lower one is mantled by relics of Oligocene and Miocene sediments (Walsh et al. 1987) which indicate the Palaeogene age of gross landscape features.

In Germany, the widespread planation model (Büdel 1977) is contradicted by the occurrence of deep karstic depressions of Oligocene age in the Rhenish Slate Mountains (Schmidt 1976), structural relief with cuestas and hogbacks in Lower Saxony (Brunotte & Garleff 1980; Spönemann 1989), the inherited nature of some escarpments in the Southern German Uplands (Bremer 1989) and the co-existence of limestone domes and plains in Frankenalb (Pfeffer 1986). Large tracts of relatively level terrain may have existed, yet they seem to have played a rather subordinate part in the overall morphology.

Conclusions

The main conclusions from this study, which may be of some value in further attempts to decipher uplift and erosional histories of European upland areas, may be summarized as follows.

1. Much of the present-day relief differentiation in elevated areas may be adequately explained as being the result of long-term selective denudation controlled by bedrock properties, chiefly of etchplanation. There is no need to invoke multiple uplift and erosion phases, for which there is usually scarce independent evidence.
2. Remnants of planation surfaces do not provide a reliable key to infer differential uplift. Surfaces of low relief may form at different altitudes and may be separated by structural escarpments. Moreover, some types of bedrock, such as granite, are unlikely to give rise to a planation surface if etching is the dominant geomorphic process.
3. More accurate reconstructions of landscapes that existed prior to the onset of uplift are necessary. Buried and exhumed landscapes have usually escaped the attention of geomorphologists and geologists, yet they may provide a vital source of information.
4. To assess the amount and pattern of uplift realistically, relief differentiation resulting from long-term denudation and characteristic landforms has first to be recognized and then the 'denudational' component may be subtracted from the total relief.

This paper is the outcome of field research carried out in the Sudetes for a number of years and extensive literature studies conducted in the School of Geography, University of Oxford, UK. Funding of the latter by The Royal Society Postdoctoral Fellowship Programme is gratefully acknowledged. Thanks are also extended to Karna Lidmar-Bergström for valuable comments on the manuscript.

References

ANDRES, W. 1989. The Central German Uplands. *Catena, Supplement*, **15**, 25–44.

BADURA, J., PRZYBYLSKI, B. & KRZYSZKOWSKI, D. 1992. Stratotype of Pleistocene deposits in Przedgórze Sudeckie: preliminary report. *Przeglad Geologiczny*, **40**, 545–551 (English summary).

BREMER, H. 1989. On the geomorphology of the South German Scarplands. *Catena, Supplement*, **15**, 45–67.

BRUNOTTE, E. & GARLEFF, K. 1980. Tectonic and climatic factors of landform development on the

northern fringe of the German Hill Country (Deutsche Mittelgebirge) since the early Tertiary. *Zeitschrift für Geomorphologie, N. F., Supplement-Band*, **36**, 104–112.

BUDKIEWICZ, M. 1971. Kaolin deposit at Kamień near Mirsk. *Kwartalnik Geologiczny*, **15**, 345–357 (English summary).

BÜDEL, J. 1977. *Klima-Geomorphologie*. Gebrüder Borntraeger, Berlin.

CIUK, E. & PIWOCKI, M. 1979. The Tertiary in the Ząbkowice Śląskie area (Lower Silesia). *Biuletyn Instytutu Geologicznego*, **320**, 27–56 (English summary).

CWOJDZIŃSKI, S. & ŻELAŹNIEWICZ, A. 1995. Crystalline basement of the Fore-Sudetic Block. *In: LXVI Annual Meeting of Polish Geological Society, Geology and Environmental Protection of the Fore-Sudetic Block*. Annales Societas Geologorum Poloniae, Special Volume, 11–28 (English summary).

CZUDEK, T., DEMEK, J., MARVAN, P., PANOŠ, V. & RAUŠER, J. 1964. Verwitterungs- und Abtragungsformen des Granits in der Böhmischen Masse. *Petermanns Geographische Mitteilungen*, **108**, 182–192.

DEMEK, J. 1975. Planation surfaces and their significance for the morphostructural analysis of the Czech Socialist Republic. *Studia Geographica*, **54**, 133–164.

—— 1984. Bohemian Massif. *In*: EMBLETON, C. (ed.) *Geomorphology of Europe*. Macmillan, London, 216–224.

DEMOULIN, A. 1988. Cenozoic tectonics on the Hautes Fagnes plateau (Belgium). *Tectonophysics*, **145**, 31–41.

DÉSIRÉ-MARCHAND, J. & KLEIN, C. 1987. Fichtelgebirge, Böhmerwald, Bayerischer Wald. Contribution à l'étude du problème des Piedmonttreppen. *Zeitschrift für Geomorphologie, N. F., Supplement-Band*, **65**, 101–138.

DON, J. & ŻELAŹNIEWICZ, A. 1990. The Sudetes – boundaries, subdivision and tectonic position. *Neues Jahrbuch für Geologie, Paläontologie usw., Abhandlungen*, **179**, 121–127.

DUMICZ, M. 1964. Geology of the crystalline massif of the Bystrzyckie Mts. *Geologia Sudetica*, **1**, 169–208.

DYJOR, S. 1975. Late Tertiary tectonic movements in the Sudety Mts. and Fore-Sudetic Block. *In: Współczesne i neotektoniczne ruchy skorupy ziemskiej w Polsce*, 1. Wydawnictwa Geologiczne, Warszawa, 121–132.

—— 1993. Stages of Neogene and Early Quaternary faulting in the Sudetes and their Foreland. *Folia Quaternaria*, **64**, 25–41 (English summary).

DZIEDZIC, K. & TEISSEYRE, A. K. 1990. The Hercynian molasse and younger deposits in the Intra-Sudetic Depression, SW Poland. *Neues Jahrbuch für Geologie and Paläontologie usw., Abhandlungen*, **179**, 285–305.

EDWARDS, R. A. & FRESHNEY, E. C. 1982. The Tertiary sedimentary rocks. *In*: DURRANCE, E. M. & LAMING, D. J. C. (eds) *The Geology of Devon*. University of Exeter, 204–237.

FEDAK, J. & NIŚKIEWICZ, J. 1979. Rudy niklu. *In*: DZIEDZIC, K., KOZŁOWSKI, S., MAJEROWICZ, A. & SAWICKI, L. (eds) *Surowce mineralne Dolnego Śląska*. Ossolineum, Wrocław, 136–142.

GAWROŃSKI, O. 1982. Rejon Świdnicy. *In: Surowce kaolinowe*. Wydawnictwa Geologiczne, Warszawa, 63–77.

GROCHOLSKI, A. 1977. The Marginal Sudetic Fault against the Tertiary volcanotectonics. *Acta Universitatis Wratislaviensis*, 378, *Prace Geologiczno-Mineralogiczne*, **6**, 89–103 (English summary).

HALL, A. M. 1986. Deep weathering patterns in northeast Scotland and their geomorphological significance. *Zeitschrift für Geomorphologie, N. F.*, **30**, 407–422.

IVAN, A. 1983. Geomorphology of Žulovska pahorkatina (Hilly land) in North Moravia. *Zpravy Geografickeho Ustavu ČSAV*, **20**(4), 49–69 (English summary).

JAHN, A. 1980. Main features of the Tertiary relief of the Sudetes Mountains. *Geographia Polonica*, **43**, 5–23.

JUNGE, H. 1987. Der Einfluß von Tektonik und eustatischen Meeresspiegelschwankungen auf die Ausbildung der Reliefgenerationen im Norden der Eifeler Nord-Süd-Zone. *Zeitschrift für Geomorphologie, N. F., Supplement-Band*, **65**, 35–84.

KLIMASZEWSKI, M. 1958. The geomorphological development of Poland's territory in the pre-Quaternary period. *Przegląd Geograficzny*, **30**, 3–43 (English summary).

KOPECKY, A. 1972. Main features of the neotectonics in Czechoslovakia. *Sbornik geologickych věd, A*, **6**, 77–155 (English summary).

KRAL, V. 1968. Geomorfologie vrcholové oblasti Krušnych hor a problem parovin y. *Rozprávy ČSAV, ř. MPV*, **78**, 1–66.

KRZYSZKOWSKI, D. (in press). The Quaternary of Western Poland. *In*: GERRARD, J. (ed.) *Encyclopaedia of Quaternary Science*.

——, MIGOŃ, P. & SROKA, W. 1995. Neotectonic Quaternary history of the Sudetic Marginal Fault, SW Poland. *Folia Quaternaria*, **66**, 73–98.

KURAL, S. 1979. Origin, age and geologic background of the kaolin in the western part of the Strzegom granitic massif. *Biuletyn Instytutu Geologicznego*, **313**, 9–68 (English summary).

LIDMAR-BERGSTRÖM, K. 1996. Long term morphotectonic evolution in Sweden. *Geomorphology*, **16**, 33–59.

MEYER, W., ALBERS, H. J., BERNERS, H. P. *et al.* 1983. Pre-Quaternary uplift in the central part of the Rhenish Massif. *In*: FUCHS, K., GEHLEN, K. VON, MÄLZER, H., MURAWSKI, H. & SEMMEL, A. (eds) *Plateau Uplift. The Rhenish Shield – A Case History*. Springer, Berlin, 39–46.

MIERZEJEWSKI, M. P. & OBERC-DZIEDZIC, T. 1990. The Izera-Karkonosze Block and its tectonic development. *Neues Jahrbuch für Geologie and Paläontologie usw., Abhandlungen*, **179**, 197–222.

MIGOŃ, P. 1996. Evolution of granite landscapes in the Sudetes (Central Europe): some problems of interpretation. *Proceedings of the Geologists' Association*, **107**, 25–37.

—— 1997. The geological control, origin and significance of inselbergs in the Sudetes, NE Bohemian Massif, Central Europe. *Zeitschrift für Geomorphologie, N. F.*, **41**, 45–66.

NIŚKIEWICZ, J. 1967. Geological structure of the Szklary Massif. *Rocznik Polskiego Towarzystwa Geologicznego*, **37**, 387–416 (English summary).

—— 1982. Geological setting of the occurrence of chryzoprase and related gem-stones in the Szklary Massif, Lower Silesia. *Geologia Sudetica*, **17**, 125–139 (English summary).

OBERC, J. 1975. Neotectonic Rozdroże Izerskie Trough. *In*: *Współczesne i neotektoniczne ruchy skorupy ziemskiej w Polsce*, 1. Wydawnictwa Geologiczne, Warszawa, 157–170.

—— & DYJOR, S. 1969. Marginal Sudetic Fault. *Biuletyn Instytutu Geologicznego*, **236**, 41–142 (English summary).

PFEFFER, K.-H. 1986. Das Karstgebiet der nördlichen Frankenalb zwischen Pegnitz und Vils. *Zeitschrift für Geomorphologie, N. F., Supplement-Band*, **59**, 67–85.

PUZIEWICZ, J. 1990. Strzegom-Sobótka granitic massif (SW Poland). Summary of recent studies. *Archiwum Mineralogiczne*, **45**, 135–154 (English summary).

RÓŻYCKI, M. 1968. Geological structure of the vicinity of Wrocław. *Biuletyn Instytutu Geologicznego*, **214**, 181–230 (English summary).

SADOWSKA, A. 1977. Vegetation and stratigraphy of Upper Miocene Coal seams of the South-western Poland. *Acta Palaeobotanica*, **18**, 87–122 (English summary).

SCHMIDT, K.-H. 1976. Strukturbedingte tertiäre Reliefgestaltung am Biespiel von Kalkgebieten am Nordrand des rechtsrheinischen Schiefergebirges. *Zeitschrift für Geomorphologie, N. F., Supplement-Band*, **24**, 68–78.

ŠEBESTA, J. 1993. The exodynamic analysis of the western part of the Bohemian Massif. *In*: VRANA, S. (ed.) *Abstracts on Geological Model of Western Bohemia in Relation to the Deep Borehole KTB in the FRG*. Czech Geological Survey, Prague.

SPÖNEMANN, J. 1989. Homoclinal ridges in Lower Saxony. *Catena, Supplement*, **15**, 133–149.

TEISSEYRE, H. 1960. Rozwój budowy geologicznej od prekambru po trzeciorzęd. *In*: TEISSEYRE, H. (ed.) *Regionalna Geologia Polski, III, Sudety*, 2. Polskie Towarzystwo Geologiczne, Kraków, 335–357.

THOMAS, M. F. 1989. The role of etch processes in landform development. *Zeitschrift für Geomorphologie, N. F.*, **33**, 129–142, 257–274.

VILIMEK, V. 1994. Quaternary development of Kateřinohorska Vault relief in the Krušné hory Mountains. *Acta Universitatis Carolinae, Geographica*, **30**, Supplement, 115–137.

WALCZAK, W. 1972. Sudety i Przedgórze Sudeckie. *In*: Klimaszewski. M. (ed.) *Geomorfologia Polski*, 1. PWN, Warszawa, 167–231.

WALSH, P. T., ATKINSON, K., BOULTER, M. C. & SHAKESBY, R. A. 1987. The Oligocene and Miocene outliers of West Cornwall and their bearing on the geomorphological evolution of Oldland Britain. *Philosophical Transactions of the Royal Society*, **A323**, 211–245.

The application of palaeolandsurface analysis to the study of recent tectonics in central Italy

ROBERTO BASILI[1], FABRIZIO GALADINI[2] & PAOLO MESSINA[2]

[1] *University 'La Sapienza', c/o CNR – Istituto di Ricerca sulla Tettonica Recente, Via Fosso del Cavaliere, 00133 Rome, Italy*
[2] *CNR – Istituto di Ricerca sulla Tettonica Recente, Via Fosso del Cavaliere, 00133 Rome, Italy*

Abstract: The analysis of remnant landsurfaces was carried out on the ranges surrounding three intermontane basins located in the central and southern Apennines. All three basins developed during the Pliocene and Quaternary, but each had a different geological, geomorphological and structural setting. Landsurfaces taken into account were those which aided in the identification of strath terraces situated on the walls of the basins. Recognition and comparison of terrace sequences aided the evaluation of the tectonic behaviour of each area by identifying sectors with differential vertical movements and by assessing the activity of the faults that caused them. The analysis of individual cases also provided useful information on fault activity that could help in devising structural schemes related to active tectonics.

The identification of the main active faults in the Apennines is hindered by its complex Quaternary tectonic setting (e.g. CNR–PFG 1987). This represents one of the most compelling problems of seismotectonic research in Italy.

The Apennine thrust belt has experienced NE–SW extension and uplift during various post-Miocene stages, forming intermontane basins which alternate with NW–SE trending ranges (Fig. 1). The extension created a large number of dip-slip normal faults (as well as strike-slip and oblique-slip faults) that are up to 20 km long, strike between NW–SE and WNW–ESE and generally lie at the boundaries between the basins and ranges. These faults are arranged in a complex structural pattern which is clearly outlined in the *Neotectonic Map of Italy* (CNR–PFG 1987). In this map the main faults and folds are classified on the basis of their age (throughout the Pliocene and Quaternary) and the territory is subdivided into sectors having different estimated uplift rates.

Another significant overview of the complexity of recent Apennine tectonics comes from the pioneering work by Bosi (1975), which represented one of the first attempts to identify active faults in Italy. This paper provides a useful inventory of active faults in central and southern Italy, classified in terms of 'probability of activity'. The complexity of the structural setting is well documented by the large number of detected faults, their length (usually no greater than a few kilometres) and their trends. These pioneering studies therefore represent a very valuable foundation for the subsequent research into the discrimination between the main active and inactive faults. However, knowledge of the behaviour of many recent tectonic structures has recently been greatly improved thanks to small-scale observations, such as kinematic and palaeoseismological studies, on specific structural elements. However, it is still difficult to distinguish between faults that have recently been active (Late Pleistocene to Holocene) and those whose activity ended some hundred thousand years ago. This aspect often leads to misinterpretations and greatly hinders a thorough understanding of the later stages of the tectonic evolution.

The main purpose of this paper is to illustrate the contribution of ancient landsurface analysis in evaluating the effects of tectonics, in particular for (i) identifying the main Quaternary faults and (ii) collecting data on their kinematic characteristics. This objective is pursued by an investigation of the relationships between large-scale, long-term geomorphic processes (e.g. extensive denudation followed by dissection producing terraced sequences), regional tectonics and small-scale, short-term, Late Pleistocene to Holocene tectonic processes (e.g. fault activity producing ground-surface displacement). After a brief summary of the general landsurface features, a landsurface analysis is performed on the ranges surrounding three intermontane basins chosen for the presence of active faults and historical earthquakes. The results obtained in evaluating the tectonic behaviour of these areas and in assessing fault activity are also discussed.

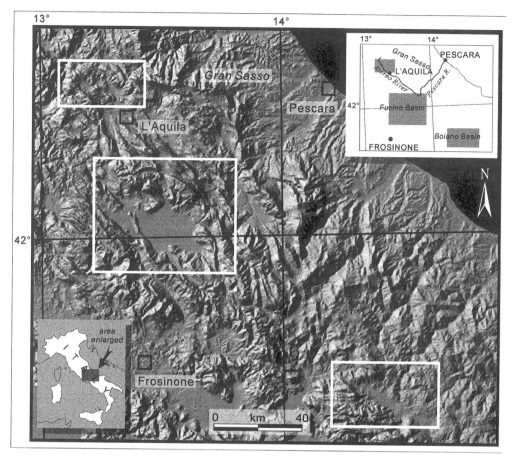

Fig. 1. Shaded relief of central and southern Apennines and location maps of the study areas (boxes).

Remnant landsurfaces

In the Apennine landscape it is common to find sub-horizontal, low-relief areas of land (generally a few square hectometres to square kilometres) carved into mountain slopes, which truncate different rock types and formations, such as Meso-Cenozoic carbonates, Miocene siliciclastics and Plio-Quaternary clastic continental deposits. These landsurfaces lie at various elevations above the present valley floors and usually occur as stepped terrace sequences in which vertical spacing varies from tens to hundreds of metres. Such surfaces were first identified by analysing aerial photos at the scale 1:33 000 and topographic maps at the scale 1:25 000. The surface inner and outer edges have been considered to delimit the terraces to be mapped and to determine the range of elevation of each terrace tread. Elevation and accordance of elevation between adjacent surfaces have been double checked in the field, whereas specific surveys have been aimed at determining presence and nature of terrace deposits.

Fill terraces and cut terraces (*sensu* Merritts *et al.* 1994) have generally been found near the floors of intermontane basins and within the valleys that feed them, whereas strath terraces (*sensu* Bucher, 1932; Merritts *et al.* 1994) exist at higher elevations in the ranges that surround the basins.

It is possible to find datable sediments within the deposits of the fill terraces and cut terraces, and as such the late evolution of a basin can generally be established with reasonable confidence and placed within a restricted chronological framework. Unfortunately, the traces of early basin evolution are usually represented by strath terraces that cannot be directly dated because of the lack of sediments. This latter kind of remnant landsurface, which is erosional in nature, can only be reconstructed and correlated on a morphological basis. The criteria to reconstruct landsurface levels and sequences used in this study were the

following: (i) spatial continuity of a surface, verified by following each surface level on aerial photographs, on topographic maps and in the field; (ii) accordance of elevation; (iii) similar vertical spacing between remnants, taking into account the possible original gradient of each surface; and (iv) equivalence of the position of a surface in more sequences, verified by comparing one sequence to another.

On this basis, correlation is reliable only for terrace sequences belonging to the same basin or for those that were controlled by the same local base-level. However, it should be taken into account that (a) the central Apennine mountain belt developed in a continental environment (mainly fluvial and lacustrine) since the Early Pliocene, (b) the presence of slope deposits attributable to the 'Brecce di Bisegna' formation indicates an Early Pleistocene age (Bosi & Messina 1991; D'Agostino et al. 1997), and (c) according to the age of the beginning of volcanic activity along the Tyrrhenian margin (Fornaseri 1985), the presence of volcanic layers within fill deposits indicates a Middle Pleistocene age or later. These are all commonly accepted chronological indicators of the region and with them it is possible to establish a correlation between the terrace sequences and the regional chronological framework. Hence, according also to estimates reported in the literature (see Bosi & Messina 1991), the ages of terraces described in the following sections range from a generic Pliocene to the Holocene.

The range of latitudes and altitudes of the central and southern Apennines permit one to assume that climate changes have affected the entire region in the same fashion and intensity. The differences found in the matching of reconstructed terrace sequences inside the same basin may thus be interpreted as being the result of tectonics.

The flat sub-horizontal landsurfaces described above, which formed during basin-and-range building, may have recorded fault activity and the resulting differential vertical movements of the crust. The ways in which landsurfaces may have recorded tectonic activity have been envisaged as follows: (a) displacement of a landsurface, when two areas of land are separated by a recognized fault-scarp; (b) displacement of landsurface sequences, when an offset between two or more sequences is observed; (c) occurrence of a different number of landsurfaces on different sequences, when the absence of unpaired terraces was verified; and (d) tilting of a landsurface. The last case is mainly theoretical because it is not possible to discriminate tilted surfaces from gently sloping surfaces carved into the bedrock.

The above procedure offers the valuable opportunity to investigate highly denuded and dissected uplands which represent the larger part of an uplifting mountain region where conventional stratigraphic methodology cannot be applied because of the scarcity of preserved deposits. However, at present this procedure is still in its development phase and some aspects need to be improved or refined. Uncertainties often arise in the geometrical reconstruction of landsurface levels because of the presence of isolated remnants, gently sloping surfaces, and surfaces with an inherent relief. As such, there is a general difficulty in determining the original elevation of surfaces and in correlating them to the correct landsurface level, bringing the possibility of incorrect estimations of elevation range and number of landsurface levels. Also, the origin and meaning of strath terrace sequences should be further investigated and clarified and their age more precisely determined. These drawbacks obviously reflect on any subsequent calculation and estimation in terms of tectonic activity using landsurfaces as reference levels across fault-lines.

Case studies from central Italy

Over the last two years research has been conducted in the Apennines to evaluate the potential of palaeolandsurface analysis in the study of recent tectonics.

This section deals with three examples from the central and southern Apennines (the Fucino basin, the uppermost reach of the Aterno River valley and the Boiano basin; Fig. 1) where strath terraces have been analysed in order to compare the landscape evolution and tectonic setting of each area. The formation of these Pliocene and Quaternary intermontane basins, which are characterized by fluvial and lacustrine deposition, was strictly related to normal faulting of the same age. Some of these faults show clear signs of recent activity (Late Pleistocene–Holocene), such as the displacement of recent slope deposits (e.g. unconsolidated gravel) in correspondence to fault scarps, and large earthquakes ($M_S = 6.5–7$) have occurred during the last millennium in all three areas.

The Fucino basin was selected as the ideal place to test the described methodology as it has already been thoroughly studied from a structural, geomorphological and stratigraphic point of view. The results obtained in this area encouraged the application of the methodology to different tectonic settings. Even though the details of the geological evolution of the Aterno

and Boiano basins have not been thoroughly studied, the analysis performed in these basins resulted in important information regarding fault behaviour and useful constraints for active tectonics.

The Fucino basin (central Apennines)

The Plio-Quaternary tectonic evolution of the Fucino basin is the result of the complex interaction of structures parallel and transverse to the general NW–SE trend of the Apennine chain (Fig. 2). The chronology and kinematics of these normal dip-slip and left-lateral strike-slip faults have been defined through structural, geomorphological and stratigraphic analyses (Bosi et al. 1995; Galadini & Messina 1994).

Recent palaeoseismological studies have precisely characterized the Late Pleistocene–Holocene activity of the NW–SE faults (Michetti et al. 1996; Galadini et al. 1997) and co-seismic faulting due to the destructive 1915 earthquake ($M_S = 6.9$; Ward & Valensise 1989) represents further evidence of this activity. The basin is a typical half-graben evolving in a graben-like structure and its evolution has clearly been driven by the NW–SE normal faults along the eastern border of the basin (Galadini & Messina 1994).

Remnant landsurface analysis has been performed over an area of $c.\,1500\,km^2$ that includes all the ranges surrounding the Fucino basin (Bosi et al. 1996). All remnants were erosional strath terraces distributed between 800 and 1950 m, whereas the elevation of the present-day basin floor is $c.\,650\,m$ (Fig. 3).

By chronologically relating the strath terraces with the previously recognized continental stratigraphy of the area, Bosi et al. (1996) deduced that (i) except for the highest ones, all the strath terraces found in the area are younger than the Pliocene formations (Bosi & Messina 1991) and (ii) the lowest strath terrace is correlated with Middle Pleistocene deposits (Bosi et al. 1995).

The spatial distribution of the terraces indicates that as many as four different terrace sequences are present in the area and that each characterizes a different sector (Bosi et al. 1996; Fig. 4). A comparison of Figs 2 and 4 shows that the boundaries of the four sectors correspond to the main faults of the area which are already known. Therefore, the four fault-bounded sectors could have had a different amount of vertical movement from each other, and from the general uplift of the chain, between the Pliocene and Holocene. In other words, while the terraces were forming these faults were producing such high vertical displacements that the development of the different terrace sequences was eventually determined.

Uppermost reach of the Aterno River (central Apennines)

The uppermost reach of the Aterno River is a straight valley that runs NW–SE; at its SE end it joins the large intermontane basin of Pizzoli–Barete. The studied part of this valley falls within the area struck by a sequence of earthquakes in 1703 (two main shocks of $M_S = 6.7$ and $M_S = 6.2$, according to Camassi & Stucchi 1996).

On the NE flank of the valley a steep scarp is clearly exposed which is related to a NW–SE striking, SW dipping and $c.\,10\,km$ long normal fault (Mount Marine fault). In some places along the scarp an offset of some metres can be detected which affects Late Pleistocene slope deposits (unconsolidated gravel).

Patches of remnant landsurfaces, ranging in size from a few square hectometres to a few square kilometres, were first identified from aerial photographs and then mapped at a scale of 1:25 000. Field surveys supplied some provisional data on clastic continental deposits in

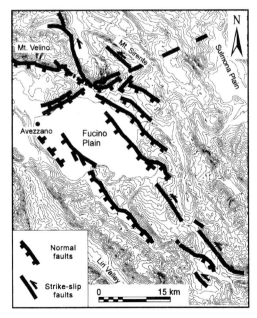

Fig. 2. Structural sketch-map of the Fucino basin (adapted from Galadini & Messina 1994).

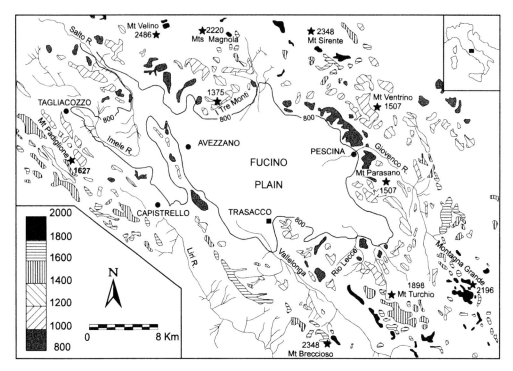

Fig. 3. Altimetric distribution (metres) of remnant landsurfaces in the Fucino area (after Bosi *et al.* 1996).

order to interpret the relative chronology of the landsurfaces. Correlation between patches of remnant landsurfaces, based on the criteria described above, led to the reconstruction of as many as seven remnant landsurface levels distributed between 700 m and 1600 m, with levels being separated from one another by slopes averaging 100 m high (Basili *et al.* 1997).

The recognized terrace sequences have been plotted in Fig. 5 from W to E as well as the range of elevations of the top surface of each terrace. The lowest level in Fig. 5 represents an undetermined number of fill terraces and cut terraces hanging at a maximum of 100 m above the modern flood-plain. The details of these terraces are beyond the scope of this paper and thus will not be addressed. In contrast, higher levels represent flights of strath terraces where only a few slope deposits are still preserved. As a whole these terrace sequences range from the Early Pleistocene to the Holocene, according to the regional chronological framework of Bosi & Messina (1991).

On the basis of the fault characteristics and the spatial distribution of the terraces, it was possible to derive a series of preliminary deductions regarding the tectonics of this area. The distribution of the strath terraces indicates that the ranges within the study area may belong to a single large block having homogeneous tectonic behaviour, i.e. continual uplift since the Early Pleistocene, because significant displacement does not exist between the terrace sequences. Within this block, a down-thrown sector is present (i.e. a depression of a few tens of square kilometres – the Pizzoli–Barete basin) in correspondence with the active Mount Marine fault (Fig. 6).

The Boiano basin (southern Apennine)

The Boiano basin is a NW–SE striking intermontane depression of the southern Apennine chain. Basin fill consists of Middle–Late Pleistocene fluvial and lacustrine sediments which crop out predominantly in the lowest portions of the basin (Russo & Terribile 1995).

The basin is bounded by two NW–SE trending normal fault-sets whose main branches displace Mesozoic carbonate units with respect to the Cenozoic siliciclastic units which outcrop extensively along both flanks of the basin. A steep scarp affecting Mesozoic carbonate formations is related to the fault along the SW slope of the basin. In places this fault also displaces slope

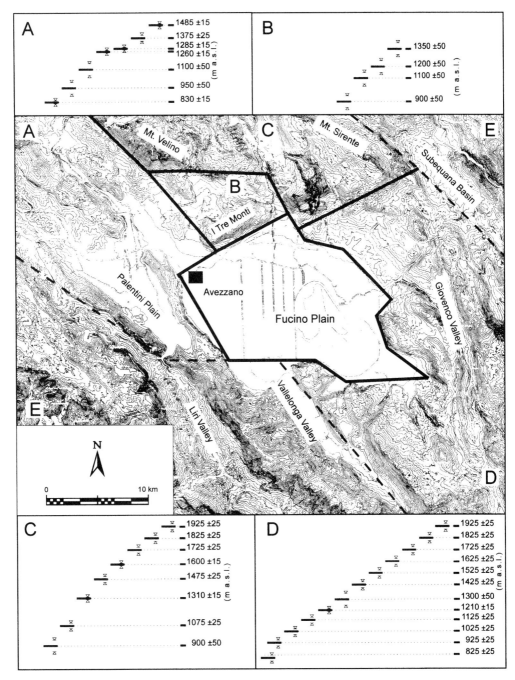

Fig. 4. Sketch-map of the different sectors of the Fucino basin as defined through the palaeolandsurface analysis. The different terraced sequences are shown in the insets and identify A, B, C and D sectors (after Bosi *et al.* 1996).

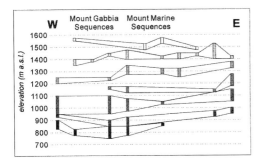

Fig. 5. Altitudinal distribution of remnant landsurfaces of the upper reach of the Aterno River valley. The lowest level represents fill terraces and cut terraces; all the others are strath terraces.

deposits which are most probably Late Pleistocene in age. In contrast, the fault is less evident on the NE wall since it mainly cuts Cenozoic siliciclastic units which are more erodible, thereby allowing the slope to recover more easily. According to Corrado *et al.* (1997) a Middle Pleistocene NE–SW extension (still active) was responsible for the evolution of this graben-like structure.

From a seismological point of view, this area is particularly important because it was affected by two strong earthquakes ($M_S > 6.5$) in 1456 and 1805 (Camassi & Stucchi 1996). As knowledge regarding the tectonics of the Boiano basin is presently insufficient to identify which fault was responsible for the high-magnitude seismicity, an analysis of the remnant landsurfaces was carried out on both walls of the basin. Several surfaces were defined between 500 and 1250 m and all represent the top of strath terraces which are estimated to have formed mainly during Early–Middle Pleistocene since they are older than the basin fill. Three different terrace

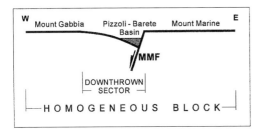

Fig. 6. Schematic cross-section showing the relationship between the activity of the Mount Marine fault (MMF) and the formation of the Pizzoli–Barete basin. The latter represents a downthrown portion inside a homogeneous block characterized by the same terrace sequences everywhere.

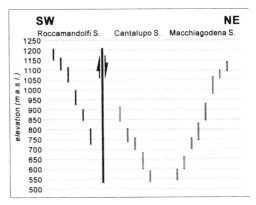

Fig. 7. Altitudinal distribution of remnant landsurfaces of the Boiano basin. Significant displacement was produced by the fault between the Roccamandolfi sequence and the Cantalupo–Macchiagodena landsurface sequences.

sequences were identified, one on the NE side of the basin (the Macchiagodena sequence) and two on the SW side (the Cantalupo and Roccamandolfi sequences).

The vertical distribution of these terrace sequences (Fig. 7) shows that a fairly good accordance exists between the Macchiagodena and Cantalupo sequences, whereas there is no accordance between these sequences and the Roccamandolfi sequence. This anomaly in the evolution of the basin landscape was interpreted as being due to the effects of the SW fault. The fault could have determined a different number of terraces on the Roccamandolfi sequence, as well as an offset of this sequence with respect to the others, while younger terraces were being formed inside the basin (probably during the Late Pleistocene).

These observations indicate that the SW fault was the primary structural element in controlling the tectonic evolution of the Boiano basin and that it is the major active structure in the area since the offset it produced is recorded in the later stages of landscape evolution. The NE fault, instead, may have contributed only during the early stages of basin formation or during the very latest stages when the basin actually became a graben. In this latter case the fault may be responsible for vertical offsets lower than the accuracy of the measurement method used.

In short, even though this analysis has not quantified the present tectonic regime (e.g. slip rates, extension rates), it has permitted the identification of the major fault of the area. As such, this preliminary study has helped to direct future research (e.g. palaeoseismological studies)

towards the NW-SE trending fault located on the SW flank of the Boiano basin.

Discussion and conclusions

The analysis of ancient landsurfaces, like those that identify strath terraces, has been applied to the ranges surrounding three Apennine intermontane basins.

This methodology offered the opportunity to investigate an environment that, because of its scarcity of deposits, is sometimes neglected in geological research which relies on the application of systematic stratigraphic analysis and dating techniques. For the same reason, unfortunately, the reconstruction of landscape evolution can only define a relative chronology. This aspect, as well as the low accuracy by which landsurfaces could be detected, represents obvious methodological limitations which at present hinder the realization of precise calculations, such as deformation rates.

Despite these drawbacks, the analysis of ancient landsurfaces at all three sites provided useful indications regarding fault activity and their effects on landscape evolution. In the first area examined (Fucino) it was found that Quaternary faults left clear signs of their activity on landscape evolution by determining the formation of different strath terrace sequences. In the second area (Aterno) the reconstruction of strath terrace sequences provides evidence of the tectonic behaviour of the area and useful elements that describe the activity of a fault. In the third area (Boiano) the main active fault was identified as being responsible for the displacement of strath terrace sequences.

The application of this methodology has thus been shown to be a very powerful way in which to retrace the geological evolution of both dissected and extensively denuded areas, as well as to assess the effects of tectonics. In more general terms this research has revealed that an extended use of palaeolandsurface analysis allows one to construct reliable hypotheses regarding Quaternary kinematic evolution, as well as contributing to the identification of the faults that played a dominant role in the recent geological history of the Apennine chain. These issues are of great importance in terms of merging information on the long-term evolution of fault-generated basins (to which the reconstruction of landscape evolution is essential) with information on the short-term history of tectonics derived from palaeoseismological studies or inferred from the historical seismological record of a region.

The results obtained in these case studies should encourage further research in this direction, as a contribution not only towards the proper reconstruction of a region's geological evolution, but also to the solution of some outstanding problems of active tectonics.

V. Bosi and P. Galli from SSN (Italian Seismic Survey) are thanked for supporting field survey in the Boiano area. Two anonymous referees are also thanked for their comments which improved the final version of the manuscript.

References

BASILI, R., BOSI, C. & MESSINA, P. 1997. La tettonica Quaternaria dell'alta valle del F. Aterno desunta dall'analisi di successioni di superfici relitte. *Il Quaternario*, **10**, 621–624.

BOSI, C. 1975. Osservazioni preliminari su faglie probabilmente attive nell'Appennino centrale. *Bollettino della Società Geologica Italiana*, **94**, 827–859.

—— & MESSINA, P. 1991. Ipotesi di correlazione fra successioni morfo-litostratigrafiche plio-pleistoceniche nell'Appennino laziale-abruzzese. *Studi Geologici Camerti*, vol. spec. (1991/2) CROP **11**, 257–263.

——, CAIAZZO, C., CINQUE, A. & MESSINA, P. 1996. Le superfici relitte della regione fucense (Appennino centrale) ed il loro possibile significato nella ricostruzione della evoluzione geologica. *Il Quaternario*, **9**, 381–386.

——, GALADINI, F. & MESSINA, P. 1995. Stratigrafia plio-pleistocenica della conca del Fucino. *Il Quaternario*, **8**, 83–94.

BUCHER, W. H. 1932. 'Strath' as a geomorphic term. *Science*, **75**, 130–131.

CAMASSI, R. & STUCCHI, M. 1996. *A parametric catalogue of damaging earthquakes in the Italian area*. GNDT, National Group for the Defence Against Earthquakes, Milano.

CNR–PFG 1987. *Neotectonic Map of Italy*. Quaderni della Ricerca Scientifica, No. 114, Vol. **4**.

CORRADO, S., DI BUCCI, D., LESCHIUTTA, I., NASO, G. & TRIGARI, A. 1997. La tettonica quaternaria della piana d'Isernia nell'evoluzione strutturale del settore molisano. *Il Quaternario*, **10**, 609–614.

D'AGOSTINO, N., SPERANZA, F. & FUNICIELLO, R. 1997. Le brecce Mortadella dell'Appennino Centrale: primi risultati di stratigrafia magnetica. *Il Quaternario*, **10**, 385–388.

FORNASERI, M. 1985. Geochronology of volcanic rocks from Latium (Italy). *Rendiconti della Società Italiana di Mineralogia e Petrografia*, **40**, 73–106.

GALADINI, F. & MESSINA, P. 1994. Plio-Quaternary tectonics of the Fucino basin and surrounding areas (central Italy). *Giornale di Geologia*, **56**, 73–99.

——, GALLI, P. & GIRAUDI, C. 1997. Geological investigations of Italian earthquakes: new paleoseismological data from the Fucino plain (central Italy). *Journal of Geodynamics*, **24**, 87–103.

MERRITTS, D. J., VINCENT, K. R. & WOHL, E. E. 1994. Long river profiles, tectonism, and eustasy: A guide to interpreting fluvial terraces. *Journal of Geophysical Research*, **99**(B7), 14031–14050.

MICHETTI, A. M., BRUNAMONTE, F., SERVA, L. & VITTORI, E. 1996. Trench investigations of the 1915 Fucino earthquake fault scarps (Abruzzo, central Italy): Geological evidence of large historical events. *Journal of Geophysical Research*, **101**(B3), 5921–5936.

RUSSO, F. & TERRIBILE, F. 1995. Osservazioni geomorfologiche, stratigrafiche e pedologiche del bacino di Boiano (Campobasso). *Il Quaternario*, **8**, 239–254.

WARD, S. N. & VALENSISE, G. R. 1989. Fault parameters and slip distribution of the 1915 Avezzano, Italy, earthquake derived from geodetic observations. *Bulletin of the Seismological Society of America*, **79**, 690–710.

An overview of Pliocene to present-day uplift and denudation rates in the Northern Apennine

CARLO BARTOLINI

Dipartimento di Scienze della Terra, Via G. La Pira 4, Firenze, Italy

Abstract: Quaternary, Holocene and present denudation rates in the Northern Adriatic basin and sedimentary supply rates, formerly to the Northern Apennine Foredeep and at present to the Adriatic Sea, are approximately consistent. The Quaternary denudation rate, based on sediment feeding to the foredeep is $0.45 \, \text{mm} \, \text{a}^{-1}$. This value is congruous with a mean exhumation rate since the Early Pliocene, deduced from apatite fission track dating, of $0.5 \, \text{mm} \, \text{a}^{-1}$ in the Apuan Alps. The Quaternary supply rate to the foredeep ($0.047 \, \text{km}^3 \, \text{a}^{-1}$) is over twice as fast as that of the Pliocene ($0.021 \, \text{km}^3 \, \text{a}^{-1}$), and reflects an increased denudation rate due to higher energy relief and a slightly larger catchment basin. The Pleistocene increase in uplift/denudation rates is emphasized by the onset of a fast prograding sedimentation pattern since 1.0 to 0.8 Ma BP on both the Padan and the Tyrrhenian side of the Apennines. Pleistocene uplift also had a considerable impact on Apennine morphology and drainage pattern evolution.

Denudation rates since Pliocene times in the Northern Adriatic watershed (i.e. Alps, Northern Apennine and Dinarides) have been examined previously from studies of the volume of accumulated sediments (Bartolini *et al.* 1996). This paper aims to:

- place the Pliocene and the Quaternary sedimentary yield to the Northern Apennine Foredeep in the context of the sedimentary history of the Foredeep in terms of depocenter sedimentation rates
- relate the denudation rates obtained by Bartolini *et al.* (1996) to available published and previously unpublished data on the timing of the Northern Apennine uplift.

Foredeep feeding rates

The history of a mountain chain is inversely written in its foredeep. The Northern Apennine foredeep (Fig. 1) is the latest evolutionary stage of a basin dating back at least to the Early Cretaceous (e.g. Boccaletti *et al.* 1990) and is now largely filled with the Adriatic Sea as a remnant feature.

In order to outline the recent history of the Northern Apennine foredeep, sedimentary feeding rates have been worked out (Fig. 2). Although less meaningful, depocentre sedimentation rates are given rather than mean rates, since the maximum thickness is a more reliable parameter when dealing with units which have been affected by tectonics and partially eroded, such as the Upper Oligocene to Upper Miocene units.

When hard rock formations are compared in terms of volume and/or thickness, with largely unconsolidated and undisturbed sedimentary units, porosity corrections have to be applied. Based on the data of Fuchtbauer (1974), Johnson & Olhoeft (1984) and Manger (1963), a porosity of 15% has been applied to the Macigno, Marnoso Arenacea and Laga Formations, which comprise sandstones and marlstones. On the basis of porosity/depth charts, worked out for the Po plain subsurface (Bartolini *et al.* 1996), rounded values of 20% and 30% respectively were assigned to the Pliocene and the Quaternary sedimentary units.

From Fig. 2, the impact of the emersion of the inner sector of the foredeep in Late Miocene is clear. At that time, the foredeep (which was previously fed by the erosion of well lithified eo-Alpine, Ligurian and Epiligurian units) began to be self-fed by the proto-Apenninic chain made up of poorly consolidated and highly erodible sedimentary units, resulting in a marked increase in sedimentation rate. However, the relatively low sedimentation rates affecting the depocenter of the Pliocene and, to a larger extent, that of the Quaternary unit, are not due to lower feeding and denudation rates (see ahead) but to the foredeep termination due to the wearing out of the Northern Apennine – Southern Alps and Dinarides gap (see for instance Castellarin & Vai 1986). During Pliocene times the foredeep was progressively replaced by a shallow basin: the Adriatic Sea.

An appraisal of the latest phases of foredeep volumetric sedimentation patterns was recently carried out by Bartolini (1993*a,b*) and by Bartolini *et al.* (1994, 1996). The present volume of Pliocene and Quaternary sediments may be reasonably inferred from maps of the base of the

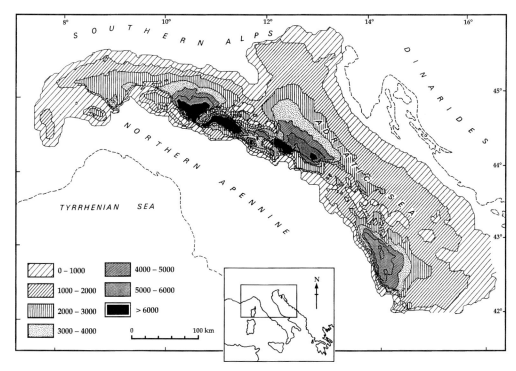

Fig. 1. Isobaths (in metres) of the base of the Pliocene–Quaternary succession (from Bartolini et al. 1996).

Pliocene and Quaternary successions (Fig. 1). It should be pointed out, however, that the base of the Quaternary is set at the base of the *Hyalinea balthica* Zone (i.e. the base of the Emilian Substage, 1.4 Ma BP). The Pliocene volume has been corrected by adding an estimate of the underthrust sediments, while no correction has been attempted for the eroded marginal deposits. The estimates of 97 000 and 95 000 km^3 as the present volumes of the Pliocene and Quaternary deposits are therefore likely to be somewhat lower than the volumes originally deposited. The volume to area ratio (i.e. the average thickness) divided by the elapsed time is the mean sedimentation rate, which ranged from 0.23 mm a^{-1} during the Pliocene to 0.66 mm a^{-1} throughout the Quaternary.

Sedimentation rates depend on the supply or feeding rate as well as on basin morphology

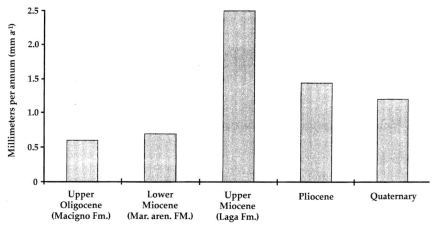

Fig. 2. Depocentre sedimentation rates in the Northern Apennine Foredeep.

and dimension. In order to remove this factor, present volumes (featuring different porosities according to grain size and depth of burial) have been transformed into net volumes in order to obtain the relative net feeding rates which are $0.021 \text{ km}^3 \text{ a}^{-1}$ (Pliocene) and $0.047 \text{ km}^3 \text{ a}^{-1}$ (Quaternary).

Exhumation and denudation rates

Data on uplift and denudation rates of the Northern Apennine catchment area were, until recently, few and of local scope (Bartolini 1984; Ciccacci *et al.* 1986; Clark & Jäger 1969; Wagner *et al.* 1977). However, apatite fission-track dating, recently carried out over different tectonic units of the Northern Apennine, has produced Pliocene–Quaternary mean exhumation rates ranging from 0.5 to 1.7 mm a^{-1} across the Apuane Alps (Abbate *et al.* 1994) and Late Miocene–Quaternary mean denudation rates of 0.3–0.4 mm a^{-1} in the Ligurian Apennine (Balestrieri *et al.* 1996). The higher rates in both areas derive from cooling rates assuming a paleogeothermal gradient of $30°\text{C km}^{-1}$. Lower rates (0.5 mm a^{-1} and a minimum of 0.2 mm a^{-1}, respectively) derive directly from the slope of the age-elevation profiles. A mean exhumation rate of 0.2–0.3 mm a^{-1} in the Ligurian Apennine since Late Miocene and of 0.5 mm a^{-1} in the Apuan Alps since Early Pliocene can be used as reference values.

Mica cooling ages obtained from borehole samples in the Larderello geothermal field, Southern Tuscany, indicate a downward migration of the brittle/ductile boundary occurring since Langhian times (i.e. since the onset of multistage extensional tectonics) at a mean rate of 0.27 mm a^{-1} (Dallmayer *et al.* 1995).

A mass-balance palaeogeographic reconstruction was carried out by Bartolini *et al.* (1996) to obtain the Quaternary average denudation rate in the catchment area. In this the rated mean pore volume of the eroded rocks ($11\,700 \text{ km}^3$) was added to the aforementioned net sediment volume ($64\,000 \text{ km}^3$). By introducing the present-day catchment area ($127\,870 \text{ km}^2$, see Fig. 3) which did not change appreciably throughout the Quaternary, a *mean denudation rate* over the entire area of 0.45 mm a^{-1} is obtained.

Direct comparison of the fission track rate and of the supply rate/denudation rate data is not possible because the former refers to single sites located right on the chain (and therefore presently affected by relatively high uplift rates) while the latter are averaged over the entire catchment basin.

A wealth of seismo-stratigraphy surveys carried out during the last decades in the western sector of the Adriatic, where most of the sedimentation occurs, allow an estimate of the net volume of Holocene sediments deposited over the last 6000 years (i.e. since the Adriatic coastline attained its present position) in the Po Plain–Adriatic Sea basin, which is rated at approximately 250 km^3, and of the relative net feeding rate ($0.042 \text{ km}^3 \text{ a}^{-1}$), which is not significantly different from the Quaternary rate ($0.047 \text{ km}^3 \text{ a}^{-1}$, see above). Due to the relatively large errors which, unavoidably, affect a volume estimate carried over a veneer of sediments such as the Holocene unit, an interpretation of these data, for instance the role of climate on erosion and feeding rates, will not be attempted. It can, however, be said that the Holocene feeding rate is consistent with the Quaternary value.

Quaternary denudation rates imply that an average theoretical thickness of some 600 m was eroded from the Northern Apennine (i.e. Northern Adriatic Sea) catchment basin during the Quaternary (Table 1). Due to the much higher erodibility of the rocks outcropping in the Apennines, the present Apennine denudation rate is several times higher than that affecting, on average, the whole catchment basin. Since the disparity was even larger through the Quaternary, because of wider outcropping of the Ligurides, it can be assumed that an average theoretical thickness in excess of 1 km was eroded in the Apennine area during this time span.

Table 1. *Erosion in the catchment area of the Northern Apennine Foredeep (Alps, N. Apennine and Dinarides)*

Interval	Time elapsed (Ma)	Theoretical average thickness (m)	Volume eroded (km³)
Quaternary	1.4	600	70 000
Pliocene	3.6	850	100 000
Pliocene and Quaternary	5.0	1450	170 000

An evaluation of the Pliocene denudation rate would require a reliable estimate of the relative source area, which was smaller than at present, due to the recent uplift of the Apennine chain. By restoring the Pliocene coastline on the eastern slope of the Northern Apennine, a Pliocene source area $\frac{1}{5}$ smaller than the present (which amounts to $128\,000\,\text{km}^2$) is suggested. It is therefore estimated that an additional average theoretical thickness of 850 m could have been removed during the Pliocene from the catchment basin of the Northern Apennine foredeep.

In conclusion, since the beginning of the Pliocene, a minimum total of $170\,000\,\text{km}^3$ and a theoretical *average* thickness of 1450 m has been eroded. This volume could be compared to the present volume of the Alps and of the Northern Apennine from the watershed to sea level in the present source area, amounting approximately to $110\,000\,\text{km}^3$. This means that with Pliocene and Quaternary denudation rates, it would take slightly over 3 Ma to denude the Alps and the Apennines, if uplifting did not occur. Due to the different erodibility of the exposed rocks in the two mountain chains, the required time would be longer in the Alps and shorter in the Apennines.

Timing of the Northern Apennine uplift

The timing of Apennine uplift, deduced from feeding and denudation rates, has not been satisfactorily elucidated. All that is known is that the Quaternary feeding rate to the foredeep ($0.047\,\text{km}^3\,\text{a}^{-1}$) is over twice as fast as the Pliocene rate ($0.021\,\text{km}^3\,\text{a}^{-1}$). Since the catchment basin of the Apennine foredeep was somewhat smaller during Pliocene times (because of the much smaller width of the Apennine chain) the denudation rate actually increased less than the ratio 0.047:0.021. A round figure of 2:1 can be assumed as a ratio closer to reality. A few points can be raised to clarify the timing of the Quaternary uplift of the Northern Apennine.

Lower–Middle Pleistocene sedimentation rates

Seismic profiles traced on the Padan and Tyrrhenian side by Argnani *et al.* (1997) point to the onset of a fast prograding sedimentation pattern from 1.0 to 0.8 Ma BP. Accordingly, since the Middle Pleistocene, coarse-grained fluvial sediments were laid down at a fast rate over the older, mostly lacustrine deposits of the Northern Apennine intermontain basins. A marked increase in sedimentary supply rate to the eastern Ligurian shelf during late Pleistocene has also been pointed out by Fanucci *et al.* (1995).

Uplifted summit areas of low relief

The Northern Apennine chain was a persistent narrow belt of low relief, shifting northeastwards from its early emersion at the end of Miocene until Early Pleistocene times. The Adriatic coastal plain shrank and widened following sea-level changes without, however, appreciably affecting the area of the catchment basin where erosion actually occurs. One may wonder whether this time interval was long enough to allow the carving of planation surfaces in the present-day chain area. The evaluation of longitudinal river profiles, frequency distributions of contour lines and the hypsographic curve obtained from a digital terrain model has led to the identification of the summit areas lying north of Lucca and Pistoia as uplifted palaeoforms (Bartolini 1980*a,b*). The lack of sedimentary deposits coeval with the paleosurfaces does not allow confirmation or direct dating of the planation and/or of the subsequent uplift. Saprolites featuring iron oxide and hydroxide mottling are, however, common features of the summit area bedrock. Such features were interpreted, by means of colorimetric and chemical analyses, as deep remnants of palaeosols antedating the uplift. Truncated palaeosols of Middle Pleistocene age, which can be correlated with the saprolites, are preserved in the continental succession laid in the intermountain basin located at the margin of the uplifted area (Magaldi *et al.* 1983; Dallan 1988). A pre-Middle Pleistocene age can thus be inferred for the aforementioned summit areas of low relief, as the planation process was interrupted in the Middle Pleistocene due to mountain uplift (Bartolini *et al.* 1984; Bartolini & Nishiwaki 1985).

The mode of Northern Apennine uplift

Middle Pleistocene uplift was neither widespread nor uniform. It mainly affected the present day chain areas and, to a much lesser extent, the outer (eastern) intermontane basins, while the inner (western) basins continued subsiding. The differential character of uplift patterns in Southern Tuscany has recently been documented by Bossio *et al.* (1995) through an evaluation of vertical displacements which affected the *Globorotalia puncticulata–G. aemiliana* interface.

Geomorphic impact of the uplift

The highly differential uplift rates affecting the Tyrrhenian side of the Northern Apennine since Middle Pleistocene led to a wide reaching rearrangement of the drainage pattern (e.g. Bartolini & Pranzini 1981), whereby upstream river sectors lying in areas of high uplift rate captured neighbouring river systems. The uplift pattern of the Lima drainage basin, as revealed by a stratigraphic appraisal of fault block mobility (Bartolini & Fazzuoli 1997), is remarkably consistent with the regional uplift pattern revealed by trend surface analysis of the already mentioned uplifted summit areas (Fig. 3).

The geodynamic framework

From a geodynamic standpoint, the Middle Pleistocene uplift phase marked a drastic decrease in tangential movements in the external belt (Castellarin & Vai 1986; Vai 1989). The effects on crustal deformation arising from the mid-Middle Pleistocene termination of compressive tectonics in the Northern Apennines has been investigated by Sabadini *et al.* (1986). Assuming that the only tectonic stress that presently remains active is induced by isostatic readjustment of crustal density anomalies, the authors outlined a model whereby the relative vertical motion between the Tyrrhenian area

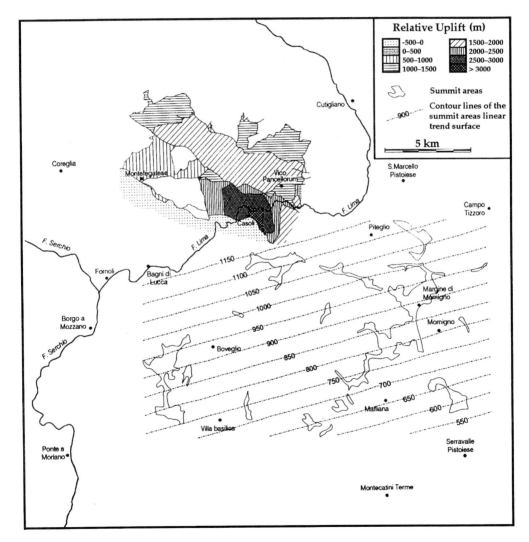

Fig. 3. Uplift pattern of the Lima Valley matched with that revealed by linear trend surface analysis of the uplifted summit areas of low relief. The Lima Valley 'high' controlled the uplift of a large area southward.

and the chain should range between 0.5 and 1.0 mm a^{-1}, thus confirming the data obtained from foredeep feeding rates.

Summary and conclusions

The final stages of the Northern Apennine foredeep sedimentary evolution have been evaluated in terms of sediment yield. The denudation rates affecting the Northern Apennine foredeep (i.e. the present-day Northern Adriatic catchment area) during Pliocene, Quaternary and Holocene times have been briefly reviewed in order to point out the rapidity with which an orogenic belt can be worn away. Differences of rock erodibilitiy in the Alps and in the Northern Apennine should, however, be noted.

The available data point to an acceleration of the uplift rate in early Middle Pleistocene times. Indeed, a present-day fast uplift rate is required to explain outcrops of erodible sandstones to largely outcrop at 1800–2000 m under relatively severe climatic conditions.

Criticism by two anonymous referees is gratefully acknowledged. The research was funded by Progetto MURST 40% 'La risposta dei processi geomorfologici alle variazioni ambientali'.

References

ABBATE, E., BALESTRIERI, M. L., BIGAZZI, G., NORELLI, P. & QUERCIOLI, C. 1994. Fission-track datings and recent denudation in Northern Apennines, Italy. *Memorie della Società Geologica Italiana*, **48**, 579–585.

ARGNANI, A., BERNINI, M., DIDIO, G. M., PAPANI, G. & ROGLEDI, S. 1997. Registrazione stratigrafica di eventi tettonici a scala crostale nei depositi Quaternari del Nordappennino. *Abstracts Convegno 'Tettonica Quaternaria del Territorio Italiano: Conoscenze, Problemi, Applicazioni'*, Parma, Feb. 25–27.

BALESTRIERI, M. L., ABBATE, E. & BIGAZZI, G. 1996. Insights on the thermal evolution of the Ligurian Apennines (Italy) through fission track analysis. *Journal of the Geological Society*, London, **153**, 419–425.

BARTOLINI, C. 1980a. Uplifted low-relief morphology of the Northern Apennines. *Abstracts 26th International Geological Congress*, Paris, July 7–17, 1980.

—— 1980b. Su alcune superfici sommitali dell'Appennino Settentrionale. (Prov. di Lucca e di Pistoia). *Geografia Fisica e Dinamica Quaternaria*, **3**, 42–60.

—— 1984a. Uplift and denudation rates in the Northern Apennines. Miocene to present. *Abstracts 25th International Geographical Congress*, Paris – Alpes, August 27–31, 1984.

—— 1984b. Rates of vertical movements over a thinning crust. *Bulletin INQUA Neotectonics Commission*, **7**, 49–53.

—— 1993a. Pliocene versus Quaternary sedimentation in the Adriatic Foredeep. *Abstracts International Symposium 'Dynamics of the Fluvial-Coastal System'*, S. Benedetto Del Tronto, June 21–24, 1993.

—— 1993b. The Apennine build up and its sedimentary budget. *Abstracts 3rd International Geomorphology Conference*, Hamilton, August 23–29, 1993.

——, BIDINI, D., FERRARI, G. & MAGALDI, D. 1984. Pedostratigrafia e morfostratigrafia nello studio delle superfici sommitali situate fra Serchio e Ombrone Pistoiese. *Geografia Fisica e Dinamica Quaternaria*, **7**, 3–9.

——, CAPUTO, R. & PIERI, M. 1994. Erosion and sedimentation rates in the Northern Apennine Foredeep. *Abstracts 77a Riunione Estiva della Società Geologica Italiana*, Bari, Sep. 23–Oct. 1, 1994.

——, —— & —— 1996. Pliocene – Quaternary sedimentation in the Northern Apennine Foredeep and related denudation. *Geological Magazine*, **133**(3), 255–273.

—— & FAZZUOLI, M. 1997. Ruolo della tettonica e della morfoselezione nell'evoluzione dell'idrografia nel bacino del Fiume Serchio. *Il Quaternario*, **10**(2),

—— & NISHIWAKI, N. 1985. Uplift model by trend analysis of an Apennine region lying south of the Lima River (Northern Tuscany). *Geografia Fisica e Dinamica Quaternaria*, **8**, 14–22.

—— & PRANZINI, G. 1981. Plio-Quaternary evolution of the Arno basin drainage. *Zeitschrift für Geomorphologie N.F., Supplementband*, **40**, 77–91.

BOCCALETTI, M., CALAMITA, F., DEIANA, G., GELATI, R., MASSARI, F., MORATTI, G. & RICCI LUCCHI, F. 1990. Migrating foredeep – thrust belt system in the Northern Apennines and Southern Alps. *Palaeogeography, Palaeoclimatology, Palaeoecology*, **77**, 3–14.

BOSSIO, A., COSTANTINI, A., FORESI, L. M., LAZZAROTTO, A., LIOTTA, D., MAZZANTI, R., MAZZEI, R., SALVATORINI, G. & SANDRELLI, F. 1995. Studi preliminari sul sollevamento della Toscana meridionale dopo il Pliocene Medio. *Studi Geologici Camerti*, Vol. Spec., **1995/1**, 87–91.

CASTELLARIN, A. & VAI, G. B. 1986. Southalpine versus Po Plain Apenninic arcs. *In*: WEZEL, F. C. (ed.) *The Origin of Arcs*. Elsevier, Amsterdam, 253–280.

CICCACCI, S., FREDI, P., LUPIA PALMIERI, E. & PUGLIESE, F. 1986. Indirect evaluation of erosion entity in drainage basins through geomorphic, climatic and hydrological parameters. *Abstracts International Geomorhology Conference*, Manchester, 1985, 33–46.

CLARK JR., S. P. & JÄGER, E. 1969. Denudation rate in the Alps from geochronological and heat flow data. *American Journal of Science*, **267**, 1143–1160.

DALLAN, L. 1988. Ritrovamento di *Alephis lyrix* nelle argille della serie lacustre di Montecarlo (Lucca) e considerazioni stratigrafiche sui depositi continentali dell'area tra il Monte Albano e il Monte Pisano. *Atti della Società Toscana di Scienze Naturali, Memorie, Serie A*, **95**, 203–219.

DALLMEYER, R. D., DECANDIA, F. A., ELTER, F. M., LAZZAROTTO, A., & LIOTTA, D. 1995. Il sollevamento della Crosta nel quadro della tettonica distensiva post-collisionale dell'Appennino Settentrionale: nuovi dati dall'area geotermica di Larderello (Toscana Meridionale). *Studi Geologici Camerti, Vol. Spec.*, **1995/1**, 337–347.

FANUCCI, F., FIRPO, M. & VETUSCHI ZUCCOLINI, M. 1995. Evoluzione delle piattaforme continentali tirreniche e tettonica verticale. *Studi Geologici Camerti, Vol. Spec.*, **1995/1**, 391–398.

FUCHTBAUER, H. 1974. Some problems of diagenesis in sandstones. *Bulletin Centre Recherche, Pau*, **8**(1), 391–403.

JOHNSON, G. R. & OLHOEFT, G. R. 1984. Density of rocks and minerals. *In*: CARMICHAEL, R. S. (ed.) *CRC Handbook of Physical Properties of Rocks*, **3**, 1–38.

MAGALDI, D., BIDINI, D., CALZOLARI, C. & RODOLFI, G. 1983. Geomorfologia, suoli e valutazione del territorio tra la piana di Lucca e il padule di Fucecchio. *Annali dell'Istituto Sperimentale per la Difesa del Suolo*, **14**, 21–108.

MANGER, G. E. 1963. Porosity and bulk density of sedimentary rocks. *U.S. Geological Survey Bulletin*, **1114**-E.

SABADINI, R., SPADA, G. & BOSCHI, E. 1987. Modelling of crustal density anomalies and vertical motions in the central Italian region. *In*: BORIANI *et al.* (eds.) The Litosphere in Italy: Advances in Earth Sciences Research. *Accademia Nazionale dei Lincei, Roma*, **1987**, 45–63.

VAI, G. B. 1989. Migrazione complessa del sistema fronte deformativo-avanfossa-cercine periferico: il caso dell'Appennino settentrionale. *Memorie della Società Geologica Italiana*, **38**, 95–105.

WAGNER, G. A., REIMER, G. M. & JÄGER, E. 1977. The cooling ages derived by apatite fission track, mica Rb-Sr and K-Ar dating: the uplift and cooling history of the Central Alps. *Memorie Istituto Geologia e Mineralogia Università di Padova*, **30**, 1–27.

Structural and stratigraphic perspectives on the uplift and erosional history of Djebel Cherichira and Oued Grigema, a segment of the Tunisian Atlas thrust front

ALASTAIR W. BAIRD[1] & ANDREW J. RUSSELL[2]

[1] *School of Geological Sciences, Kingston University, Penrhyn Road, Kingston upon Thames, Surrey, KT1 2EE, UK*
[2] *Department of Earth Sciences, University of Keele, Keele, Staffordshire, ST5 5BG, UK*

Abstract: Djebel Cherichira, a segment of the Tunisian Atlas thrust front, was structurally uplifted by at least 1650 m and possibly more than 2000 m by late Tortonian and post-Messinian thrusting. After late Tortonian structural uplift and erosion, Djebel Cherichira formed a topographic high with Messinian, arid to semi-arid, debris-flow fan deposits mantling the relict topography. Oued Cherichira formed in Serravallian to early Tortonian times, prior to the growth of the thrust front, as a foreland basin fluvial system ahead of the thrust structures within the thrust belt. It maintained its antecedent drainage pattern during rapid tectonic uplift of the thrust front at Djebel Cherichira. Movement on the Kredija Strike-slip Fault records the latest tectonic event in the thrust belt. This probable Early Pliocene deformation did not displace the drainage pattern, implying that much syn-tectonic erosion and desert pediment formation had already occurred.

Oued Grigema, a tributary of Oued Cherichira, whilst appearing to be antecedent with respect to Djebel Cherichira, is a post-tectonic, consequent stream, where at least some of the fluvial erosion and deposition is of Holocene age. Many of the present-day processes of sedimentation and erosion in Oued Grigema are analogous to those interpreted from the Messinian strata of Djebel Cherichira.

The Atlas Mountains, which form part of the Alpine–Mediterranean–Atlas orogenic belt (Dewey *et al.* 1973; Biju-Duval *et al.* 1977) extend for 2100 km along North Africa from Morocco in the west, through Algeria to Tunisia in the east. The orogenic belt continues eastwards into Sicily and then northwards to form the Calabrian Arc and the Apennines in Italy (Fig. 1, inset).

In Tunisia the Atlas thrust front forms the eastern margin of a range of N–S trending hills in northern Tunisia termed the North–South Axis (Fig. 1). The thrust front extends 300 km southwards from Cap Bon towards Gafsa before trending westwards towards and across Algeria where it is termed the South Atlas Front (Fig. 1, inset). Previously it had been argued by Caire (1978), Perthuisot (1981), Snoke *et al.* (1988), Dewey *et al.* (1989) and others, that the Atlas thrust front lay considerably further to the north at what is now termed the southern margin of the Tellian Atlas (Fig. 1). Ironically, whilst the thrust geometry of the structures forming the N–S Axis has been recognized for some time (Castany 1951; Ouali 1985), it is only recently that the eastern margin of the N–S Axis has been widely accepted as the thrust front of the Tunisian Atlas (Boccaletti *et al.* 1988, 1990; Baird 1991; Burollet 1991; Creuzot *et al.* 1993; Outtani *et al.* 1995; Anderson 1996) and even so, the age and structural evolution of the thrust front are still the subjects of debate.

The flat plains of the Sahel extend eastwards from the hills of the N–S Axis to the Mediterranean coast and form the foreland to the thrust belt. Djebel Cherichira lies 35 km SSW of the city of Kairouan (Fig. 1) and is an 8 km long WSW–ENE trending hill rising abruptly to a height of 462 m (Fig. 2) which forms a small part of the mountain front of the Tunisian Atlas. A comparison of Figs 2 and 3 shows that the trace of the southeasternmost thrust plane coincides with the mountain front.

In this study the tectonic evolution and structural uplift of Djebel Cherichira are described and temporally constrained in relation to topographic uplift and erosion by fluvial and other processes in arid to semi-arid environments. The antecedent Oued Cherichira drainage system is a major ephemeral wadi system which flows eastwards, draining and eroding the hills of the N–S Axis further west, before drying up in the vicinity of Kairouan. The preservation of its base-level during formation of the mountain front facilitates quantification of the erosion which accompanied the structural uplift at the thrust front. Oued Grigema is a smaller wadi system which rises on the northern slopes of

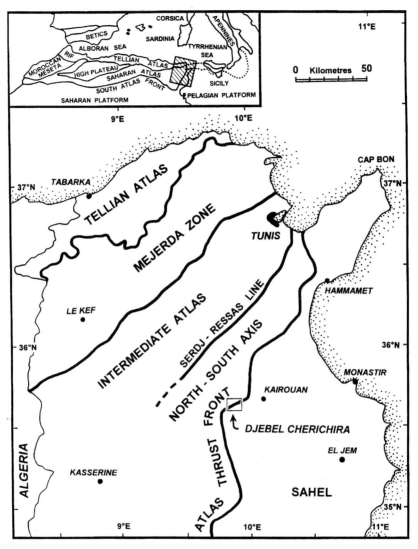

Fig. 1. Map to show the main structural zones of northern Tunisia and the location of Djebel Cherichira on the Atlas thrust front. The inset shows northern Tunisia in the context of the Atlas, Betic and Apennine orogenic belts and the western and central Mediterranean Sea.

Djebel Cherichira and cuts eastwards through the Djebel before joining Oued Cherichira (Fig. 2). The present-day sedimentary and erosional processes in Oued Grigema shed considerable light on the depositional environment of the area in Messinian times.

Structural uplift, topographic uplift and erosion in thrust belts

In any consideration of uplift and erosion it is important to distinguish clearly between topographic or surface uplift, the uplift of rocks by tectonic processes and the exhumation of rocks by erosive processes (England & Molnar 1990). In Djebel Cherichira, as at all thrust fronts in orogenic belts, even when erosion is assumed not to have taken place, the amount of structural uplift of the complex thrust structure above the local, non-elevated portions of thrust flats is greater than the overall increase in topographic elevation. Over-thickening of the crust by thrust stacking leads to additional lithospheric loading which causes regional flexural subsidence and consequently the topographic elevation of

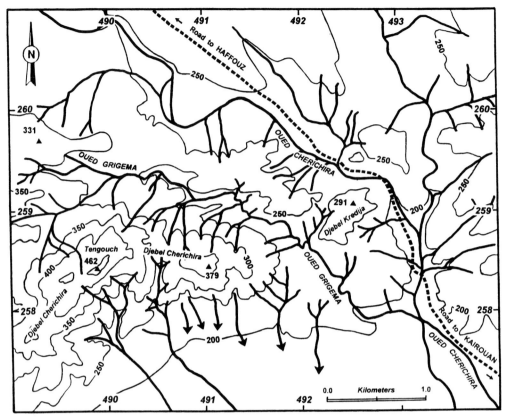

Fig. 2. A topographic map of Djebel Cherichira and the ephemeral drainage systems of Oued Cherichira and Oued Grigema. The map grid is part of the Tunisian National Grid. The location of Djebel Cherichira is indicated in Fig. 1. All heights are in metres.

the thrust belt is lower than might be expected and the adjacent foreland displays subsidence. Beaumont (1981) and Jordan (1981) have modelled thrust belt and foreland basin formation and related the amount of uplift in a thrust belt to lithospheric flexure and subsidence. Although they assumed different mechanisms of flexure and used a range of values of lithospheric flexural rigidity, a simplistic view of their results is that one might expect approximately a 5–25% reduction of topographic elevation as a result of regional lithospheric flexural subsidence, with the greatest subsidence occurring when the lithosphere has low flexural rigidity and is probably very thin. In Tunisia, where the N–S Axis has a crustal thickness of 32 km and the lithospheric thickness is in excess of 100 km (Buness et al. 1992) the reduction of topographic elevation related to lithospheric flexural subsidence is probably less that 10–15% (Beaumont 1981, fig. 9). In real-life situations the calculation of topographic uplift is much more difficult because even if the additional mass of the thrust load can be determined, its height and width are not necessarily known and furthermore much of the additional load is no longer present because of syn- and post-tectonic erosion.

Aspects of the stratigraphy of Djebel Cherichira

The evolution of Triassic-cored Djebels in northen Tunisia has previously been related to diapirism (Perthuisot 1981) but here the structure, stratigraphy and sedimentology of the Triassic strata are used to refute this model of structural evolution when applied to Djebel Cherichira. The age, sedimentological nature and distribution of pre-, syn- and post-tectonic fluvial, arid and semi-arid desert sediments are used to constrain the structural uplift, topographic uplift and erosion in the area.

On a large scale, a coherent pattern of strike of bedding is visible within the Triassic strata, even though the combination of gypsiferous

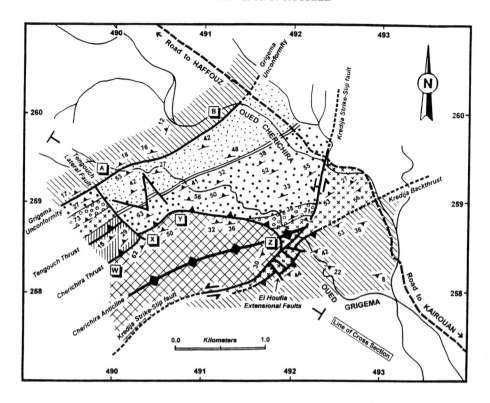

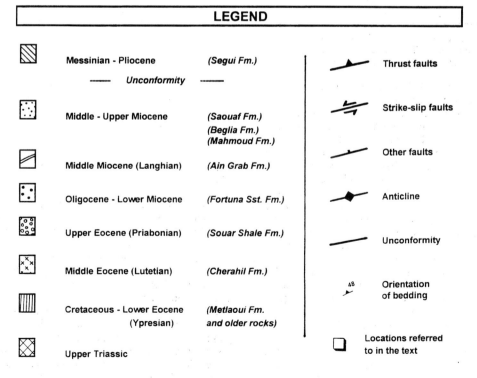

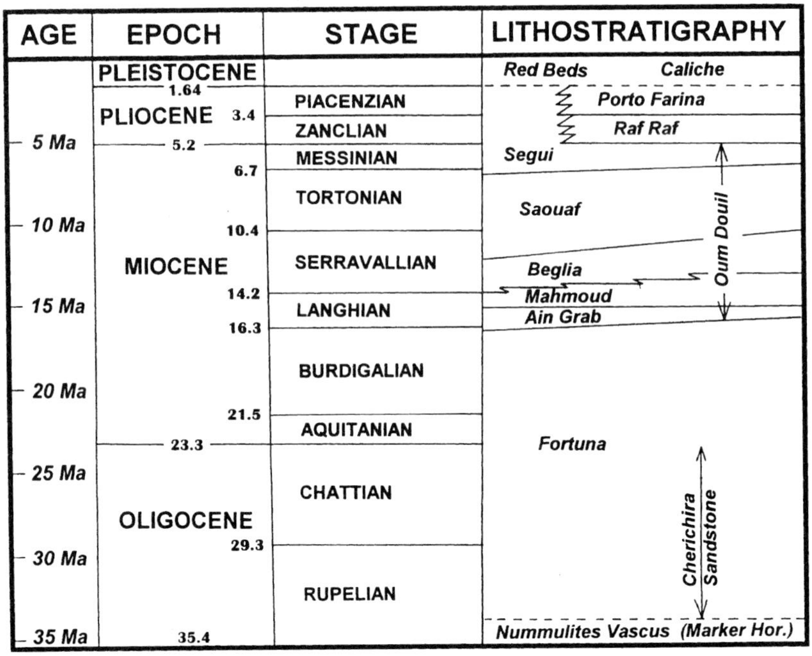

AGE	EPOCH	STAGE	LITHOSTRATIGRAPHY
	PLEISTOCENE 1.64		Red Beds --- Caliche
–5 Ma	PLIOCENE 3.4 5.2	PIACENZIAN	Porto Farina
		ZANCLIAN	Raf Raf
		MESSINIAN 6.7	Segui
–10 Ma	MIOCENE	TORTONIAN 10.4	Saouaf
		SERRAVALLIAN 14.2	Beglia
–15 Ma		LANGHIAN 16.3	Mahmoud / Ain Grab
		BURDIGALIAN	
–20 Ma		21.5	
		AQUITANIAN 23.3	Fortuna
–25 Ma	OLIGOCENE	CHATTIAN 29.3	
–30 Ma		RUPELIAN	Cherichira Sandstone
–35 Ma		35.4	Nummulites Vascus (Marker Hor.)

(Oum Douil spans Ain Grab through Saouaf)

Fig. 4. A correlation of Oligocene to Recent lithostratigraphic units in northern and central Tunisia. Modified from Ben Ferjani *et al.* (1990) using the timescale of Harland *et al.* (1989).

evaporite, dolomites and cross-bedded sandstones and siltstones often show tectonic disruption at a small scale. There is no evidence for halite in the succession and no suggestion of the Triassic strata being diapiric (cf. Ben Ferjani *et al.* 1990). All of the observed contacts between Triassic and younger sediments are tectonic. The Oueds Cherichira and Grigema cut through a sequence of well exposed Upper Eocene to Pliocene sediments in which the Oligocene to Miocene rocks form the type section for the general area of central to northern Tunisia (Castany 1951; Ben Ferjani *et al.* 1990; Yaich 1992*a,b*). The lithostratigraphy of Oligocene to Recent strata is summarized in Fig. 4. Sedimentologically varied and distinctive Upper Cretaceous to Palaeogene sediments have been thrust onto Triassic rocks along much of the northern and western margin of the Triassic outcrop and are juxtaposed against Oligo-Miocene sediments further to the east, across the NW–SE trending Tengouch Lateral Fault (Fig. 3).

The Grigema Unconformity is the principal key to understanding the structure and uplift of the area. The sediments east of the Tengouch Lateral Fault and below the Grigema Unconformity have been mapped by Castany (1951) and Yaich (1992*b*) as a continuous Oligocene to Upper Miocene succession in which no stratigraphic gaps were recorded (Fig. 4). The Saouaf Formation, which lies immediately below the Grigema Unconformity in Djebel Cherichira, has been dated regionally as Tortonian and possibly up to early Messinian age (Ben Ferjani *et al.* 1990). Erosion below the Grigema Unconformity must have occurred in a very short period of time because the Segui Formation which rests on the erosion surface (Figs 3 and 4) is of Messinian and perhaps Early Pliocene age (Ben Ferjani *et al.* 1990). However, since the Grigema Unconformity is an angular unconformity and some erosion of the underlying Saouaf Formation has occurred, it is here assumed that the Saouaf Formation is of early Tortonian age and that the Grigema Unconformity and much of the structural development occurred in late Tortonian times. It is acknowledged that detailed palynological studies are required to constrain this chronology more precisely.

Above the Grigema Unconformity the Messinian Segui Formation dips fairly uniformly away from Djebel Cherichira and it could be

Fig. 3. A map to show the structure and stratigraphy of Djebel Cherichira. The chrono- and lithostratigraphic sub-divisions are based on Ben Ferjani *et al.* (1990). The map grid is part of the Tunisian National Grid. The location of Djebel Cherichira is indicated in Fig. 1.

thought that this formation is simply a post-tectonic drape on the underlying structure, an assumption which was made by Anderson (1996) who then concluded that all thrust-related deformation occurred in the Middle Miocene. Near the headwaters of Oued Grigema (Fig. 3, locality A) the basal Segui deposits dip 15° to 20° NW and are a combination of arid or semi-arid, matrix-supported debris-flow fan deposits (Cooke *et al.* 1993; Miall 1996), soil horizons and occasional fluviatile channel deposits in which imbricated pebbles show palaeocurrents towards the NW. These sediments still retain their depositional dips. After late Tortonian thrusting and erosion, the Tengouch area still possessed some of its structurally generated, topographic elevation, with Segui Formation fan deposits and some ephemeral streams flowing off the topographic high towards the NW. However, along-strike from this locality, in Oued Cherichira (Fig. 3, locality B), extensive cross-stratification in an ephemeral, braided, fluviatile sequence of silts with interbedded muds showing desiccation cracks and mud curls, indicates deposition of the basal Segui Formation in a SE flowing drainage system. Oued Cherichira maintained its base-level and continued to flow southeastwards through the topographically elevated thrust front. The Messinian strata presently dip at 15° to 20° NW as a result of post-Messinian deformation.

The antecedent Oued Cherichira drainage system in Djebel Cherichira must have formed at some time after the deposition of the marine limestones of the early Langhian, Ain Grab Formation (Fig. 4). Ben Ferjani *et al.* (1990) noted that the shales of the overlying Langhian (to early Serravallian?) Mahmoud Formation contain a rich marine microfauna whilst the succeeding Beglia and Saouaf Formations have fluvial and deltaic affinities respectively. Thus Oued Cherichira formed during or after early Serravallian times (14 Ma), prior to the growth of the thrust front at Djebel Cherichira which commenced in the late Tortonian (8.5 to 6.7 Ma). Oued Grigema, whilst superficially appearing to be antecedent to the uplift of Djebel Cherichira, is actually a much younger consequent drainage system.

Tectonic evolution and structural uplift of the thrust front at Djebel Cherichira

The structure of the thrust front at Djebel Cherichira has evolved as the result of complex and polyphase motion on a series of thrusts, strike-slip and extensional faults. The terminology used to describe these structures has been summarized by Butler (1982*a*) and McClay (1992) and some of the key geometries are illustrated in Fig. 5.

Tengouch Thrust

The oldest thrust in the area is the Tengouch Thrust (Fig. 3) which has placed Lower Eocene (Ypresian) limestones in a hangingwall flat (Fig. 5a) over a footwall ramp of Ypresian limestones. There has been about 50 m of structural uplift on the thrust plane but the amount of horizontal transport may have been much greater. Because of the geometrical similarities of the Tengouch Thrust to the later thrusts, it is considered to be the earliest in a sequence of structures which commenced in late Tortonian times.

Tengouch Lateral Fault

The Tengouch Lateral Fault (Fig. 3) developed after the Tengouch Thrust and juxtaposed Upper Cretaceous and Palaeogene rocks against Oligocene to lower Tortonian rocks. Approximately 800 m of the Lower Oligocene to lower Tortonian succession which are present to the east of the Tengouch Lateral Fault are missing from the western side. There is no good sedimentological evidence in the Oligo-Miocene rocks east of the fault to suggest that the Tengouch Lateral Fault was a basin-bounding fault throughout this period, therefore it is assumed that the sediments were deposited on both sides of the fault prior to at least 800 m of structural uplift and erosion of the western side in the late Tortonian. All movement on the Tengouch Lateral Fault occurred before the deposition of the Messinian Segui Formation.

The Tengouch Lateral Fault probably formed synchronously with the Cherichira Thrust as a hangingwall drop fault (Butler 1982*a,b*) which was required to accommodate differential uplift above a vertically orientated, lateral thrust ramp on the Cherichira Thrust (Fig. 6b). The dip of bedding to the west of the Tengouch Lateral Fault (Fig. 3) reflects the geometry of the hangingwall anticline above the Cherichira Thrust which is schematically illustrated in Fig. 6c, area E, at a stage prior to later tectonic events. There is a possibility that the Tengouch Lateral Fault pre-dated the Cherichira Thrust, in which case the position of the 'hangingwall anticline' to the west of the lateral fault is purely fortuitous.

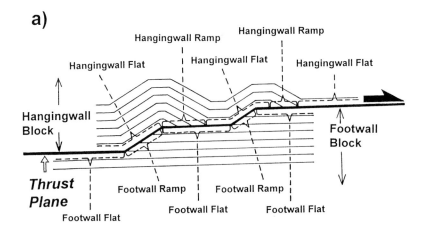

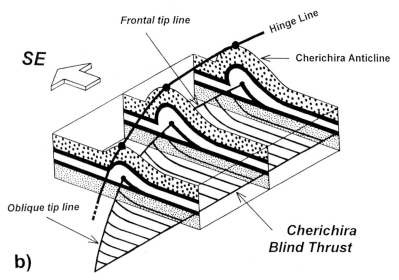

Fig. 5. Sketches to illustrate aspects of thrust geometry discussed in the text.

Cherichira Thrust

The Cherichira Thrust has transported the Tengouch Thrust and the Tengouch Lateral Fault southeastwards onto the footwall flat Triassic rocks of Djebel Cherichira. Segment XY of the Cherichira Thrust (Fig. 3) possesses footwall and hangingwall flats, and transport on the thrust flat was not associated with any structural uplift (Fig. 6c, area D). The 'young-on-old' stratigraphic relationship across the thrust can be explained in two ways. If one assumes an initially complete 'layer-cake' stratigraphy, the Oligocene rocks can be emplaced on top of Triassic rocks using a complex series of 'out-of-sequence thrusts' (Butler 1987). This mechanism has been applied to the N–S Axis by Anderson (1996) and elsewhere in the Atlas Mountains by Morley (1988) even though it requires extensive transport on numerous thrusts which are unexposed at present (Anderson 1996), and it assumes 'layer-cake' stratigraphy (Creuzot et al. 1993) although regionally there are known to be extremely rapid lateral variations of stratigraphic

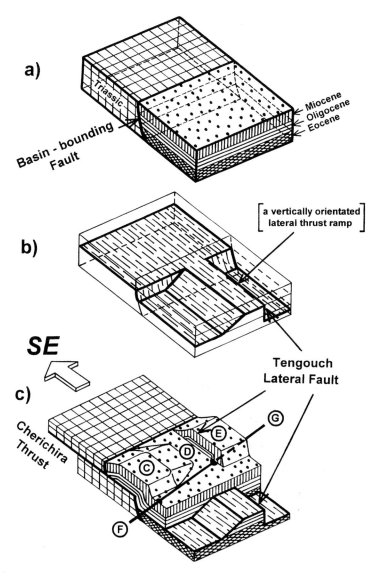

Fig. 6. Block diagrams to illustrate the evolution of the Cherichira Thrust. (**a**) Pre-thrust geometry showing the basin-bounding fault and elevated Triassic rocks in its footwall. (**b**) The geometry of the Cherichira Thrust surface. (**c**) The geometry of the folds and faults produced during southeasterly directed thrusting. Note the thrusting of younger onto older strata.

thickness (Ben Ferjani *et al.* 1990; Clayton & Baird 1997). If the Cherichira Thrust is part of an 'out-of-sequence thrust' system, the amounts of structural uplift and erosion associated with this system cannot be constrained.

Presented herein is an alternative and simpler explanation for the evolution of the Cherichira Thrust in which Oligocene sediments have been thrust southeastwards out of a basin, across its basin-bounding fault and onto the Triassic rocks which occupied a stratigraphically elevated position in the footwall of the basin-bounding fault (Fig. 6a). Similarly small, fault-controlled, Oligo-Miocene basins have been described in the sub-surface of the foreland, 30 to 70 km further east (Bedir *et al.* 1992, 1996) and noted further NW within the Tunisian Atlas (Clayton & Baird 1997).

The surface of the Cherichira Thrust was not planar within the Tertiary basin to the NW of the Triassic horst block (Fig. 6b). Thrust transport has elevated and placed approximately

450 m of Eocene to Oligocene strata onto segment YZ of the Cherichira Thrust (Fig. 3 and Fig. 6c, area C) which are not present further west above segment XY of the thrust plane. The hangingwall above segment WX of the Cherichira Thrust contains an elevated 350 m succession of Upper Cretaceous to Palaeogene strata which is unexposed above segment YZ further east, therefore at least 800 m (450 m + 350 m) of structural uplift occurred to the west of the Tengouch Lateral Fault (Fig. 6c, area E) whilst none occurred immediately east of the Tengouch Lateral Fault (Fig. 6c, area D).

Below the Grigema Unconformity 800 m of the Oligo-Miocene succession west of the Tengouch Lateral Fault has been eroded. Approximately the same stratigraphic level of the Saouaf Formation is exposed below the Grigema Unconformity from the Tengouch Lateral Fault eastwards to Oued Cherichira (Fig. 3) which requires that equal amounts of erosion have occurred all along the presently exposed Saouaf Formation. To satisfy this geometry, the present-day outcrop trace of the Grigema Unconformity would lie along line FG in Fig. 6c.

Cherichira Blind Thrust

Evidence for the nature of the thrust structures which fold and deform the Cherichira Thrust is seen in geometry of the Triassic strata below the thrust plane. Here, despite tectonic disruption, especially adjacent to thrust contacts, a coherent pattern of strike with dips of 30° to 55° defines a large-scale antiform (Fig. 3). Triassic rocks form a footwall flat to the Cherichira Thrust and were flat-lying during transport on the Cherichira Thrust (Fig. 6b). They have obtained their folded nature and relatively steep dips as the result of later deformation. The propagation of the underlying Cherichira Blind Thrust folded the Cherichira Thrust and formed the Cherichira Anticline (Fig. 3) as a fault-propagation fold (Mitra 1990). Anderson (1996) described his equivalent of the Cherichira Thrust as an 'out-of-sequence thrust' (Anderson 1996, fig. 11, thrust 2) which is not folded by the underlying thrust; however, his description is contradicted by his own field observations which show his equivalent of the Cherichira Thrust to be folded (Anderson, 1996, fig. 7).

The Cherichira Anticline can be traced eastwards through the Triassic and into Upper Eocene strata before it is truncated by the Kredija Strike-slip Fault. The eastern end of the Cherichira Anticline has an easterly plunge whereas the western portion of the fold is sub-horizontally hinged. The thrust has a frontal tip line geometry (Boyer & Elliott 1982) in the west of the area which becomes a more oblique tip line further east (Fig. 5b). The tilted upper limb of the fault-propagation fold has a horizontal width of at least 800 m (Fig. 7) and formed with a dip of 25° to 30° NW. Tilting of this long limb has generated at least 450 m of structural uplift along the crest of the fault-propagation fold at its western end. The growth of the Cherichira Anticline structurally uplifted both sides of the Tengouch Lateral Fault by the same amount.

Further east towards the lateral tip of the Cherichira Blind Thrust, where the dip of the thrust may have been lower, the amount of uplift generated by fault propagation was less, in the order of 300 m.

Reactivation of the Tengouch Lateral Fault

The Cherichira Thrust has been displaced during reactivation of the Tengouch Lateral Fault (Fig. 3) and since the Grigema Unconformity is not displaced, the reactivation must be late Tortonian in age. The apparent downthrow of 50 m to the west on the Tengouch Lateral Fault is in the opposite sense to its earlier large displacement and is difficult to relate to the growth of the Cherichira Anticline which preferentially uplifted the western end of the fold. Reactivation probably reflects the differential loading of the Triassic strata below the Cherichira Thrust when the hangingwall block of the Cherichira Thrust was in excess of 800 m thicker on the western side than on the eastern side of the Tengouch Lateral Fault.

It is possible that reactivation of the Tengouch Lateral Fault may be unrelated to other fault movements and a significant amount of transport on the 'lateral fault' parallel to the Cherichira Thrust plane may have occurred if the fault motion was strike-slip.

Cherichira Sole Thrust

During thrust propagation, it has been noted by Boyer & Elliott (1982) and Jamison (1987) that the dip of bedding towards the hinterland does not exceed that of the ramp angle of the underlying thrust (Fig. 5b). These authors observed that the most common values of ramp angles in naturally occurring thrust structures are between 10° and 40°. In northern and central Tunisia, thrust ramp angles have been observed by the

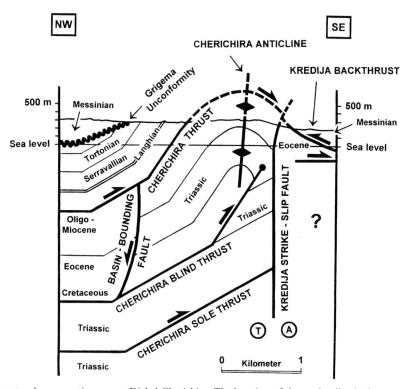

Fig. 7. Structural cross-section across Djebel Cherichira. The location of the section line is shown on Fig. 3.

authors to be almost invariably 30 ± 2°. Thus it is highly unlikely that the steep hinterland dips to the NW of up to 55° to 60° in the area NW of the crest of the Cherichira Anticline (Fig. 3) have been generated solely by the formation of a fault-propagation fold. There must be another unexposed thrust, the Cherichira Sole Thrust, lying below the Cherichira Blind Thrust. Transport up the ramp of the Cherichira Sole Thrust has back-steepened the 25° to 30° NW dipping long limb of the Cherichira Anticline to average dips of 45° to 60° NW in the rocks below the Grigema Unconformity (Fig. 7). The short foreland-dipping limb of the Cherichira Anticline fault-propagation fold has also been rotated by the Cherichira Sole Thrust. This has the effect of reducing its formerly steep dip of 60° SE to 30° SE and giving the fault-propagation fold a more symmetrical appearance at exposure level. The Cherichira Sole Thrust has back-steepened rocks across a horizontal NW–SE width of at least 1.5 km from at least as far SE as the Kredija Strike-slip Fault to the Grigema Unconformity at locality A (Fig. 3) and probably further NW in the sub-crop of the unconformity. This requires at least 750 m of uplift on the ramp of the Cherichira Sole Thrust (Fig. 7). At locality A (Fig. 3) the Messinian Segui debris-flow fan deposits retain their depositional dips and rest unconformably on the Saouaf Formation which dip 45° to 55° NW. Therefore some, if not all, of the 750 m of uplift and back-steepening on the ramp of the Cherichira Sole Thrust must have occurred in late Tortonian Times.

Post-Messinian transport on the Cherichira Sole Thrust has back-steepened the unconformable, originally gently SE dipping fluviatile Segui Formation (Fig. 3, locality B) to dips of 15° to 20° NW, whilst at the same time tilting the underlying Saouaf Formation to dips of 40° to 45° NW. The back-steepened fluviatile Segui Formation crops out for at least 200 m NW of locality B (Fig. 3) which requires transport of Messinian strata onto the base of the Cherichira Sole Thrust ramp followed by at least 250 m of transport up it which would have generated at least 100 m of uplift in post-Messinian times.

On a section through Oued Cherichira, the Cherichira Sole Thrust ramp extends from SE of the Kredija Strike-slip Fault to at least 200 m NW of the Gregima Unconformity. The ramp

has a horizontal width of at least 1.8 km and assuming a ramp angle of 25° to 30°, thrusting in late Tortonian and post-Messinian times has uplifted the hangingwall rocks by at least 900 m, with at least 100 m of the uplift being of post-Messinian age.

Thus interpretation of the field evidence demonstrates that the Cherichira Sole Thrust experienced polyphase movement, generating uplift in late Tortonian and post-Messinian times.

Kredija Backthrust

The Kredija Backthrust is so termed because of its SE dip and NW transport direction towards the orogenic hinterland. Apparently micropalaeontologically barren, fluviatile sediments in the hangingwall block have no sedimentological affinities to the Beglia and Saouaf Formations (Fig. 4), which locally pre-date most, if not all, of the thrusting. The hangingwall sediments are probably part of the Segui Formation and related to thrust front elevation and the formation of a foreland basin. They have been backthrusted onto previously folded and thrusted Middle Eocene limestones lying south of the Kredija Strike-slip Fault, a geometry which is not uncommon at thrust fronts (Vann et al. 1986). At least 250 m of structural uplift occurred above the SE dipping Kredija Backthrust, a post-Messinian structure younger than, or related to, the post-Messinian activity on the Cherichira Sole Thrust.

El Houfia Extensional Faults

Some of the NW–SE trending El Houfia Extensional Faults formed prior to the Kredija Backthrust and are truncated by it; however, many of the El Houfia Extensional Faults have been reactivated and have displaced the Kredija Backthrust (Fig. 3). The age of these extensional faults relative to the structures north of the Kredija Strike-slip Fault cannot be ascertained because the extensional faults cannot be traced north of the Kredija Strike-slip Fault.

Kredija Strike-slip Fault

The Kredija Strike-slip Fault, the youngest structure known to affect the area, truncates the El Houfia Extensional Faults and the Cherichira Thrust. At least 800 m of sinistral displacement has occurred on the fault which has highly variable dip along its length, ranging from sub-vertical to as low as 20° to 30° NW. The authors have not been able to demonstrate that fault motion has created local structural uplift, erosion and the development of topography and have only been able to constrain its age to post-Messinian.

The Kredija Strike-slip Fault, together with the suite of slightly older thrusts in Djebel Cherichira, may be analogous to the structures at the northern end of the N–S Axis, where Boccaletti et al. (1988) have described the N–S Axis as a Late Miocene–Early Pliocene zone of thrusting and left-lateral strike-slip faulting. However, Creuzot et al. (1993) working further south have noted that Tortonian thrust planes have been reactivated as transpressional structures in which sinistral strike-slip displacement is predominant and they have suggested, without stratigraphical justification, that the reactivation is of Villafranchian (Early Pleistocene) age.

Topographic uplift and erosion of Djebel Cherichira

In the previous section it was demonstrated that at least 2250 m of structural uplift has occurred in Djebel Cherichira. At least 100 m of structural uplift on the Cherichira Sole Thrust and 250 m on the Kredija Backthrust occurred in post-Messinian times but the majority, probably in excess of 1900 m, occurred in late Tortonian times.

The late Tortonian structural uplift generated some topographic uplift in the area north of Tengouch. Here the Segui Formation was deposited in an arid to semi-arid climate and is a matrix-supported, debris-flow deposit containing evidence for ephemeral stream flow to the northwest (Fig. 3, locality A) which blanketed a surface which was a northwesterly dipping topographic reflection of the uplifted thrust belt. In general, debris slopes in deserts are highly stable at dips of up to about 26° (Cooke et al. 1993, p. 135–137), although slopes up to 38° have been recorded and average slopes may range from about 12° to 18° (Abrahams et al. 1985).

In the vicinity of Oued Cherichira at least 1650 m of structural uplift was generated by thrust transport on the Cherichira Thrust (450 m), and uplift above the Cherichira Blind Thrust (300 m) and the Cherichira Sole Thrust (900 m). However, there was no topographic elevation because the Oued Cherichira antecedent fluvial system maintained its base-level, so that more than 1650 m of erosion occurred

in the area southeast of the Grigema Unconformity. The thrust-generated uplift increased progressively southeastwards from the Grigema Unconformity and would have been at its greatest above the hinge line of the Cherichira Anticline, about 1.5 km SE of the trace of the Grigema Unconformity prior to the sinistral displacement on the Kredija Strike-slip Fault.

At locality B (Fig. 3) in Oued Cherichira, the Segui Formation was a southeasterly flowing, braided, silty and muddy fluvial deposit. Such deposits in present-day deserts tend to have dips of 1° to 6° (Cooke et al. 1993). At locality A the Segui Formation is a matrix-supported, debris-flow deposit which has retained its depositional dip of 15–20° NW. Localities A and B (Fig. 3), which are separated by a horizontal distance of 1300 m, are at the same stratigraphic level and show a gradual change of facies from A to B. If the separation was in the dip direction, the maximum difference in elevation between them at the time of deposition could have been up to approximately 650 m. However, the two localities lie along the NW face of Djebel Cherichira, with locality A being 45 m higher than locality B. They lie on an ENE–WSW trending line which is almost at right angles to the NW dipping, debris-fan depositional slope at locality A and to the SE dipping, fluviatile depositional slope at locality B. The present-day 45 m difference in elevation is the combination of two factors: firstly an original difference in elevation at the time of deposition of the Segui Formation, and secondly differential, post-Messinian, tectonic uplift during the reactivation of the Cherichira Sole Thrust. The thrust reactivation elevated locality A more than locality B because thrusting transported locality A further up the underlying thrust ramp whereas locality B was transported along the thrust flat of the Cherichira Sole Thrust and onto the basal slopes of the thrust ramp to back-steepen the bedding. Therefore locality A was less than 45 m higher than the antecedent river valley floor of locality B in Messinian times. Consequently the Messinian Segui debris-fan deposits of locality A formed virtually at the base of the NW side of the topographic mass of Djebel Cherichira which was structurally uplifted in the order of 2000 m in late Tortonian times. In the main body of Djebel Cherichira it is not possible to quantify the relationship between the amount of structural uplift and topographic elevation, because one cannot quantify erosion in areas where the antecedent drainage has not been preserved. However, since Oued Cherichira managed to maintain its base-level during rapid structural uplift of up to 1650 m by eroding through a relatively young and generally poorly indurated sand and mudstone sequence, it is likely that the consequent streams which flowed off Djebel Cherichira were also highly erosive.

If one considers flexural subsidence of the thrust belt, in the absence of any erosional activity, the structural uplift of approximately 2 km in Djebel Cherichira would have generated about 1600 to 1800 m of topographic uplift. At present Djebel Cherichira is only 250 m higher than the base-level of Oued Cherichira, therefore approximately 1.5 km of erosion must have occurred in the area, mostly during syntectonic episodes.

The evolution of the Cherichira and Grigema drainage systems

Prior to the late Tortonian structural development of Djebel Cherichira, the Oued Cherichira fluvial system formed in Serravallian to early Tortonian times and flowed southeastwards towards the foreland of the Atlas Thrust Belt. The system probably developed to drain the tectonically and topographically elevating Djebel Serdj thrust system, 40 km to the northwest (Fig. 8). Oued Cherichira maintained its antecedent drainage pattern during the late Tortonian and post-Messinian periods of structural uplift of Djebel Cherichira.

The drainage pattern of Oued Cherichira has not been displaced by the Kredija Strike-slip Fault which, using Middle Eocene strata as a marker, has a sinistral displacement of at least 0.8 km and which, by analogy with the northern end of the N–S Axis (Boccaletti et al. 1988), was probably active in the Early Pliocene. During Early Pliocene propagation of the strike-slip fault it is likely that the area lying immediately south of the fault had already been heavily eroded so that Oued Cherichira continued to flow directly southeastwards and was not laterally displaced by the strike-slip fault. Conversely, if Oued Cherichira had been flowing in a deeply incised valley then it is likely that the drainage pattern would have been laterally displaced across the strike-slip fault.

Oued Grigema, a smaller system which drains much of the eastern mass of Djebel Cherichira, contains numerous fluvial deposits and river terraces which have similarities to those in Oued Cherichira. It was initially thought that Oued Grigema was antecedent with respect to Djebel Cherichira because it rises on the northern slopes of Djebel Cherichira and flows southeastwards. In its upper reaches Oued Grigema flows towards and cuts through the resistant limestone ridge of the Langhian Ain Grab Formation. If

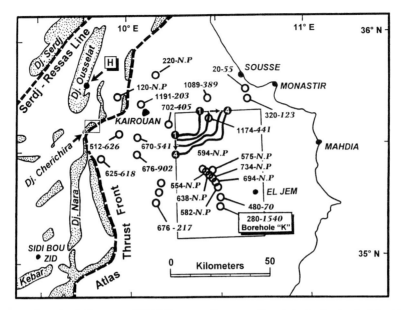

Fig. 8. A map to show the thicknesses of Mid-Miocene to Recent sediments in the foreland of the Atlas Thrust Belt. In each borehole the first number is the preserved thickness (in metres) of Messinian and younger strata; the second number is the preserved thickness (in metres) of Mid-Miocene to lower Tortonian strata. N.P – not preserved. The rectangle defines the area of the North Kairouan Permit which was studied by Jennings *et al.* (in press). Lines labelled 1 to 4 show the positions of four onlapping seismic reflectors of Serravallian to early Tortonian age which mark the southeastern distal margin of the foreland basin at this time. Messinian foreland basin deposits covered the whole of the North Kairouan Permit before local erosion on the crests of reactivated extensional and transtensional faults.

the drainage pattern was indeed antecedent, the Messinian Segui deposits at locality A (Fig. 3) should be fluviatile with depositional dips towards the SE into the putative Messinian Oued Grigema drainage system. However, it has already been noted that the Segui Formation at this locality is a debris-flow fan deposit with steep depositional dips to the NW. Thus the extensive, polyphase, interbedded alluvial deposits, soil horizons and river terraces preserved on the sides of Oued Grigema are post-tectonic, implying that many of the later fluvial features of Oued Cherichira are also relatively young. The discovery of an ampura pot handle in one of the youngest interfluvial soil horizons indicates that at least some of the formation of Oued Grigema is of Holocene age. Rognon (1987), Fontes & Gasse (1989) and White *et al.* (1996) have dated and discussed the formation of Late Pleistocene to Holocene palaeosols and alluvial fans in southern Tunisia and Steinmann & Bartels (1982) working in Oued Ain Djelloula, 22 km north of Djebel Cherichira (Fig. 8, locality H), have obtained radiocarbon dates of 18740 ± 230 and 745 ± 55 years BP from similar humic soils and fine alluvial sediments. In Oued Grigema work continues on the sedimentological and geomorphological evolution of the fluvial deposits and terrace systems and will be presented elsewhere.

Thrust front and foreland basin development

The timing, geometrical evolution and denudation of the thrust front can be constrained by examining data from the foreland. Bedir *et al.* (1992, 1996) have studied a large area of the foreland in the Sahel, but their work concentrated on pre-Messinian seismic sequences. To consider the evolution of the Djebel Cherichira and the Atlas thrust front, data supplied by KUFPEC (Tunisia) Ltd from 26 exploration oil wells and 370 km of regional seismic lines have been integrated with a study by Jennings *et al.* (1997, in press) of over 1200 km of seismic lines and borehole data from an area within the North Kairouan Permit (Fig. 8). In the well log data only the stratigraphic 'tops' of the Messinian Segui Formation, the lithologically distinctive limestones of the lower Langhian Ain Grab Formation and older formations had been defined. The lower Langhian to Messinian interval spans two of the seismic sequences defined by Jennings *et al.* (in press), who concluded that

the lower sequence (Langhian to lower Serravallian) formed during strike-slip basin reactivation and the younger sequence (lower Serravallian to Messinian) formed during foreland basin development. This interpretation contrasts with that of Anderson (1996) who assumed that increased rates of sedimentation in Langhian times reflected foreland basin development. Jennings *et al.* (in press) have identified four seismic reflectors in the upper Serravallian to lower Tortonian portion of the upper sequence. The reflectors show progressive onlap southeastwards within the permit (Fig. 8) and define the distal margin of the foreland basin at this time. The thickness of the lower Langhian to Messinian strata at each borehole (Fig. 8) is a combination of the original thicknesses of each of the two sequences, the erosive nature of the intervening sequence boundary and later, tectonically induced erosion. For example, in borehole 'K' (Fig. 8) there is an unusually thick, 1540 m succession of lower Langhian to Messinian strata, even though the Serravallian to lower Tortonian portion of the succession was not deposited in the area. The borehole lies just on the southern and downthrown side of a major strike-slip, basin-bounding fault which was active in early Langhian to early Serravallian times and contains thick deposits of this lower sequence.

The detailed studies of Jennings *et al.* (in press) show that the Messinian Segui Formation was deposited throughout the N Kairouan Permit. The distal margin of the foreland basin shifted from its position within the N Kairouan Permit in Serravallian to early Tortonian times, by at least 50 km to the SE before deposition of the Messinian Segui Formation. It is likely that the thrust front propagated a similar distance southeastwards throughout this period, advancing from Djebel Serdj to form Djebel Cherichira.

Conclusions

The Tunisian Atlas thrust front at Djebel Cherichira was formed by late Tortonian and post-Messinian thrusting. The youngest structure in the area is the sinistral Kredija Strike-slip Fault which is probably Early Pliocene, but perhaps as young as Early Pleistocene in age.

At least 2000 m of structural uplift was generated in the Tengouch area of Djebel Cherichira during thrusting, the majority of it during the earlier late Tortonian event. Topographic uplift was associated with the structural uplift and Djebel Cherichira formed a topographic high in Messinian times, but because of extensive syntectonic erosion of the relatively young strata it is believed that the creation of topography was minimal in comparison to the amount of structural uplift.

Oued Cherichira, which formed in Serravallian to early Tortonian times, was a fluvial drainage system within the foreland of the Atlas thrust belt when the thrust front lay further to the NW at Djebel Serdj. It preserved its antecedent drainage pattern during more than 1650 m of structural uplift at the Atlas thrust front in Djebel Cherichira in late Tortonian and post-Messinian times. The distal edge of the foreland basin moved at least 50 km to the SE between early Tortonian and Messinian times, corresponding to the thrust front propagation from Djebel Serdj to its present position at Djebel Cherichira.

In the Early Pliocene or later, when the Kredija Strike-slip Fault moved at least 800 m laterally, Oued Cherichira was not displaced sinistrally along the trace of the strike-slip fault. This implies that most of the topography flanking the Oued Cherichira drainage channel had been denuded and Oued Cherichira flowed across the strata on the NW side of the fault and continued to flow directly southeastwards.

Oued Grigema, which superficially appears to have an antecedent drainage pattern with respect to Djebel Cherichira, is a consequent stream which has been actively eroding in Holocene and perhaps Late Pleistocene times. The present-day erosional and depositional environment of Oued Grigema is closely analagous to that of Djebel Cherichira in Messinian times.

In general terms it has been demonstrated that an integrated structural, sedimentological and geomorphological study of this portion of the Tunisian Atlas thrust front has yielded valuable detailed information about the timing of thrust front formation, the amount of structural uplift and its relationship to the amount of topographic uplift and erosion associated with thrust-front growth. The integrated analysis has also shed light on the palaeogeography of the region and on the sequential nature of the growth of thrust-generated mountains within the thrust belt. This style of integrated analysis is clearly applicable to other thrust fronts around the world.

We are grateful to KUFPEC (Tunisia) Ltd, and in particular R. Bally, for access to confidential seismic and borehole data and for discussions about the evolution of the foreland. We thank the staff and final-year Earth Sciences students of Kingston University who survived several years of fieldtrips to the area. We particularly acknowledge discussions with C. J. Clayton, J. S. Jennings, J. M. Baird, J. D. Rolls, K. Lacey and J. A. Holmes. Finally, we would like to thank the two anonymous reviewers whose constructive comments improved the manuscript.

References

ABRAHAMS, A. D., PARSONS, A. J. & HIRSH, P. J. 1985. Hillslope gradient-particle size relations: Evidence for the formation of debris slopes by hydraulic processes in the Mojave Desert. *Journal of Geology*, **93**, 347–357.

ANDERSON, J. E. 1996. The Neogene structural evolution of the western margin of the Pelagian Platform, central Tunisia. *Journal of Structural Geology*, **18**, 819–833.

BAIRD, A. W. 1991. The geometry and evolution of the Tunisian Atlas Thrust and Foreland Basin System: theoretical modelling vs. inheritance of structure. *AAPG Bulletin*, **75**, 1402.

BEAUMONT, C. 1981. Foreland basins. *Geophysical Journal of the Royal Astronomical Society*, **65**, 291–329.

BEDIR, M., ZARGOUNI, F., TLIG, S. & BOBIER, C. 1992. Subsurface geodynamics and petroleum geology of transform margin basins in the Sahel of Mahdia and El Jem (Eastern Tunisia). *AAPG Bulletin*, **76**, 1417–1442.

——, TLIG, S., BOBIER, C. & AISSAOUI, N. 1996. Sequence stratigraphy, basin dynamics, and petroleum geology of the Miocene from Eastern Tunisia. *AAPG Bulletin*. **80**, 63–81.

BEN FERJANI, A., BURROLET, P. F. & MEJRI, F. 1990. *Petroleum Geology of Tunisia*. Entreprise Tunisienne d'Activitiés Pétrolières Tunis, Tunis.

BIJU-DUVAL, B., DERCOURT, J. & LE PICHON, X. 1977. From the Tethys Ocean to the Mediterranean Seas: a plate tectonic model of the evolution of the western Alpine system. *In*: BIJU-DUVAL, B. & MONTADERT, L. (eds), *Structural History of the Mediterranean Basins. Proceedings of an International Symposium*. Technip, Paris, 143–164.

BOCCALETTI, M., CELLO, G. & TORTORICI, L. 1988. Structure and tectonic significance of the north–south axis of Tunisia. *Annales Tectonicae*, **2**, 12–20.

——, —— & —— 1990. First order kinematic elements in Tunisia and the Pelagian block. *Tectonophysics*, **176**, 215–228.

BOYER, S. E. & ELLIOTT, D. 1982. Thrust Systems. *AAPG Bulletin*, **66**, 1196–1230.

BUNESS, H. ET AL. 1992. The EGT'85 seismic experiment in Tunisia: a reconnaissance of the deep structures (Research Group for Lithospheric Structure in Tunisia). *Tectonophysics*, **207**, 245–267.

BUROLLET, P. F. 1991. Structures and tectonics of Tunisia. *Tectonophysics*, **195**, 359–369.

BUTLER, R. W. H. 1982a. The terminology of structures in thrust belts. *Journal of Structural Geology*, **4**, 239–245.

—— 1982b. Hangingwall strain as a function of duplex shape and footwall topography. *Tectonophysics*, **88**, 235–247.

—— 1987. Thrust sequences. *Journal of the Geological Society of London*, **144**, 619–634.

CAIRE, A. 1978. The central Mediterranean mountain chains in the Alpine Orogenic Environment. *In*: NAIRN, A. E. M., KANES, W. H. & STEHLI, F. G. (eds) *The Ocean Basins and Margins, Vol. 4B. The Western Mediterranean*. Plenum, New York, 201–256.

CASTANY, G. 1951. Étude géologique de l'Atlas tunisien oriental. *Thèse, Annales des Mines et de la Géologie*, **8**.

CLAYTON, C. J. & BAIRD, A. W. 1997. Fluid-flow, Pb–Zn mineralization, hydrocarbon maturation and migration in the Tunisian Atlas. *In*: HENDRY, J. P., CAREY, P. F., PARNELL, J., RUFFELL, A. H. & WORDEN, R. H. (eds) *Geofluids II*. The Geological Society, London, 457–460.

COOKE, R. U., WARREN, A. & GOUDIE, A. S. 1993. *Desert Geomorphology*. UCL, London.

CREUZOT, G., MERCIER, E., OUALI, J. & TRICART, P. 1993. La tectogenèse atlasique en Tunisie centrale: Apport de la modelisation géometrique. *Eclogae Geologicae Helvetiae*, **86**, 609–627.

DEWEY, J. F., PITMAN, W. C. III, RYAN, W. B. F. & BONIN, J. 1973. Plate tectonics and the evolution of the Alpine systems. *Geological Society of America Bulletin*, **84**, 137–180.

——, HELMAN, M. L., TURCO, E., HUTTON, D. H. W. & KNOTT, S. D. 1989. Kinematics of the western Mediterranean. *In*: COWARD, M. P., DIETRICH, D., & PARK, R. G. (eds) *Alpine Tectonics*. Geological Society, London, Special Publication, **45**, 265–283.

ENGLAND, P. C. & MOLNAR, P. 1990. Surface uplift, uplift of rocks, and exhumation of rocks. *Geology*, **18**, 1173–1177.

FONTES, J. C. & GASSE, F. 1989. On the ages of humid Holocene and Late Pleistocene phases in North Africa – remarks on 'Late Quaternary Climatic reconstruction for the Maghreb (North Africa)' by P. Rognon. *Palaeogeography, Palaeoclimatology, Palaeoecology*, **70**, 393–398.

HARLAND, W. B., ARMSTRONG, R. L., COX, A. V., CRAIG, L. E., SMITH, A. G. & SMITH, D. G. 1989. *A Geological Time Scale*. Cambridge University, Cambridge.

JAMISON, W. R. 1987. Geometric analysis of fold development in overthrust terranes. *Journal of Structural Geology*, **9**, 207–219.

JENNINGS, J. S., BAIRD, A. W. & CLAYTON, C. J. 1997. Basin growth and deformation in an orogenic foreland during thrust belt propagation: an example from the Tunisian Atlas. *Terra Nova*, **9** (Abstracts Supplement No. 1), 321.

——, —— & —— (in press) Late Mesozoic and Tertiary basin systems and their deformation in the foreland of the Tunisian Atlas: implications for 'Atlassic' and 'Alpine' thrusting. *AAPG Bulletin*.

JORDAN, T. E. 1981. Thrust loads and foreland basin, Cretaceous, western United States. *AAPG Bulletin*, **65**, 291–329.

MCCLAY, K. R. 1992. Glossary of thrust tectonics terms. *In*: MCCLAY, K. R. (ed.) *Thrust Tectonics*. Chapman & Hall, London, 419–433.

MIALL, A. D. 1996. *The Geology of Fluvial Deposits*. Springer, Berlin.

MITRA, S. 1990. Fault-propagation folds: geometry, kinematic evolution, and hydrocarbon traps. *AAPG Journal*, **74**, 921–945.

MORLEY, C. K. 1988. Out-of-sequence thrusts. *Tectonics*, **7**, 539–561.

OUALI, J. 1985. Structure et évolution géodynamique du chaînon Nara-Sidi Khalif (Tunisie centrale). *Bulletin des Centres de Recherches, Exploration-Production, Elf-Aquitaine*, **9**, 155–182.

OUTTANI, F., ADDOUM, B., MERCIER, E., FRIZON DE LAMOTTE, D. & ANDRIEUX, J. 1995. Geometry and kinematics of the South Atlas Front, Algeria and Tunisia. *Tectonophysics*, **249**, 233–248.

PERTHUISOT, V. 1981. Diapirism in northern Tunisia. *Journal of Structural Geology*, **3**, 231–235.

ROGNON, P. 1987. Late Quaternary Climatic Reconstruction for the Maghreb (North Africa). *Palaeogeography, Palaeoclimatology, Palaeoeoclogy*, **58**, 11–34.

SNOKE, A. W., SCHAMEL, S. & KARASEK, R. M. 1988. Structural evolution of Djebel Debadib Anticline: A clue to the regional tectonic style of the Tunisian Atlas. *Tectonics*, **7**, 497–516.

STEINMANN, S. & BARTELS, G. 1982. Quartärgeomorphologische Beobachtungen aus Nord-Und Südtunesien. *Catena*, **9**, 95–108.

VANN, I. R., GRAHAM, R. H. & HAYWARD, A. B. 1986. The structure of mountain fronts. *Journal of Structural Geology*, **8**, 215–227.

WHITE, K., DRAKE, N. A., MILLINGTON, A. C. & STOKES, S. 1996. Constraining the timing of alluvial fan response to Late Quaternary climatic changes, southern Tunisia. *Geomorphology*, **17**, 295–304.

YAICH, C. 1992a. Dynamique des faciès détritiques oligo-miocènes de Tunisie. *Journal of African Earth Sciences*, **15**, 35–47.

—— 1992b. Sédimentologie, tectonique (et variations relatives du niveau marin) dans les formations du Miocene inférieur à moyen. Tunisie centrale et orientale. *Géologie Méditerranéenne*, **19**, 249–264.

Landscape evolution and the preservation of tectonic landforms along the northern Yammouneh Fault, Lebanon

R. W. H. BUTLER[1] & S. SPENCER[2]

[1] *School of Earth Sciences, The University of Leeds, Leeds LS2 9JT, UK*
[2] *Department of Geology, American University of Beirut, Bliss Street, Beirut, Lebanon*

Abstract: The Yammouneh Fault is commonly considered to be the principal active strand of the Dead Sea Transform in Lebanon – an inference reached primarily from interpretations of the geomorphological expression of the fault on satellite images. However, new geological field observations show the Yammouneh Fault to be sealed stratigraphically by the Homs Basalt, dated using new K-Ar ages at 5.2–6.5 Ma. Drainage systems which link to the pre-Homs Basalt palaeosurface show evidence of fault disruption. Those valleys incised into the basalt show no evidence for transcurrent offsets. The inferred left-lateral displacement of c. 45 km on the Dead Sea Transform that post-dates the Homs Basalt is presumed to have bypassed to the west of Mount Lebanon. These linked geological and geomorphological studies indicate that landscape evolution can be exceptionally slow in northern Lebanon. Faceted spurs, poljes and offset drainage along the Yammouneh Fault across Mount Lebanon, evident on satellite images, are interpreted as being of Miocene age and are not indicative of Plio-Quaternary displacements on the fault. Much of the Lebanese tectonic landscape has thus remained stable for many millions of years, although locally incised during large-scale uplift of the Mount Lebanon range. Presumably landscape insensitivity reflects the arid climatic conditions together with inhibited run-off due to the regional karst system.

The interpretation of landscape underpins almost all studies of neotectonics. These researches have increased greatly over the past 20 years with the widespread availability of satellite images, providing opportunities for tectonic analyses in physically inaccessible regions. Arguably the most spectacular examples of this approach have come from Tibet where offset drainage, fans, moraines and other landforms have been used both to estimate fault slip-vectors and, by assuming ages for landforms, to evaluate long-term slip rates. Peltzer *et al.* (1988, 1989) analysed the Altyn Tagh Fault using SPOT images to deduce very rapid neotectonic slip rates (>2 mm a^{-1}). Armijo *et al.* (1989) obtained similar rates for the Beng Co Fault in southern Tibet. However, these values are only underpinned by an assumption that the Tibetan landscape was formed during the last Wurm glaciation (Armijo *et al.* 1989). Elsewhere in central Asia other researchers have been less assertive as to the age of landforms imaged on SPOT and LANDSAT datasets (e.g. Thomas *et al.* 1996) and hence have not attempted such precise determinations of slip rates.

The purpose of this contribution is to examine part of an active continental plate boundary system to assess the activity of a major transcurrent fault and to evaluate the longevity of landforms which appear to be controlled by its displacements. Our example is the Yammouneh Fault, part of the Dead Sea Transform in Lebanon (Fig. 1). The region has recently become accessible again following the civil war. From the mid-1970s, assessment of displacement magnitudes and activity, using offset landscape features, has relied on LANDSAT images and only very rapid bouts of localized fieldwork (e.g. Garfunkel *et al.* 1981; Girdler 1990). Here we describe new field observations which demonstrate that at least the northern segment of the Yammouneh Fault has been inactive since Miocene times. The corollary is that landscape features, particularly small basins and apparently deflected drainage systems observable in satellite images, have been preserved with little modification for at least 5 million years. After describing field and geochronological evidence for the lack of any map-scale transcurrent offsets on the Yammouneh Fault during the Plio-Quaternary, we go on to discuss landscape evolution in the region where the Yammouneh Fault emerges northwards from Mount Lebanon. The general aim is to show that understanding rates of landscape evolution is often as important as structural analysis in any appraisal of neotectonics.

Geological setting of the Yammouneh Fault

For the most part the Dead Sea Transform is an obvious and well studied plate boundary. In its type area and south to the Red Sea, it is clearly

From: SMITH, B. J., WHALLEY, W. B. & WARKE, P. A. (eds) 1999. *Uplift, Erosion and Stability: Perspectives on Long-term Landscape Development*. Geological Society, London, Special Publications, **162**, 143–156. 1-86239-030-4/99/ $15.00 © The Geological Society of London 1999.

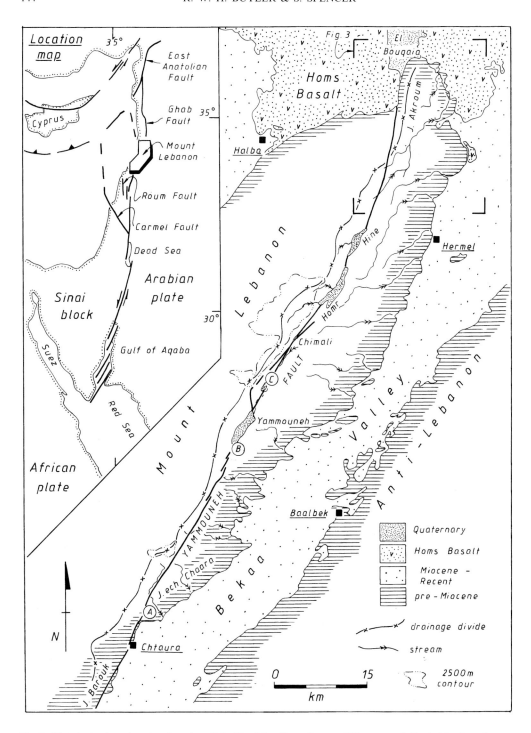

Fig. 1. Plate tectonic and regional setting of the Dead Sea Transform and Yammouneh Fault. The boxed area is the location of Fig. 3. Inset: Location map for detailed sites along the Yammouneh Fault.

constrained to lie within the Dead Sea valley (e.g. Garfunkel et al. 1981; Girdler 1990). Offsets of pre-tectonic markers suggest c. 105 km of left-lateral displacement (Quennell 1984). Tracing this amount of displacement northwards across Lebanon has long been a problem for structural geology (e.g. Quennell 1984). Upon entering Lebanon from the south (Fig. 1), the plate boundary apparently splits into at least five major fault zones (Beydoun 1977; Beydoun & Habib 1995; Walley 1988). Only one fault has been recognized to link northwards into the Syrian segment of the plate boundary (Garfunkel et al. 1981). This structure, mapped as a discontinuous array of fault segments by Dubertret (1955) and clearly evident on satellite images (e.g. Garfunkel et al. 1981; Girdler 1990) is called the Yammouneh Fault (Fig. 1). It appears to connect with the Ghab Fault in NW Syria to link with the inferred transform–transform–trench triple junction with the Tethyan collision belt in SE Turkey (Fig. 1). Consequently it is this fault that is generally considered to be the principal focus of relative plate motion and is shown as such on almost all maps of plate boundaries (e.g. Hempton 1987). Central to the argument is the relationship between the fault and the Homs Basalt in northernmost Lebanon, as documented by Dubertret (1955). He inferred that the basalt was Early Pliocene in age. His mapping shows the southern outcrop trace of the basalt to be deflected by about 11 km across the fault, implying a time-averaged displacement rate of about $2\,\text{mm}\,\text{a}^{-1}$ (Chaimov et al. 1990). Although a factor of three slower than that for the plate boundary in the south, it is still a significant value.

The view that the Yammouneh Fault accommodates a substantial portion of the bulk relative plate motion between Arabia and the Eastern Mediterranean during the Plio-Quaternary has been questioned by Girdler (1990). He points out that seismicity is strongly focused on a different structure, the Roum Fault, which bypasses the Yammouneh to the west (Fig. 1). Indeed there have been very few earthquakes recorded instrumentally in northern and central Lebanon. There is, however, historical evidence for earthquakes in the northern part of the Bekaa valley (Ambraseys & Barazangi 1989), albeit well off the trace of the Yammouneh Fault. The absence of suitable archaeological and historical records from the sparsely populated slopes of Mount Lebanon may be reflected in this lack of certainty (e.g. Ambraseys 1997) although these types of data rarely give good constraints on the three-dimensional orientation and kinematics of faulting. Consequently evaluating the longer time perspective provided by geomorphological data becomes critical. An important aspect for the landscape evolution of Lebanon is that almost all the strata that existed prior to deformation are carbonates. The region has a well developed cave and karst system by which much of the precipitation on upland areas is drained.

Geomorphological characteristics of the Yammouneh Fault

The Yammouneh Fault occupies a range of topographic positions in the landscape along its length (Fig. 1). In the south it generally defines the western boundary of the intermontane basin of the Bekaa valley. To the north the fault lies within the upland region of Mount Lebanon and it emerges from the hills just south of the Syrian border. It is difficult to correlate geomorphology across the fault in the south though the high-standing western side of the fault (Jebal Barouk, Fig. 1) does apparently contain truncated drainage patterns (e.g. Garfunkel et al. 1981). However, as a landscape feature, bounding the Jebal Barouk range, it is particularly striking.

In the vicinity of Chtaura (Fig. 1), the Yammouneh Fault passes northwards into the flank of Mount Lebanon. Here its eastern side contains a truncated spur (Jabel ech Chaara, Fig. 2a), particularly prominent on satellite images (e.g. Girdler 1990). This spur has a prominent west-facing facet which has no obvious correlative on the west side of the fault, suggesting substantial lateral motions. However, for the most part there are no systematic offsets of drainage pathways across the Yammouneh Fault within the Mount Lebanon range, although there are few significant valleys. Garfunkel et al. (1981) suggests that the lack of consistent deflections in drainage pathways is indicative of displacements greater than the lateral separation of streams (> 10 km). However, because the Yammouneh Fault runs immediately to the east of the drainage divide on Mount Lebanon (Fig. 1), few streams cross the fault and this makes this type of landscape feature inappropriate for establishing neotectonic kinematics. To the west of the fault, particularly the segment which lies west of the town of Hermel (Fig. 1), the map patterns of streams suggest evidence of capture, making it difficult to use stream deflections alone to infer tectonic processes.

The most striking geomorphological feature on the Yammouneh Fault comes from its type area at Yammouneh village (Fig. 1). Here the

fault forms the western margin of an active Quaternary depocentre (Fig. 2b). Indeed there is a modern lake within a small sedimentary basin. The basin continues parallel to the fault for about 5 km and Hancock & Atiya (1979) interpreted the basin as forming at a leftward step-over on the Yammouneh Fault, clearly suggesting that the fault system was still active and that continuing fault slip controlled the basin formation and its continuing sedimentary fill. However, Dubertret (1955) considered the basin simply to be a polje, temporarily filled by an ephemeral karst lake.

The Yammouneh polje is the largest such landscape feature along the fault. There are others at Chimali, Homr and Hine (Fig. 1), each of which captures drainage flowing directly from the watershed of Mount Lebanon. There are no major valleys which cross or are deflected by the trace of the fault. Further south in Mount Lebanon, the Yammouneh Fault lies a few kilometres to the east of the drainage divide (Fig. 2c). However, at the northern end of the mountains the topographic crest and drainage divide cross the fault. The drainage divide is a composite feature and cannot be used to infer fault offsets. The apparent dextral offset of the drainage divide across the Yammouneh Fault (Fig. 3) is an artefact. In contrast, the northern flank of the Mount Lebanon massif shows an apparent left-lateral offset across the Yammouneh Fault of about 11 km (Fig. 3). The east side of the fault is marked by the Jabel Akroum range while to the west is the relatively low-lying ground containing the Homs Basalt. It is from this northernmost outcrop domain of the Yammouneh Fault that critical evidence comes for establishing its activity in Plio-Quaternary times.

Apart from the large-scale landscape features described above, there is little evidence for active faulting along the Yammouneh Fault. The recently deposited lacustrine and marginal deposits of the Yammouneh polje show no pressure ridges or other deformation features. It should be noted, however, that to date there have been no sub-surface investigations or trenches cut that might unequivocally demonstrate this point. Apart from the four poljes along the Yammouneh Fault in the northern part of Mount Lebanon, for the most part the fault cuts a landscape dominated by very thin soils or frost-shattered limestone – terrain hardly conducive to trenching. Moreover, there are no signs of fresh fault scarps on aerial photographs or satellite images.

Evidence against Plio-Quaternary transcurrent faulting in northern Lebanon

The use of the apparent left-lateral offset of the northern margin of Mount Lebanon to demonstrate young displacements on the Yammouneh Fault is critically dependent upon the age of this landscape. Crucial evidence for the Pliocene and younger displacements on the northern part of the Yammouneh Fault comes from the relationship between the fault and the Homs Basalt (Fig. 3). Here we describe new field observations and present new K-Ar geochronological data from the basalt outcrops at the fault. The fault itself is clearly evident (Fig. 4a), roughly following the valley of Wadi Chadra. This valley has basalt on its western flank and Cretaceous carbonates on the east. The relationship between these two rock types may be observed in side-valleys which provide natural trenches through the Yammouneh Fault.

Homs Basalt and the Yammouneh Fault: field observations

In many places along Wadi Chadra, the Cretaceous limestones to the east of the Yammouneh Fault show widespread fracturing and brecciation. This cataclasis, presumably a product of seismogenic faulting, increases into a zone, up to 20 m wide, of intensely milled fault gouge. When seen in outcrop, this rock flour is exclusively composed of limestone fragments. Thus

---

Fig. 2. Geomorphological expression of the Yammouneh Fault in central Lebanon. (**a**) The truncated spur of Jabel ech Chaara, viewed to the north (A on Fig. 1). The Jebel forms the right (E) skyline. The right (west)-facing facet has no correlative on the west side of the fault. The Yammouneh Fault runs down the valley to the west (L) of the Jabel and then continues as a broad zone of cataclasite on the slopes above (L) of the village. The fresh, bright outcrops are of cataclasite, preferentially quarried as aggregate. The visible topographic relief from Jabel ech Chaara to the village is about 600 m. (**b**) The Yammouneh Fault at its type area, looking north (B on Fig. 1). The fault apparently bounds the high ground of the drainage divide on Mount Lebanon (visible elevation *c.* 2700 m) from the relatively low-lying polje and lacustrine basin in the foreground (elevation *c.* 1400 m). Yammouneh village is visible at the foot of the scarp at the extreme left of the photograph.
(**c**) Looking north along the Yammouneh Fault, with the village of Ainata in the middle distance (C on Fig. 1). The visible topographic relief is *c.* 700 m, with the fault following the base of the prominent scarp and gully line running diagonally from the skyline in mid-view to bottom left of the photograph.

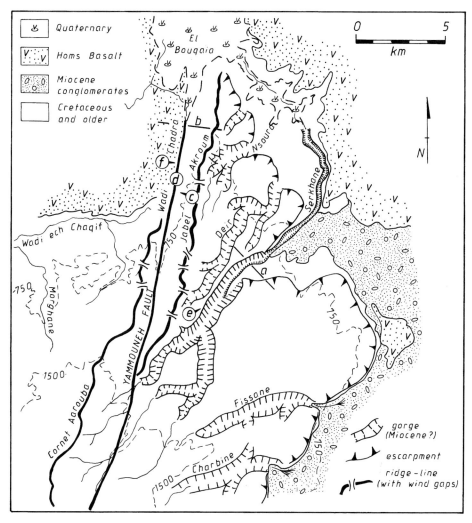

Fig. 3. Simplified geomorphological/geological map of the northern part of Mount Lebanon (see Fig. 1 for location). The arrow-heads on the escarpment symbol point upslope. Locations a & b indicate the topographic profiles of Fig. 6; c, viewpoint for Fig. 4a; d, viewpoint for Figs 4b, c and the profile of Fig. 5; e, viewpoint of Fig. 7a; f, viewpoint of Fig. 7b and location of samples R96–7 and R96–8.

there is a spectacular deformation gradient into the Yammouneh Fault. Bedding, where visible in the limestones, is folded into a N–S trending antiform. The mapping by Dubertret (1955) shows this fold to have a core of older Cretaceous sediments and to be offset sinistrally along the Yammouneh Fault. The fold has no obvious expression within the landscape. These

Fig. 4. Field aspects of the geology at Wadi Chadra. (**a**) Looking across Wadi Chadra (running diagonally through the centre of the photograph) from Jabel Akroum onto the trap morphology of the Homs Basalt to the west (taken from location c on Fig. 3). The prominent sub-horizontal flow structure in the lavas is evident in centre-left of view. These exposures lie at the head of the side-valley which contains location d on Fig. 3. (**b**) Undeformed pillow lavas at the deepest exposure level of Homs Basalt at Wadi Chadra, location of samples R96–9 and R96–10 (location d on Fig. 3 and b on Fig. 5). The white material just behind the outcrop is unstructured carbonate fault gouge of the Yammouneh Fault. (**c**) Undeformed subaerial lavas of the Homs Basalt in Wadi Chadra, location of sample R96–12 (location d on Fig. 3 and c on Fig. 5). The columnar jointed flow is about 4 m high.

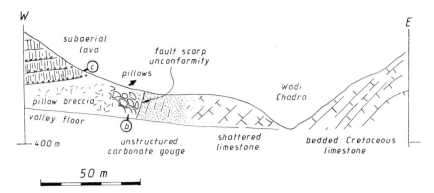

Fig. 5. Sketch section across the Yammouneh Fault in Wadi Chadra (location d on Fig. 3). The locations of field photographs in Fig. 4b and c are indicated by appropriate letters. The samples of pillow lava (R96–9 and R96–10) came from location b on this profile and the columnar-jointed flow sample (R96–12) came from location c.

relationships imply that the folding pre-dated at least the later stages of slip on the Yammouneh Fault. The deformation gradient as indicated by a simple, westward increase in cataclasis, suggests that, within the preserved eastern side of the fault, the vast majority of fault slip was accommodated in the narrow zone of gouge along the contact with the Homs Basalt.

In contrast to the carbonates, exposures of Homs Basalt show very little fracturing, save for jointing. Critical relationships between the lavas and the highly shattered carbonates along the fault may be found in a side-valley to Wadi Chadra (location d on Fig. 3; Fig. 5). Here the structurally deepest parts of the lavas crop out (Fig. 4b). They are pillow basalts with hyaloclastites and thin volcaniclastic sediments. The pillow shapes are very well preserved adjacent to the carbonate-derived fault gouge. They show no evidence of tectonic fracturing subsequent to eruption. The immediate fault gouge contains no fragments of basalt. These relationships are indicative of an unconformity (Fig. 5). The Homs Basalt rests upon a deformed landscape, locally a fault scarp, but is not itself deformed. The shapes of pillow lavas indicate that the subaqueous flow banked against the fault scarp.

Structurally and stratigraphically overlying the pillow basalts and hyaloclastites is a pile of subaerial basalts (Fig. 4c). They form 2–5 m thick flows, commonly separated by thin red soils, and reach a total thickness in excess of 200 m. Neither along the side of Wadi Chadra nor to its west do these lavas show any sign of faulting. The sole exception to this is a small linear scarp developed in the lavas, facing east and trending N–S, which is evident on air photographs on the SW margin of the El Bouqaia depression. We interpret this feature as a very minor normal fault, perhaps generated by differential compaction of the Miocene sediments which underlie the basalts.

Field relationships indicate that the Homs Basalt banks against the scarp of the Yammouneh Fault. The basalts are not deformed by the faulting. Thus we infer that the Homs Basalt post-dates all significant transcurrent displacements upon the Yammouneh Fault. Although our observations are only directly applicable to Wadi Chadra, pinning the fault here effectively precludes transcurrent displacements of more than a few tens of metres in adjacent areas after basalt eruption. Thus the Yammouneh Fault is not the active segment of the Dead Sea Transform in Lebanon. The most pressing question now is: how far back in the geological record does this situation pertain?

Age of the Homs Basalt

To answer this question the Homs Basalt has been dated using the whole-rock K-Ar method. Samples were collected from the pillow basalts adjacent to the Yammouneh Fault in Wadi Chadra. However, to provide a geological test of the results, samples were also collected from the bottom and top of the subaerial basalts which overlie the pillows. A synopsis of the results is presented in Table 1.

The basalts are of latest Miocene age, erupted at about 6 Ma. The two subaerial sample sites give results consistent with their relative stratigraphic position. Furthermore the chemistries of these samples are appropriate for reliable K-Ar

Table 1. K-Ar data from the Homs Basalt

	Sample	K (%)	Volume of radiogenic ^{40}Ar ($\times 10^{-5}$ cm^3 g^{-1})	Radiogenic ^{40}Ar (%)	Age (Ma)	Error
Columnar flow, top of section, Aandquat	R96–7	1.013 / 0.0202	0.0211	11.3 / 24.1	5.2	0.2
Columnar flow, top of section, Aandquat	R96–8	1.086 / 0.0228	0.0238	23.1 / 23.3	5.5	0.2
Pillow lava, Wadi Chadra	R96–9	0.671 / 0.0143	0.0154	9.8 / 6.5	5.7	0.7
Pillow lava, Wadi Chadra	R96–10	0.799 / 0.0204	0.0209	21.0 / 21.7	6.7	0.2
Lowest columnar flow, Wadi Chadra	R96–12	1.607 / 0.0401	0.0411	47.8 / 44.6	6.5	0.2

All sample analyses were duplicated. Location of samples is shown in Figs 3 and 5. Samples R96–9 and R96–10 were both collected from the same outcrop of pillow lavas in Wadi Chadra (grid reference 20522948). R96–12 came from the columnar-jointed subaerial basalts immediately overlying the pillow lavas, also in Wadi Chadra (GR 20512948). R96–7 and R96–8 both come from the top of the basalt pile at about 520 m elevation from the top of the Wadi Chadra valley side, about 2 km N of Aandqat village (location f on Fig. 3). Neither was *in situ* but are presumed to have been transported no more than a few metres.

dating, with relatively high concentrations of ^{40}K and of radiogenic ^{40}Ar. Consequently the error on the ages is only a few per cent. In contrast, one pillow lava sample (R69–9) had rather low ^{40}K and ^{40}Ar concentrations. The sample's age (5.7 Ma) therefore has a rather larger error (>10%). The other sample (R69–10), which has better concentrations of ^{40}K and ^{40}Ar, yields an analytically robust age of 6.7 ± 0.2 Ma. The oldest subaerial lava overlying the pillows and hyaloclastites yields a similarly robust age of 6.5 ± 0.2 Ma. This is apparently in accord with stratigraphic relationships but, given the resolution of the method, the ages of the pillows and subaerial flows are indistinguishable (c. 6.5 Ma). However, to provide a stratigraphic test of the geochronology, the youngest preserved basalt flows, about 150 m above Wadi Chadra (RB96–7, RB96–8), were dated. These yielded ages of 5.2 ± 0.2 Ma and 5.5 ± 0.2 Ma. It can be concluded that the duration of eruption of the Homs Basalt in the Wadu Chadra area was about one million years. In stratigraphic terms, these lavas were erupted during the Messinian. Thus the Yammouneh Fault, where it emerges from the northern slopes of Mount Lebanon, has not accommodated any significant transcurrent displacement during Plio-Quaternary times.

Drainage evolution and landscape correlations across the northern Yammouneh Fault

In order to provide further tests of our geological interpretations of the relationship between the Homs Basalt and displacements on the Yammouneh Fault, the geomorphological evolution of the area was examined. If the fault accommodated significant transcurrent slip since the latest Miocene (c. 1 km) then landscape features younger than the basalt should show offsets. Critical to this is an understanding of valley system evolution. This is represented here using two selected lateral topographic profiles (Fig. 6), illustrating the geomorphology using two field photographs (Fig. 7).

The ridge of Jabel Akroum, on the eastern side of the Yammouneh Fault, is crossed by a series of deeply incised valleys which drain north and eastwards into the northern Bekaa valley or the El Bouqaia depression. (Fig. 3). The most prominent of these is the NE–SW trending valley which contains the modern Wadi Serkhane. This incised valley has its headwaters on the western flank of the Yammouneh Fault and they show no apparent offset from the downstream sectors. Wadi Serkhane continues downstream to cut deeply beneath the Homs Basalt and clearly links onto the base-level of the El Bouqaiaa basin (Fig. 3). However, the modern Wadi Serkhane is incised into a much broader valley (Figs 6 and 7). This ancestral valley links to a gentle palaeosurface which dips NE away from the Jabel Akroum ridge (Fig. 8). This surface is overlain by subaerial conglomerates inferred to be late Miocene in age (Dubertret 1955), which are in turn overlain by the Homs Basalt. Thus drainage from Jabel Akroum shows a distinct, two-stage evolution.

A suite of valleys may be traced eastwards from the Yammouneh Fault down to the late

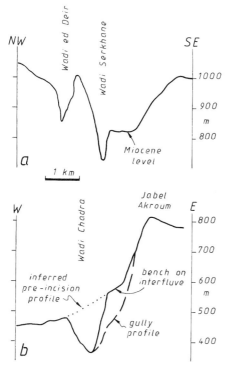

Fig. 6. Topographic profiles across selected composite valleys adjacent to Jabel Akroum. (**a**) Profile across Wadi Serkhane (location a on Fig. 3); (**b**) profile across Wadi Chadra (location b on Fig. 3).

Miocene palaeosurface in the northern Bekaa. This surface terminates to the west at a prominent erosional scarp (Fig. 3) which represents the ancestral boundary to the Mount Lebanon uplift. The valleys of Wadis Fissane and Charbine, which run to the south of the Serkhane valley, emerge through the scarp and preserve Miocene conglomeratic fills in their lower reaches (Fig. 3). These valleys show no indications of significant post-Miocene incision, presumably because their headwaters have largely been captured by the karstic system and because there is little surface run-off to allow gully-head erosion. To the north of Wadi Serkhane the complex valley systems of Wadis ed Deir and Nsoura show a two-stage incision history. The lower reaches are ravines that link onto the Quaternary base-level of the El Bouqaia depression. However, the knick-points associated with these have only migrated part-way up valleys which breach the late Miocene mountain edge. These wadis are therefore inferred to have been initiated in Miocene times.

Although drainage post-dating the Homs Basalt shows no indication of offset or of truncated headwaters across the Yammouneh Fault, the earlier valley systems do. The ridgeline of Jabel Akroum shows a series of windgaps, particularly above the valleys of Wadi ed Deir and Wadi Nsoura (Fig. 3). These windgaps form part of the Miocene drainage network. Thus offsets and truncation of these landscape features by the Yammouneh Fault could have happened at any stage since their formation in Miocene times. There is no evidence for post-Miocene movement. Since the eruption of the Homs Basalt some drainage basins have been incised, with incision only strongly developed along Wadi Serkhane which has significant headwaters. The other valleys remain dry for much of the year, with precipitation presumably entering the well developed karstic system of dolines along the Jabel Akroum ridge.

Drainage on the western flank of Jabel Akroum links strongly into the modern level of Wadi Chadra (Fig. 3). However, many of the interfluves show a marked bench at about the elevation of the top of the Homs Basalt (Figs 6 and 7). This suggests that the western flank of the Jabel formed a prominent scarp tailing off onto the early Pliocene palaeosurface which caps the Homs Basalt. Rather more speculatively, the ridge line of Jabel Akroum contains perched valleys which drain westwards directly opposite the upper valley systems of Wadi ech Chaqif, on the west side of the Yammouneh Fault (Fig. 3). These linked palaeolandscape features correlate without lateral offset across the Yammouneh Fault. However, they have been strongly modified by incision along the modern valley system of Wadi Chadra (Fig. 8). Thus the headwaters of Wadi ech Chaqif, and much of those of Wadi Serkhane, have been captured and now drain northwards. In summary, there are no landscape features in the study area post-dating the Homs Basalt that show offset or tectonic truncation by the Yammouneh Fault. Truncations and offsets are only shown by those features which are part of the pre-Homs landscape.

Preservation of landscapes

The northern part of the Yammouneh Fault preserves landscape features which pre-date eruption of the Homs Basalt, and others which appear to post-date it but pre-date incision by the modern Wadi Chadra (Fig. 8). Preservation of these features seems to be critically dependent upon the karstic nature of the region. Those valleys with high headwaters located on the main ridge of Mount Lebanon appear to have

Fig. 7. Views across composite valleys adjacent to Jabel Akroum (see Fig. 6). (**a**) The modern, incised valley of Wadi Serkhane and the broader palaeovalley, seen looking NE from location e on Fig. 3. The late Miocene base-level is evident as the prominent bench, overlain in the distance by conglomerates and the Homs Basalt. The slope in mid-right of picture is the inferred Miocene escarpment that bounded the ancestral NE Mount Lebanon. This escarpment is approximately 250 m high. (**b**) The view looking east across Wadi Chadra (hidden) from the palaeosurface which caps the Homs Basalt (foreground) onto the western flank of Jabel Akroum. The grassy slope in mid-picture represents the perched terrace on an interfluve. The ridge-line is approximately 400 m above the viewpoint. Taken from location f on Fig. 3.

maintained the ability to incise, as typified by Wadi Chadra and Wadi Serkhane. These valleys are therefore able to capture drainage from those valley systems whose headwaters percolate into the karst system. In these latter situations, landscape evolution has been greatly retarded.

More strikingly, if the northern segment of the Yammouneh Fault has not accommodated transcurrent offsets since Miocene times, it is necessary to re-evaluate the landforms along this fault further south. Note that the fault could still accommodate minor vertical movements to the south of Wadi Chadra. However, the apparent 'step-over' basin at its type area and the truncated spur of Jabel ech Chaara have presumably been inherited from these pre-Pliocene times. Garfunkel *et al.*'s (1981) interpretation of the lack of systematic offset of minor streams crossing the Yammouneh Fault need not imply fault displacements greater than the lateral separation of these streams. Rather, it is consistent with the notion that the fault has not been active since these streams were established within the landscape.

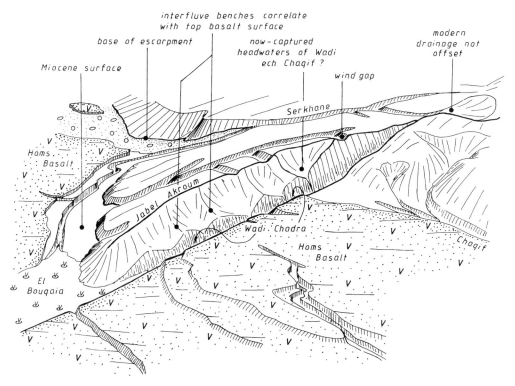

Fig. 8. Schematic block diagram showing the relationships between various landscape features at Jabel Akroum and Wadi Chadra, viewed towards SE (see Fig. 3 for locations, key to geological units and spatial separation between locations). The Yammouneh Fault runs close to the course of Wadi Chadra but is buried by Quaternary lacustrine deposits and, presumably, the Homs Basalt in the depression of El Bouqaia. The truncated spurs and benches along the slopes of Jabel Akroum running down to Wadi Chadra reflect incision of this Wadi. The landsurfaces correlate from Jabel Akroum to the top surface of the Homs Basalt on the west side of the Wadi. Drainage networks on the eastern side of Jabel Akroum (shown in very simplified form here), which are apparently truncated by the Yammouneh Fault, link to the Miocene palaeosurface upon which was erupted the Homs Basalt.

Discussion

The northern margin of the Mount Lebanon massif shows a left-lateral offset across the Yammouneh Fault of about 11 km. The use of this relationship to infer active displacement on the fault is ambiguous because the uplift of Mount Lebanon was initiated during Miocene times (e.g. Beydoun 1977; Walley 1988). Thus the offset of the massif's northern margin could have occurred at any stage thereafter. The original mapping of Dubertret (1955) implies that the northern margin of the massif existed as a geomorphological feature with associated depositional fans and drainage systems in the Miocene and Pliocene. This margin also acted as the southern limit to the Homs Basalt. Our fieldwork in northern Lebanon has confirmed these relationships and clarified them in one main regard. The outcrops in Wadi Chadra show that the Homs Basalt unconformably lies against the Yammouneh Fault and has not been deformed by it. The apparent offset on the southern outcrop edge of the basalt is a topographic artefact caused by the restricted eruption of lava on the relatively low-lying ground around Jabel Akroum (Fig. 8). This represents a palaeolandscape, overlain by the Homs Basalt. Drainage systems which form part of the pre-basalt landscape show evidence for disruption by the Yammouneh Fault. They may be linked morphologically to the pre-basalt palaeosurface. However, drainage systems which incise into the basalt show no evidence of lateral offset. Indeed, landforms along the flanks of Wadi Chadra correlate across the fault without offset (Fig. 8). They are dismembered now because of incision along the valley. Although we can only date these younger landforms as being post-5 Ma, the age of the youngest flows of the Homs

Basalt, their lack of offset is at least consistent with our main conclusion – that the Yammouneh Fault ceased activity as a major strike-slip structure prior to the eruption of the Homs Basalt.

The relative stratigraphy of landscape and fault activity outlined above may be given some precision by the K-Ar ages on the Homs Basalt. The earliest dated lavas are about 6.5 Ma old, with the lavas erupted until about 5.2 Ma. Thus our analysis of tectonic geomorphology relies not only on landscape description but also on gaining ages of palaeosurfaces, bracketed by the basalt ages. In the absence of an active Yammouneh Fault for the past 5 Ma, other structures are needed that could have accommodated the expected $c.45$ km of left-lateral displacement on the Dead Sea Transform in Lebanon (e.g. Hempton 1987; Chaimov et al. 1990). As the Homs Basalt and its associated palaeosurface effectively seal faults in the whole of northern Lebanon, it is likely that the principal active transcurrent structure is the Roum Fault in SW Lebanon (Fig. 1, Butler et al. 1997). It is certainly the focus for much of the historical and instrumentally recorded seismicity (e.g. Girdler 1990; Ambraseys 1997).

Our results raise serious issues for the preservation and stability of tectonic geomorphology elsewhere along the Yammouneh Fault. The faceted spurs, poljes and disrupted drainage systems along the fault were presumably formed in Miocene times. These are extinct landforms (Fig. 2) and are not indicative of active transcurrent faulting (cf. Garfunkel et al. 1981). However, we infer that the poljes in particular continued to accumulate sediment after the termination of displacement along the Yammouneh Fault. Clearly future work should be directed at investigating the state of deformation in these sediments.

The southern segments of the Yammouneh Fault, along the eastern flank of Jabel Barouk in particular, may indeed be active. Uplift of the Mount Lebanon range has continued through Plio-Quaternary times, as indicated by the array of exposed marine terraces and benches along its western slopes (e.g. Beydoun 1977). It is probable that the Wadi Chadra area has been uplifted, as indicated by the rejuvenation of valley incision after the eruption of the Homs Basalt and the subaerial exposure of the pillow lavas at its base. This uplift may be controlled in part by dip-slip faulting, possibly locally reactivating the Yammouneh Fault, but with different kinematics to that of its earlier transcurrent history.

We finish with some general observations relating to the stability of tectonic landforms. Our conclusions indicate that large-scale tectonic landforms may be preserved for many millions of years with only minor modification by surface processes. However, this stability relies on very low rates of landscape evolution. In the arid to semi-arid environment of the eastern slopes of Mount Lebanon, where surface run-off is especially inhibited by the karstified limestone subsubstrate, landscape longevity might be expected. It is clearly unwise for the same investigative methods as adopted for elevated regions prone to Quaternary glaciation to be used in these environments. Peltzer et al. (1989) imply that slip rates of greater than $c.1$ mm a^{-1} during Holocene times are amenable to analysis using SPOT images. Their appraisal is based on the inferred rate of landscape evolution and stability of tectonic landforms on the Tibetan plateau. In less sensitive landscapes, where erosion may be far slower, the use of remote sensing alone to deduce slip rates may be misleading. An evaluation of the rates of landscape evolution, tied to dated palaeosurfaces, is as important as the recognition of the landforms themselves in the study of neotectonics through geomorphology.

This work was carried out while R.W.H.B. held a Nuffield Foundation Science Research Fellowship. Preliminary fieldwork was supported by the British Council. We are indebted to the late Z. Beydoun. We also thank H. Griffiths, K. Khair and C. Walley and we are grateful for the comments of two anonymous referees on an earlier draft of this paper. The K-Ar dating was carried out by P. Guise in Leeds.

References

AMBRASEYS, N. 1997. The earthquake of 1 January 1837 in Southern Lebanon and Northern Israel. *Annali di Geofisica*, **40**, 923–935.

—— & BARAZANGI, M. 1989. The 1759 earthquake in the Bekaa valley: implications for earthquake hazard assessment in the eastern Mediterranean region. *Journal of Geophysical Research*, **94**, 4007–4013.

ARMIJO, R., TAPPONNIER, P. & HAN, T. 1989. Late Cenozoic right-lateral strike-slip faulting across southern Tibet. *Journal of Geophysical Research*, **94**, 2787–2838.

BEYDOUN, Z. R. 1977. The Levantine countries: the geology of Syria and Lebanon (maritime regions). *In*: NAIRN, A. E. M., KANES, W. H. & STHELI, F. G. (eds) *The Ocean Basins and Margins, 4A, the Eastern Mediterranean*. Plenum, New York, 319–353.

—— & HABIB, J. G. 1995. Lebanon revisited: new insights into Triassic hydrocarbon prospects. *Journal of Petroleum Geology*, **18**, 75–90.

BUTLER, R. W. H., SPENCER, S. & GRIFFITHS, H. M. 1997. Transcurrent fault activity on the Dead

Sea Transform in Lebanon and its implications for plate tectonics and seismic hazard. *Journal of the Geological Society, London*, **153**, 757-760.

CHAIMOV, T. A., BARAZANGI, M., AL-SAAS, D., SAWAF, T. & GEBRAN, A. 1990. Crustal shortening in the Palmyride Fold Belt, Syria, and implications for movement along the Dead Sea Fault System. *Tectonics*, **9**, 1369-1386.

DUBERTRET, L. 1955. *Carte géologique du Liban au 1/200,000, avec notice explicative*. Ministère des Travaux Publics, Beyrouth.

GARFUNKEL, Z., ZAK, I. & FREUND, R. 1981. Active faulting in the Dead Sea Rift. *Tectonophysics*, **80**, 126

GIRDLER, R. W. 1990. The Dead Sea transform fault system. *Tectonophysics*, **180**, 1-13.

HANCOCK, P. L. & ATIYA, M. S. 1979. Tectonic significance of mesofracture systems associated with the Lebanese segment of the Dead Sea transform fault. *Journal of Structural Geology*, **1**, 143-153.

HEMPTON, M. R. 1987. Constraints on Arabian plate motion and extension history of the Red Sea. *Tectonics*, **6**, 687-705.

PELTZER, G., TAPPONNIER, P., GAUDEMER, Y., MEYER, B., GUO, S., YIN, K., CHEN, Z. & DAI, H. 1988. Offsets of late Quaternary morphology, rate of slip and recurrence of large earthquakes on the Change Mar Fault (Gansu, China). *Journal of Geophysical Research*, **93**, 7793-7812.

——, —— & ARMIJO, R. 1989. Magnitude of late Quaternary left-lateral displacements along the northern edge of Tibet. *Science*, **246**, 1285-1289.

QUENNELL, A. M. 1984. The western Arabian rift system. *In*: DIXON, J. E. & ROBERTSON, A. H. F. (eds) *The Geological Evolution of the Eastern Mediterranean*. Geological Society, Special Publications, **17**, 775-788.

THOMAS, J. C., COBBOLD, P. R., WRIGHT, A. & GAPAIS, D. 1996. Cenozoic tectonics of the Tadzhuk depression, central Asia. *In*: YIN, A. & HARRISON, M. (eds) *The Tectonic Evolution of Asia*. Cambridge University, 191-207.

WALLEY, C. D. 1988. A braided strike-slip model for the northern continuation of the Dead Sea Fault and its implications for Levantine tectonics. *Tectonophysics*, **145**, 63-72.

Influence of crustal movements on landforms, erosion and sediment deposition in the Irangi Hills, central Tanzania

MATS G. ERIKSSON

Department of Physical Geography, Stockholm University, S-106 91 Stockholm, Sweden

Abstract: The influence of crustal movements on landforms, erosion and sediment deposition has been analysed in the severely eroded Irangi Hills in Kondoa District, central Tanzania. Field observations, in combination with satellite imagery, aerial photographs, and topographical and geological maps have been used. The Irangi Hills, which are uplifted relative to the vast Maasai Plain, consist of several tectonic blocks, with each block being tilted to the west. Frequent earthquakes indicate that relative uplift, faulting and tilting of the Irangi Hills are still taking place. The drainage pattern of the area and the deposition of river sediments are directly influenced by the morphotectonics. Where fault scarps occur, large rivers have cut into the bedrock across the fault scarps and follow antecedent stream courses, while small streams have been dammed, forming swamps or lakes. The denudation rate for a small catchment (33.4 km^2) has been calculated to 2.7 mm a^{-1} for the years 1836–1992 by using a dated sediment record. This is considered to be an accelerated denudation rate. The soil erosion of the Irangi Hills is a long-term natural stripping of regolith, accelerated by anthropogenic causes, probably in combination with crustal uplift and the tilting of tectonic blocks.

The Irangi Hills of Kondoa District are one of the most severely eroded areas in Tanzania. The problem of land degradation in the area has been acknowledged since early colonial times (Kannenberg 1900), and several soil conservation programmes have been implemented. The latest included the prohibition of grazing in an area covering 1256 km^2 in 1979, and is one of the most controversial steps against soil erosion ever effected in Tanzania (Östberg 1986).

This study is part of the multidisciplinary research programme 'Man–Land Interrelations in Semi-arid Tanzania' (Christiansson *et al.* 1991; Christiansson 1996), in which one of the main objectives is to describe the landscape history of the Irangi Hills, with special reference to soil erosion. When describing the physical characteristics and environmental history of an area located in the East African Rift Zone, the effects of tectonic activity cannot be overlooked. Modern, man-induced erosion needs to be separated from the results of geological processes that have operated over a much longer time scale.

The purpose of this paper is to analyse the relationship between tectonically influenced landforms, drainage patterns and sediment deposition in the Irangi Hills, and to discuss the effect of crustal movements on soil erosion. In addition, the recent denudation rate of a small catchment has been calculated and evaluated in relation to long-term denudation rates. Anthropogenic factors influencing erosion in the area, such as changes in land use, demography and socio-economic characteristics, have been examined by others (e.g. Mung'ong'o 1995; see also Christiansson & Kikula 1996).

Geology and climate

The Irangi Hills in north central Tanzania are located within the eastern branch of the East African Rift System (Fig. 1) (Quennell *et al.* 1956; Grove 1986). This area has been uplifted relative to the vast Maasai Plain to the east and is separated by a major escarpment (Fig. 2). The escarpment began to form in the Miocene and has continued to develop into recent times (James 1956; Veldkamp 1996). The uplifted area is characterized by hills, the summits of which have been stripped of saprolite, and which are separated by broad valleys with thick layers of regolith. It can be classified as a partly stripped etchsurface developed according to a model with 'double surfaces' as described by Büdel (1957, 1982) and Thomas (1994). From an altitude above 2000 m (Tomoko Hills) at the edge of the escarpment in the northeast, the Irangi Hills decline to around 1500 m in the southwest.

The bedrock is of Precambrian age and consists mainly of feldspathic gneiss and schist (Fozzard 1963; Selby & Mudd 1965). The upper hillslopes are dominated by quartz gravels and bare rock. Soils occur in catenary sequences in which soil types change gradually downhill (Payton *et al.* 1992; Payton & Shishira 1994). Most pediments in the Irangi Hills are dissected

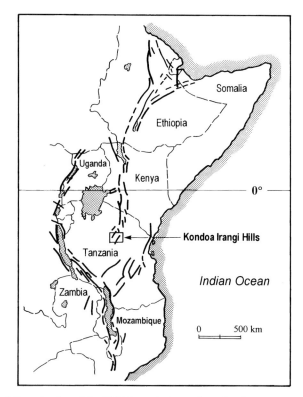

Fig. 1. East Africa and the extension of the Rift Zones (redrawn from Somi 1993, p. 11). Outlined box shows the location of the study area.

by numerous gullies, up to 20 m deep, and badlands are common (Fig. 3). In addition, sheet erosion has removed up to 2–3 m of topsoil (Payton & Shishira 1994).

Lake Haubi (Fig. 2), which was a seasonally inundated swamp before the turn of the century, is located within a severely degraded catchment of 33.4 km². Here, extensive gully systems and badlands feed directly into large sandfans, which terminate at the edge of the lake. The sandfans and the existence of the water body have facilitated a thorough sorting of sediments according to particle size. Therefore the swamp/lake sediment consists of more than 80% clay, while the fans are composed of sandy material.

The climate of the Irangi Hills is semi-arid, with an annual average precipitation ranging from c. 600 mm in the southwest to c. 900 mm in the more elevated northeast (Yanda 1995). Annual potential evaporation exceeds 2000 mm (Woodhead 1968).

Natural vegetation consists of 'Miombo' woodland, with *Brachystegia* the most common woody genus (Backéus et al. 1994). Extensive deforestation has taken place on several occasions in the past, for instance along caravan routes in the nineteenth century (Christiansson 1981, 1988). Clearing of Miombo woodland also occurred over large areas in Kondoa District during the 1930s and 1940s in order to combat the tsetse fly (Mung'ong'o 1995, p. 74). In the Tomoko Hills (Fig. 2), a dry Afromontane forest grows (Lyaruu et al. 1997) on a highly leached and weathered soil, rich in organic matter (Yanda, pers. comm.). If this soil had been exposed to the intensive rains, it would probably have been washed away, and its continued existence suggests that the Tomoko Hills have never been deforested. The forest, which is rapidly diminishing at present (Lyaruu et al. 1997), is likely to be a remnant of a once larger forest covering the more elevated parts of the Irangi Hills.

Methods

The mapping of fault lines, rivers and sand deposits was made from topographical (1963) and geological (1965) maps, aerial photographs (1:40 000, July 1960; 1:25 000, June 1987) and

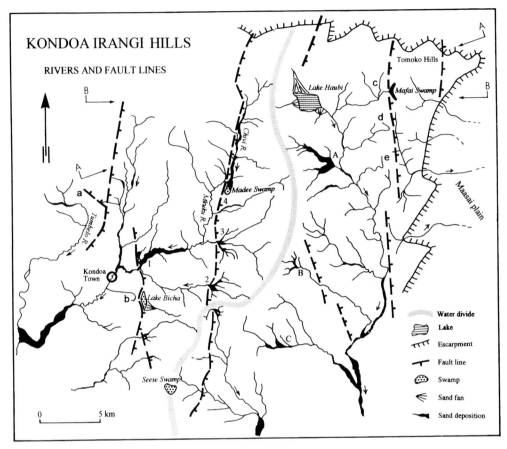

Fig. 2. Map showing fault lines, rivers, lakes and swamps in the Kondoa Irangi Hills. Numbers 1–4 indicate locations where streams have cut antecedent courses concurrently with uplift. Increased sand deposition occurs in these places. Letters A, B and C indicate where enhanced sand deposition occurs due to the westward tilting of the tectonic blocks. Letters a–e indicate where reversal of headwaters has occurred. The location of the topographical profiles shown in Fig. 4 are indicated with A–A and B–B.

satellite imagery (SPOT, October 1987). Topographical profiles were drawn in order to interpret the location of individual tectonic blocks and their effect on the morphology. Two of these are presented in Fig. 4. Interpretations were checked during several field visits.

A 361 cm deep sediment profile covering the period from c. 1836 to 1988 was retrieved from Lake Haubi in order to obtain information on the recent sedimentation and soil erosion history. The sediment record was divided into three stratigraphical units: A, B and C (Fig. 5). Unit A is a swamp sediment, unit B a transitional layer and unit C a laminated lake sediment. A chronology for the sediment record has been established by analyses of ^{210}Pb and ^{137}Cs (El-Daoushy & Eriksson 1998). The sediment record is used here for the calculation of sedimentation (Fig. 5) and denudation rates of the catchment. Additional analyses of the sediment record are presented elsewhere (Eriksson & Christiansson 1997; Eriksson & Sandgren in press; Erikssen et al. in press).

Depths from the soil surface to the bedrock in the Haubi basin have been measured using geoelectrical depth sounding equipment. The measurements, carried out on sandfans close to the lake shore, indicate a bedrock surface inclined to the west with soil depths of between 30 and 60 m.

Tectonics and drainage systems

Several faults, aligned primarily N–S, can be distinguished within the Irangi Hills (Fig. 2).

Fig. 3. Large parts of the slopes of the Kondoa Irangi Hills are dissected by gullies up to 20 m deep, and badlands are common. Photo by M. G. Eriksson, August 1992.

They divide the area into a number of tectonic blocks, each tilted up to 3° towards the west (Fig. 4). The summit levels of some of the tectonic blocks reveal an inclined surface, suggesting that a planation surface was developed before the tilting took place (Fig. 4A). Generally, the eastern parts of the block units are dominated by bare rock and regolith, while the western parts are often covered by regolith and sediments.

Ephemeral rivers drain the area, which forms part of the 10 000 km² Bubu River catchment,

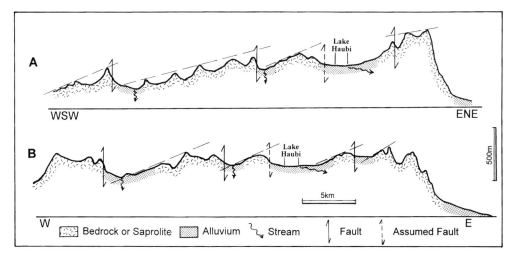

Fig. 4. Topographical transects across the Kondoa Irangi Hills. Both profiles run across the Lake Haubi basin (for location, see Fig. 2). The thin broken lines indicate the summit levels of surfaces which are assumed to have been formed by deep weathering before the later rise and inclination of the Irangi Hills took place.

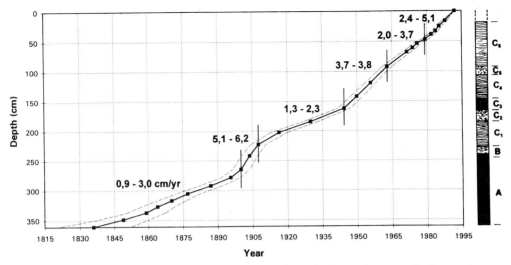

Fig. 5. Apparent sedimentation rate in the Haubi basin since the early nineteenth century. Dating was done through the analyses of ^{210}Pb, with each black dot in the diagram representing one analysis. The margin of error is indicated by dashed lines. The formation of Lake Haubi took place around the turn of the century as interpreted from the sediment record (unit B). The high sedimentation rate at that time was partly due to high biogenic production in combination with poor decomposition of organic matter within the swamp/lake.

eventually ending in an internal drainage basin, the Bahi Swamp. The drainage system is divided into a western and an eastern system (Fig. 2). Much of the drainage pattern has been modified by the fault scarp development and block tilting. In the western drainage system, north–south streams follow the fault lines until they are diverted to the west, forming an angulate drainage pattern. The eastern drainage system runs in a southerly and southeasterly direction.

Due to the continuous supply of eroded material from numerous gullies, sandbed rivers are a prominent feature in the Irangi Hills. The relation between the drainage systems and the faults explains the location of some of the extremely wide sandbed rivers, swamps and lakes. In rivers flowing westwards, large streams have been powerful enough to cut down concurrently with local uplift and now follow narrow antecedent stream courses across the north–south fault scarps (1–4 in Fig. 2 and Fig. 6). Several tributaries converge here on their way westwards and the water is dammed, which causes enhanced sediment deposition (Figs 6 and 7). Small streams have either been dammed against the fault scarps to become lakes or swamps (Lake Bicha, Madee Swamp, Mafai Swamp) (Fig. 2), or have been deflected to join adjacent larger streams. Whether a stream channel has been deflected by a crustal displacement or has eroded it, is determined by the stream power, i.e. the overall effect of discharge and gradient.

In the eastern river system, tilting of the bedrock has caused increased sand deposition in streams flowing eastwards. Summerfield (1991, p. 412) states that a slight tilt of a landsurface, perhaps only by a fraction of a degree, will cause pronounced changes in the drainage with river disruption or reversal as a result. As the tectonic blocks are tilted towards the west and the streams flow to the east, the stream gradient is lowered and stream velocities decrease, causing sedimentation. If tilting continues, stream velocities will decrease so much that all drainage is cut off and a swamp or lake begins to form. Lake Haubi, which existed as a permanent lake from the early nineteenth century to 1994, and otherwise as a seasonally inundated swamp, might have been formed partly in this way (Fig. 4). The formation was also aided by growing sand-fan deposits, which blocked the swamp outlet (Eriksson & Christiansson 1997). South of Lake Haubi, several very wide stream beds are found which could be prospective lakes (A–C in Fig. 2) about to form in the same way. Tilting of the bedrock has also affected some of the river headwaters. The rivers marked a–e in Fig. 2 have developed new stream courses subsequent to the westward tilting.

Earthquakes, uplift and natural erosion

The earthquake diagram for the Irangi Hills and its vicinity (latitude 3.5–6° S, longitude

Fig. 6. Aerial photograph showing an ephemeral river following an antecedent stream course which cuts through a N–S oriented fault line. Here the stream velocity decreases, resulting in enhanced sand deposition. Small streams have not been able to maintain their courses concurrently with tectonic uplift and fault scarp formation; instead they converge to follow the main river. The light areas in the upper part of the stream catchments are badlands and gullied areas.

34.5–37° E) (Fig. 8) shows that tectonic activity is frequent, although the intensity varies over time. At the end of the 1960s numerous earthquakes were recorded, with a maximum of 28 events in 1966, after which activity declined in the 1970s. Earthquake data exist for Tanzania from 1846 onwards (Iranga 1991). However, it should be noted that the instruments used for recording earthquakes have improved over time. Instruments used after 1954 are sensitive enough to detect most earthquakes, whilst only the major earthquakes are likely to have been recorded before 1954 (Iranga 1992). Most of the events recorded (Fig. 8) reach magnitude 4 to 5 on the Richter scale.

It is shown (Figs 2, 6 and 7) that crustal movements have markedly affected the drainage pattern and the spatial distribution of sediment accumulation within the Kondoa Irangi Hills. Earthquake data indicate that crustal movements still occur (Fig. 8) and it is likely that they affect the erosional processes in the area. When land is elevated and tilted relative to a base-level, the potential energy and hence the erosional power of running water increase in the uplifted area (Burbank *et al.* 1996). The tilting may also create a biased distribution of soil and regolith cover in the landscape (Fig. 9), which in turn speeds up movement of soil within the tilted area until a more stable situation is reached. Many authors have noted how the number and size of incised stream courses increase with uplift in tectonically active areas

Fig. 7. Schematic diagram showing the relationship between fault scarps and rivers in the Irangi Hills. Large streams have cut antecedent stream courses concurrently with uplift and caused enhanced sedimentation upstream of these locations. Small streams have been dammed against the fault scarps to become lakes or swamps. The lightly shaded area indicates the places where sediments generally accumulate.

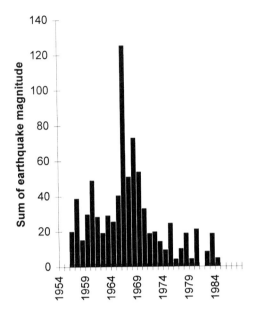

Fig. 8. Seismic energy release per year between 1954 and 1988 in central Kondoa District (i.e. lat. 3.5–6° S, long. 34.5–37° E). Very few earthquakes were recorded in the area between 1846, when registration began, and 1954, because the instruments were not sensitive enough at that time. Data compiled from Iranga (1991).

They estimated the average denudation rate to 0.0084 mm a^{-1}, equivalent to a yield of 22 t km^{-2} a^{-1} for the Tertiary, using a rock density of 2.65 g cm^{-3}, and 0.029–0.075 mm a^{-1} (77–200 t km^{-2} a^{-1}) for the Pliocene and Quaternary. Dunne (1979) estimated the denudation rate for humid Kenya to be 20–30 t km^{-2} a^{-1} (0.008–0.01 mm a^{-1}) when he used sediment yields from forested catchments in tectonically stable areas. A similar study in semi-arid regions in Australia (Douglas 1967) gave a peak sediment yield of 115 t km^{-2} a^{-1} (0.043 mm a^{-1}).

Recent sedimentation and denudation rates

Since 1836, the apparent sedimentation rate of the Haubi basin (Fig. 5) has increased from about 1 cm a^{-1} to more than 5 cm a^{-1}. High sedimentation rates occurred at the turn of the century and have also occurred since the 1940s, indicating a recent acceleration of erosion. However, the comparatively low sedimentation rates during the first half of the nineteenth century are also regarded as being accelerated compared to natural conditions (see below).

Depth to the bedrock beneath Lake Haubi can be interpolated to about 50 m. Part of this ought to be saprolite, although it could not be detected from geoelectrical soundings. If we simplify and assume the whole depth to be sediments, the time needed to fill up the Haubi basin can be calculated by extrapolating back in time the earliest (and lowest) sedimentation rate obtained (c. 1 cm a^{-1}). The time needed would thereby be only c. 5000 years, which is a very short time period, bearing in mind that the relative uplift and formation of the Irangi Hills started in the Miocene. The sedimentation rate has therefore probably been considerably lower in the past and the lowest recorded sedimentation rate in the Haubi basin is regarded as being accelerated, while the long-term rate of sedimentation remains unknown.

The dated sediment record of the Haubi basin has enabled the recent denudation rate for the catchment to be calculated. From the dry bulk density of the Haubi sediment the weight of the accumulated material in Lake Haubi was determined. The sediment consists of between 80 and 100% clay. The extremely high clay content in the sediment has enabled the clay content in soils from the pediment slopes of the Irangi Hills (Payton & Shishira 1994; Yanda 1995) to be used in order to calculate the amount of soil lost from the catchment since 1836. Lixisols (FAO-Unesco 1988) are the dominant soil type

(Ollier & Pain 1981; Burbank et al. 1996), which indicates that there is an interrelationship between uplift and gully incision. Similarly, crustal uplift is likely to have a direct effect on the intensity of soil erosion in the Kondoa Irangi Hills.

Data on uplift rates of the African Rift Zones are scarce but some authors indicate that the rates could be around 1 mm a^{-1} (Saggerson & Baker 1965; Baker & Wohlenberg 1971; Summerfield 1991). Much lower figures (0.1–0.3 mm a^{-1}) were obtained through fission track dating of apatites in a study of the uplift of the Sinai Peninsula prior to the opening of the Red Sea (Kohn & Eyal 1981). Although uplift rates may be relatively low in a long-term perspective, single earthquakes may cause displacements of several metres (Summerfield 1991, p. 375). In order for a net surface uplift to take place in an area, the long-term denudation rate cannot be higher than the crustal uplift rate. For the Irangi Hills, this means a probable denudation rate of less, or much less, than 1 mm a^{-1} following the figures referred to above. Estimations of the long-term geological erosion rate in Kenya, using differences in elevation of dated Cenozoic surfaces, were made by Dunne et al. (1978).

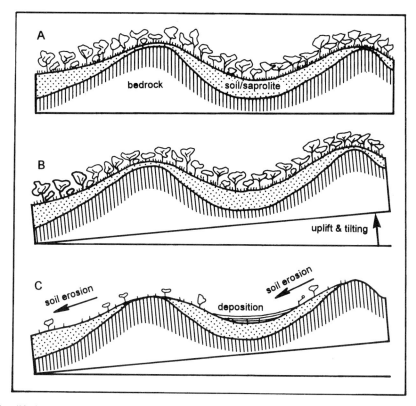

Fig. 9. Simplified model of how crustal uplift in combination with deforestation might have accelerated erosion in the Kondoa Irangi Hills. (**A**) Kondoa Irangi Hills protected against soil erosion by dense natural forest cover. (**B**) Uplift and tilting occur which cause increases in slope instability and in the energy gradient of surface flow. However, dense vegetation still stabilizes the regolith and protects the area from soil erosion. The tectonic movements have taken place over a considerable period of time. (**C**) Vegetation cover removed by human activities. Soil erosion is initiated and becomes more severe than if the landscape had not been destabilized by tectonic movements.

of the pediments in the catchment. Average clay content in lixisols, based on 35 particle size analyses from 14 soil profiles, amounted to 37%. It was found that 5.2×10^6 t of *clay* had been deposited during 156 years (1836–1992) in the lake basin (2.1 km^2), corresponding to a *soil* loss of 2700 t km^{-2} a^{-1}. This figure is equivalent to a catchment denudation rate of 2.7 mm a^{-1}, using a soil bulk density of 1 g cm^{-3} (Dunne *et al.* 1978). This rate, which is calculated on a catchment basis, including both erosion and accumulation areas, is among the highest recorded when compared with figures from similar environments (Table 1). If the soil loss is allocated to the erosion area solely (23.9 km^2), the erosion rate will amount to 3750 t km^2 a^{-1} (3.75 mm a^{-1}).

Measurements of tree root exposures in central Tanzania (Strömquist & Johansson 1978), and in southern Kenya (Dunne *et al.* 1978), show erosion rates of around 3 mm a^{-1}, which are similar to those recorded in the Haubi basin. However, due to the limited areal extent of the study sites, the above values can be compared to those obtained from plot experiments, which generally show higher values than do measurements from small catchments. For example, plot experiments in a badland area in the Ebro Basin, semi-arid Spain, gave erosion figures varying between 6 and 25.8 mm a^{-1} (Benito *et al.* 1992; Sirvent *et al.* 1997), and plot studies at Mpwapwa, Tanzania gave figures of ≤ 9.8 mm a^{-1}, with the variations explained by differences in the vegetation cover (Temple 1972; Christiansson 1981, p. 130ff). Generally, plot experiments tend to indicate high erosion rates and/or highly variable rates, both in time and space. They are therefore not suitable for extrapolations to larger areas or longer time periods (Rydgren 1993).

Table 1. *Recent denudation rates from catchments in semi-arid regions, mainly East Africa*

Location	Catchment area (km^2)	Rainfall (mm a^{-1})	Method	Denudation rate (mm a^{-1})	Sediment yield (t km^{-2} a^{-1})	Reference
Haubi, Tanzania	33.4	~890	Lake sediment	2.7	2700	This paper
Ikowa, Tanzania	640	~570	Reservoir sediment	0.1–0.38	100–380	Christiansson 1981
Matumbulu, Tanzania	18.1	~570	Reservoir sediment	0.44–0.63	440–630	Christiansson 1981
Msalatu, Tanzania	8.7	~570	Reservoir sediment	0.46–0.64	460–640	Christiansson 1981
Imagi, Tanzania	1.5	~570	Reservoir sediment	0.5–0.75	500–750	Christiansson 1981
Kisongo, Tanzania	9.3	~570	Reservoir sediment	0.45–0.64	446–640	Murray-Rust 1972
Morogoro, Tanzania	19.1	890	Suspended river sediment	0.39	390	Rapp *et al.* 1972*a*
Mtera, Tanzania	–	~450	Tree root measurement	2–4	2000–4000	Strömquist & Johansson 1978
Southern Kenya	–	~700	Tree root measurement	3	3000	Dunne *et al.* 1978
Cape Verde Islands	3–71	Semi-arid	Reservoir sediment	0.08–1.65	80–1650	de Brum Ferreira 1996

The figures are based on sediment studies of lakes and reservoirs, except for two studies, which are based on measurements of tree root exposures. Erosion plot studies are not taken into consideration, as they generally are incomparable with estimations of denudation rates on catchment basis.

A number of studies on erosion and sedimentation in degraded catchments subjected to intense grazing have been carried out in central, semi-arid Tanzania (Murray-Rust 1972; Rapp et al. 1972a,b; Christiansson 1981). Denudation rates for these catchments were calculated to around $0.5\,\mathrm{mm\,a^{-1}}$. Suspended-sediment yields in the rivers draining large catchments in East Africa show even lower figures as the physiography, land use and vegetation become more heterogeneous, and as the filtering effect of local sediment basins increases with increasing catchment size. Walling (1984) took a number of studies from large catchments into consideration when he estimated the present average suspended sediment yield in East Africa to $100-1000\,\mathrm{t\,km^{-2}\,a^{-1}}$, corresponding to a denudation rate of $0.1-1\,\mathrm{mm\,a^{-1}}$. In light of the above figures, which concern landscapes affected by humans, Haubi stands out as a catchment that has been characterized by highly accelerated soil erosion since at least 1836.

Discussion

Soil and sediment studies of the Haubi basin indicate that soil erosion has been severe in this area for at least the last 200 years, and probably much longer (Eriksson & Christiansson 1997). The Haubi area has been intensively utilized for cultivation during this period (Mung'ong'o 1991, 1995), an activity which increases soil erosion. Still, few areas utilized intensively by man suffer from such extreme soil erosion as the Irangi Hills. An additional process, or specific natural condition, must be taken into consideration to explain the extent of erosion in the Irangi Hills. A high rate of relative uplift and tilting could be such a process. However, if a dense vegetation cover existed (as indicated by the remaining vegetation on the Tomoko Hills) during the time of active uplift, effective soil erosion would have been restricted until vegetation removal. Hence, the result is more severe soil erosion than would have been the case if erosion had taken place successively (Fig. 9).

The importance of climate and its variation for soil erosion has been stressed by numerous authors. However, as the extreme soil erosion is concentrated mainly in the Irangi Hills, rather than being regional, large-scale changes in climate are unlikely to have been a major cause of the severe erosion. However, on a smaller scale, the uplift of land within the rift zones in East Africa has locally resulted in increased orographic precipitation (Partridge et al. 1995), which may have contributed to a more efficient stripping of the regolith. In that case, again, uplift is the ultimate cause of the increased erosion.

Conclusions and implications

The Irangi Hills belong to a tectonically active area where earthquakes indicate that crustal movements occur. The drainage pattern and the deposition of alluvial sediments are influenced by uplift, faulting and tilting. The larger ephemeral rivers flowing westwards have cut down into the bedrock and follow antecedent stream courses across fault scarps. The smaller rivers have been diverted along fault lines or have been dammed against scarps to form swamps or lakes. Increased sediment deposition on river beds caused by energy decrease of the stream, is found in west-flowing streams upstream from fault scarps because of damming. In east-flowing streams, sediments are spread over wide valley bottoms because of the tilting of tectonic blocks towards the west. Tilting, which has been observed to be up to $3°$, accounts for the fact that within each tectonic block there are generally extensive stripped surfaces in the east and more sediment-laden surfaces in the west.

The soil erosion is part of a long-term natural stripping of regolith and the recent (1836–1992) denudation rate for the Haubi basin was calculated to $2.7\,\mathrm{mm\,a^{-1}}$ ($2700\,\mathrm{t\,km^{-2}\,a^{-1}}$). This is much accelerated compared to the long-term geologic erosion rate which is much lower, although unknown. It is probable that relative crustal uplift and tilting of tectonic blocks have been contributing factors to accelerated soil erosion in the Irangi Hills. The effects of crustal movements must not be overlooked when studying soil erosion and sediment deposition in tectonically active areas.

In this study a wider geomorphological analysis of the Irangi Hills is combined with an estimation of the denudation rate of a small catchment. A study of only one or the other would not provide a sufficient understanding of soil erosion and its relation to morphotectonics in the Irangi Hills. However, by combining broad geomorphological studies with studies of key local sites, a more comprehensive understanding of the land-forming processes in an area is likely to be achieved. In this way, for instance, erosion plot studies, which are unsuited for extrapolation purposes, could become useful for wider assessments of erosion and denudation rates.

This study has been made possible through funding by the Swedish Agency for Research Cooperation with Developing Countries (SAREC), the Swedish Society

for Anthropology and Geography (SSAG), Lillemor and Hans W:son Ahlmann's Fund, and Axel Lagrelius' Fund. I am grateful to Karna Lidmar-Bergström, Carl Christiansson and Gunhild Rosqvist for improvements to the manuscript. Thanks are also extended to Cliff Ollier and two anonymous referees for critically reading the manuscript and to Annika Dahlberg and Jessie Karlén for improving the language. Many thanks are also directed to the staff at the Institute of Resource Assessment at the University of Dar es Salaam for great help during my field work. The help provided by A. C. Mbegu and C. Sianga and other staff at Hifadhi Ardhi Dodoma (HADO) is also gratefully acknowledged.

References

BACKÉUS, I., RULANGARANGA, Z. K. & SKOGLUND, J. 1994. Vegetation changes on formerly overgrazed hill slopes in semi-arid Tanzania. *Journal of Vegetation Science*, 5, 327–336.

BAKER, B. H. & WOHLENBERG, J. 1971. Structure and evolution of the Kenya Rift Valley. *Nature*, 229, 538–542.

BENITO, G., GUTIÉRREZ, M. & SANCHO, C. 1992. Erosion rates in badland areas of the central Ebro Basin (NE-Spain). *Catena*, 19, 269–286.

BÜDEL, J. 1957. Die doppelten Einebenungsflächen in den feuchten Tropen. *Zeitschrift für Geomorphologie, N.F.*, 1, 201–288.

—— 1982. *Climatic Geomorphology*. Princeton, New Jersey.

BURBANK, D. W., LELAND, J., FIELDING, E., ANDERSON, R. S., BROZOVIC, N., REID, M. R. & DUNCAN, C. 1996. Bedrock incision, rock uplift and threshold hillslopes in the northwestern Himalayas. *Nature*, 379, 505–510.

CHRISTIANSSON, C. 1981. *Soil erosion and sedimentation in semi-arid Tanzania: Studies of environmental change and ecological imbalance*. Scandinavian Institute of African Studies, Uppsala.

—— 1988. Degradation and rehabilitation of agropastoral land – Perspectives on environmental change in semiarid Tanzania. *AMBIO*, 17, 144–152.

—— 1996. Linking landscapes and societies: glimpses from ongoing studies in the 'Man–Land Interrelations' Research Programme. *In*: CHRISTIANSSON, C. & KIKULA, I. (eds) *Changing Environments. Research on Man–Land Interrelations in Semi-Arid Tanzania*. Report No. 13, Regional Soil Conservation Unit, Swedish International Development Cooperation Agency, Nairobi, 7–14.

—— & KIKULA, I. S. (eds). 1996. *Changing Environments. Research on Man–Land Interrelations in Semi-Arid Tanzania*. Report No. 13, Regional Soil Conservation Unit, Swedish International Development Cooperation Agency, Nairobi.

—— KIKULA, I. S. & ÖSTBERG, W. 1991. Man–Land Interrelations in Semiarid Tanzania: a multi-disciplinary research program. *AMBIO*, 20, 357–361.

DE BRUM FERREIRA, D. 1996. Water erosion in the Cape Verde Islands: factors, characteristics and methods of control. *In*: SLAYMAKER, O. (ed.) *Geomorphic Hazards*. Wiley, Chichester, 111–124.

DOUGLAS, I. 1967. Man, vegetation and sediment yields of rivers. *Nature*, 215, 925–928.

DUNNE, T. 1979. Sediment yield and land use in tropical catchments. *Journal of Hydrology*, 42, 281–300.

——, DIETRICH, W. E. & BRUNENGO, M. J. 1978. Recent and past erosion rates in semi-arid Kenya. *Zeitschrift für Geomorphologie N. F., Supplementband*, 29, 130–140.

EL-DAOUSHY, F. & ERIKSSON, M. G. 1998. Radiometric dating of recent lake sediments from a highly eroded area in semiarid Tanzania. *Journal of Palaeolimnology*, 19, 377–384.

ERIKSSON, M. G. & CHRISTIANSSON, C. 1997. Accelerated soil erosion in Central Tanzania during the last few hundred years. *Physics and Chemistry of the Earth*, 22, 315–320.

—— & SANDGREN, P. in press. Mineral magnetic analyses of sediment cores recording recent soil erosion history in central Tanzania. *Palaeogeography, Palaeoclimatology, Palaeoecology* (in press).

——, BONNEFILLE, R. & LAFON, S. in press. Recent lake level variations in Lake Haubi, central Tanzania, interpreted from pollen and sediment studies. *Journal of Paleolimnology* (in press).

FAO-UNESCO. 1988. *FAO-Unesco Soil Map of the World*, revised legend. FAO, Rome.

FOZZARD, P. M. H. 1963. *The general geology of south Masailand including the Kondoa and Babati area*. Report **PMHF/19**, File 2904, Geological Survey of Dodoma.

GROVE, A. T. 1986. Geomorphology of the African Rift System. *In*: FROSTICK, L. E., RENAUT, R. W., REID, I. & TIERCELIN, J. J. (eds) *Sedimentation in the African Rifts*. Geological Society, London, Special Publications, 25, 9–16.

IRANGA, M. D. 1991. *An Earthquake Catalogue for Tanzania, 1846–1988*. Report No. **1–91**, Seismological Department, Uppsala University, Sweden.

—— 1992. Seismicity of Tanzania: distribution in time, space, magnitude, and strain release. *Tectonophysics*, 209, 313–320.

JAMES, T. C. 1956. *The Nature of Rift Faulting in Tanganyika*. First Meeting, East-Central Regional Committee for Geology, Dar es Salaam, 81–94.

KANNENBERG, H. 1900. Reise durch die hamitischen Sprachgebiete um Kondoa. *Mittheilungen von Forschungsreisenden und Gelehrten aus den Deutschen Schutzgebieten*, 13, 144–172.

KOHN, B. P. & EYAL, M. 1981. History of uplift of the crystalline basement of Sinai and its relation to opening of the Red Sea as revealed by fission track dating of apatites. *Earth and Planetary Science Letters*, 52, 129–141.

LYARUU, H. V. M., ELIAPENDA, S., MWASUMBI, L. B. & BACKÉUS, I. 1997. *The Afromontane forest at Mafai in Kondoa Irangi Hills, central Tanzania. A proposal to conserve a threatened ecosystem*. EDSU Working paper No. 37, School of Geography, Stockholm University.

MUNG'ONG'O, C. 1991. Socioecological processes and the land question in the Kondoa Irangi Hills, Tanzania. *AMBIO*, 20, 362–365.

—— 1995. *Social processes and ecology in the Kondoa Irangi Hills, central Tanzania*. Meddelanden series **B 93**, Department of Human Geography, Stockholm University.

MURRAY-RUST, D. H. 1972. Soil erosion and reservoir sedimentation in a grazing area west of Arusha, northern Tanzania. *Geografiska Annaler*, **54A**, 325–343.

OLLIER, C. D. & PAIN, C. F. 1981. Active gneiss domes in Papua New Guinea. New tectonic landforms. *Zeitschrift für Geomorphologie N.F.*, **25**, 133–145.

ÖSTBERG, W. 1986. *The Kondoa Transformation – Coming to grips with soil erosion in Central Tanzania*. Research report No. **76**, Scandinavian Institute of African Studies, Uppsala.

PARTRIDGE, T. C., BOND, G. C., HARTNADY, C. J. H., DE MENOCAL, P. B. & RUDDIMAN, W. F. 1995. Climatic effects of late Neogene tectonism and volcanism. *In*: VRBA, E., DENTON, G., BURCKLE, L. & PARTRIDGE, T. C. (eds) *Paleoclimate and Evolution, with Emphasis on Human Origins*. Yale University, 8–23.

PAYTON, R. W. & SHISHIRA, E. K. 1994. Effects of soil erosion and sedimentation on land quality: Defining pedogenetic baselines in the Kondoa District of Tanzania. *In*: *Soil Science and Sustainable Land Management in the Tropics*. CAB, Wallingford, 88–119.

——, CHRISTIANSSON, C., SHISHIRA, E. K., YANDA, P. & ERIKSSON, M. G. 1992. Landform, soils and erosion in the north-eastern Irangi Hills, Kondoa, Tanzania. *Geografiska Annaler*, **74A**, 65–79.

QUENNELL, A. M., MCKINLAY, A. C. M. & AITKEN, W. G. 1956. *Summary of the Geology of Tanganyika. Part 1: Introduction and Stratigraphy*. Memoir No. **1** (reprinted 1986).

RAPP, A., AXELSSON, V., BERRY, L., & MURRAY-RUST, D. H. 1972a. Soil erosion and sediment transport in the Morogoro river catchment, Tanzania. *Geografiska Annaler*, **54A**, 110–124.

—— MURRAY-RUST, D. H., CHRISTIANSSON, C., & BERRY, L. 1972b. Soil erosion and sedimentation in four catchments near Dodoma, Tanzania. *Geografiska Annaler*, **54A**, 255–318.

RYDGREN, B. 1993. *Environmental impacts of soil erosion and soil conservation. A Lesotho case study*. UNGI Report **85**, Institute of Earth Sciences, Uppsala University.

SAGGERSON, E. P. & BAKER, B. H. 1965. Post-Jurassic erosion-surfaces in eastern Kenya and their deformation in relation to rift structure. *Quarterly Journal of the Geological Society of London*, **121**, 51–72.

SELBY, J. & MUDD, G. C. 1965. *Brief Explanation of the Geology of Quarter Degree Sheet 104 (KONDOA)*. Geological Survey Division, Dodoma (map).

SIRVENT, J., DESIR, G., GUTIERREZ, M., SANCHO, C. & BENITO, G. 1997. Erosion rates in badland areas recorded by collectors, erosion pins and profilometer techniques (Ebro Basin, NE-Spain). *Geomorphology*, **18**, 61–75.

SOMI, E. J. 1993. *Paleoenvironmental changes in central and coastal Tanzania during the Upper Cenozoic*. Doctoral Thesis, Department of Geology and Geochemistry, Stockholm University.

STRÖMQUIST, L. & JOHANSSON, D. 1978. Studies of soil erosion and sediment transport in the Mtera Reservoir Region, Central Tanzania. *Zeitschrift für Geomorphologie N. F., Supplementband*, **29**, 43–51.

SUMMERFIELD, M. A. 1991. *Global Geomorphology*. Longman, Essex.

TEMPLE, P. H. 1972. Runoff and soil erosion at an erosion plot scale with particular reference to Tanzania. *Geografiska Annaler*, **54A**, 203–220.

THOMAS, M. F. 1994. *Geomorphology in the Tropics*. Wiley, Chichester.

VELDKAMP, A. 1996. Late Cenozoic landform development in East Africa: The role of near base level planation within the dynamic etchplanation concept. *Zeitschrift für Geomorphologie N. F., Supplementband*, **106**, 25–40.

WALLING, D. E. 1984. The sediment yields of African rivers. *In*: WALLING, D. E. (ed.) *Challenges in African Hydrology and Water Resources*. IAHS Publication No. **144**, 265–283.

WOODHEAD, T. 1968. *Studies of potential evaporation in Tanzania*. East African Common Services Organization, and Water Development and Irrigation Division, Ministry of Lands, Settlement and Water Development, Government of Tanzania.

YANDA, P. Z. 1995. *Temporal and Spatial Variations of Soil Degradation in Mwisanga Catchment, Kondoa, Tanzania*. Department of Physical Geography, Stockholm University, Dissertation Series No. 4.

Uplift and landscape stability at Taroko, eastern Taiwan

DAVID N. PETLEY & SHARON REID

Department of Geology, University of Portsmouth, Burnaby Road, Portsmouth, PO1 3QL, UK

Abstract: The Taroko area of Taiwan represents an extremely dynamic environment, of rapid geological and geomorphological change. The location of the area on the collision zone between the continental Eurasian plate and the oceanic Philippine Sea plate leads to a high frequency of seismic activity, intense internal deformation and high rates of tectonic uplift. The rate of uplift, which is believed to be one of the highest measured, is approximately 5.5 mm per annum, with higher rates inferred for the last million years. In consequence, the area, which consists predominantly of relatively weak, intensely deformed marble and schist, has a high potential energy providing ideal conditions for rapid landform evolution.

In addition to high rates of uplift, high rates of denudation are also favoured by the extreme climatic conditions. In particular, the area is seasonally affected by tropical cyclones (typhoons); an average of 16 of these occur in the northwest Pacific per annum, about three of which affect Taiwan. The passage of a typhoon is marked by heavy rainfall which may reach intensities of 100 mm per hour and 1500 mm in 24 hours. The high rainfall intensities generate intense fluvial activity, including extensive overland flow, high peak discharges, the mobilization of debris channels, landslides and rockfalls.

Intense tectonic and geomorphological activity provide the ideal environment for studying the rapid evolution of landscapes. In this paper, uplift data are combined with daily hydrological records from a major river for the last fifteen years. The hydrological records, which include discharge, mean flow rates and sediment volumes, are used to estimate the sediment budget for the river and hence to estimate denudation rates. It is demonstrated that average aerial denudation rates approximately equal uplift rates. These data are then used to generate a simple computer model for the evolution of Taroko Gorge, which demonstrates that current landforms may be created without substantial changes in the rate of uplift.

The evolution of landscapes remains a complex and poorly understood topic, but one that is of greatest interest and relevance to geomorphologists and geologists. The examination of this subject allows the detailed, local-scale investigations that form the backbone of geomorphology to be related to the regional or global scale and, as such, represents a holistic approach to the subject. Neotectonic areas have a major role to play in the understanding of geomorphology as in these environments processes frequently occur at sufficiently high rates to allow the evolution of the landscape to be studied. This paper presents an examination of the evolution of such an area, the Taroko area of the eastern side of Taiwan.

Geological setting

Taiwan is an island located on the eastern margin of continental Asia at approximately 121° E and 24° N (Fig. 1). It is spindle-shaped with a maximum length of 385 km and a width of 143 km at the widest point, and a surface area of 36 000 km². The island is believed to have formed as the result of the highly complex interaction of the Philippine Sea oceanic plate and the Eurasian continental plate (Fig. 2) (Ho 1987). To the east of the island the collision zone is oriented approximately east–west and is represented by the Ryuku Arc subduction system, in which the Philippine Sea plate is being underthrust northwards beneath the Eurasian plate. The collision zone has all of the 'text-book' features of an oceanic–continental plate collision zone (Ho 1987), including a deep trench (the Ryuku Trench), a sub-oceanic accretionary complex, an island arc (the Ryuku Arc) and a back-arc basin (Fig. 2). The trench, volcanic arc and the Okinawa Trough are all seen to be in contact with the northeast margin of Taiwan island (Fig. 2).

Collision to the south of Taiwan is represented by the Luzon Arc (Fig. 2), which is oriented approximately N–S. Here the plate interaction is rather unusual as the Manila Trench represents the location of subduction of thin crust attached to the Eurasian plate beneath the thicker crust of the Luzon Arc (Ho 1987). It is possible to trace this system northwards to the eastern margin of Taiwan. Along the eastern coast there is a clear, linear, narrow valley (Fig. 3), known as the Eastern Longitudinal Valley. To the east of this feature is the Coastal Range, which consists of a small mountain range of

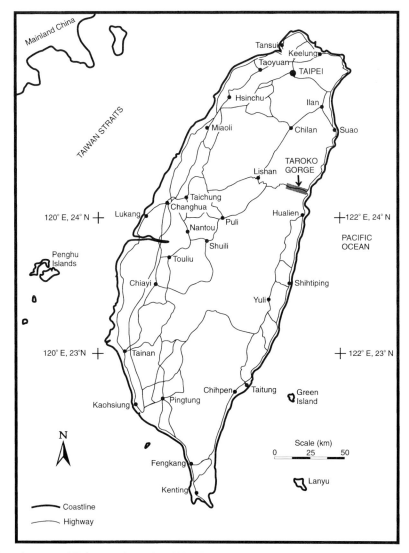

Fig. 1. Location map of Taiwan and associated islands.

Neogene volcaniclastic and turbiditic (including mélange) sediments lying on igneous and sedimentary rocks that represent island arc material. To the west is the eastern margin of the Central Range, consisting of a fold-thrust belt of Permian basement. Here the rocks are predominantly marbles and schists, with some isolated bodies of gneiss. To the west the rocks become progressively younger, comprising slates, shales and sandstones. It is generally believed that the Coastal Range represents an extension of the Luzon volcanic arc, with the Longitudinal Valley being the remnants of the suture between the arc and the Eurasian plate. The suture has now been filled to just above sea-level by sediments eroded from the adjacent mountain ranges and deposited by fluvial activity. However, it remains difficult to reconcile the different types of collision in the vicinity of Taiwan. To the northeast the Philippine Sea plate is undergoing subduction, whereas to the south it is over-riding the Eurasian plate. The east coast of Taiwan represents the division between these two collisions, and it has thus been proposed that along the margin of Taiwan there is no direct subduction occurring (Ho 1986, 1987; Ma & Liu 1996); instead strain is accommodated in a broad zone of strike-slip

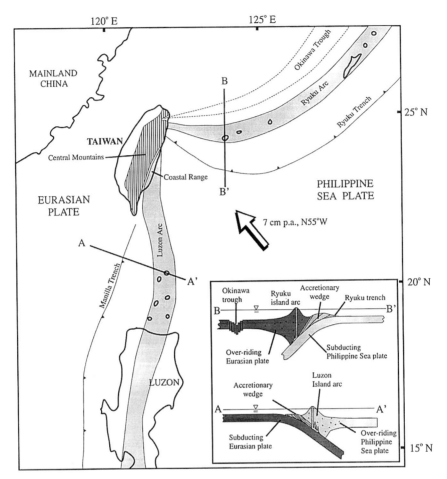

Fig. 2. The tectonic setting of Taiwan (after Ho 1987).

faulting and compressive deformation. This is supported by seismic tomography which has suggested that there is no subducting slab along the eastern coast of Taiwan (Ma & Liu 1996), probably because the crust associated with the continental margin of Taiwan and the northern extension of the Luzon volcanic arc is too thick.

The current collision event was initiated to the north of Taiwan island at approximately 4 Ma BP, although there is evidence of earlier collision events. The collision propagated southwards along the Luzon Arc at about 90 km Ma^{-1}. Thus the landscape in the north of the island has evolved over a maximum of about 4 million years, whereas to the south it may have had only a million years or less.

As a result of the continuing plate collision, Taiwan is experiencing extremely high rates of tectonic uplift. Measured uplift rates for Taiwan island have included 5.0 ± 0.7 m ka^{-1} for the Coastal Range over the last 9000 years (Lin 1991), 2 m ka^{-1} for north Taiwan for 5.5–1.5 ka BP and 5.3 m ka^{-1} for 8.5–5.5 ka BP (Peng et al. 1977), and 1.4–4.8 m ka^{-1} for south Taiwan (Bonilla 1975). Wang & Burnett (1991) measured rates of uplift of 4.7 to 5.3 m ka^{-1} for the Eastern Coastal Range and 3.3 to 3.5 m ka^{-1} for the south of Taiwan. Brunsden & Lin (1991) proposed that, as a consequence of the tectonic setting, Taiwan can effectively be divided into two tectonic zones, based on the analysis of uplift data for the last 3 Ma (Table 1). To the north rates of uplift are dominated by the Ryuku collision event in which active subduction is occurring. As a result rates of uplift are relatively low (c. 2 m ka^{-1}). The remainder of west and central Taiwan is within the Luzon Arc regime, in which crustal thickening associated with the active collision is occurring. In

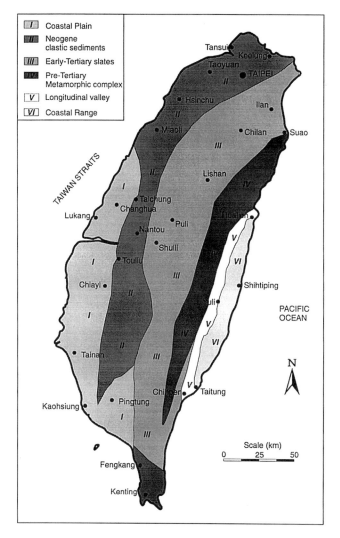

Fig. 3. A simplified map of the main geological units of Taiwan (after Ho 1982).

Table 1. *Uplift and denudation rates for Taiwan based on fission track measurements*

	Time (Ma BP)	Estimated uplift rate (mm a^{-1})	Estimated denudation rate (mm a^{-1})	Uplift − Denudation (mm a^{-1})
Central Mountains	3.0	c. 0.1	c. 0.1	0.0
	2.0	c. 1.0	0.384	0.616
	1.0	7.0	3.1	3.9
	0.5	7.0	5.0	2.0
	present	c. 5.5	5.0 ± 0.7	0.5 ± 0.7
Ryuku rate	1.0	2.0	2.6	0.6
(N Taiwan)	present	2.0	2.6	0.6

After Brunsden & Lin (1991)

these areas the rate of uplift is much greater ($c.\ 5\,\mathrm{m\,ka^{-1}}$).

The continuing collision also generates a high frequency of seismic activity, which may be significant in the evolution of the landforms of Taiwan. An analysis of historical records suggested that an average of 269 significant earthquakes are detected annually throughout Taiwan island (Hsu 1975), with the highest frequency on the east coast in the vicinity of the city of Hualien, which experiences an annual average of 116 events. The frequency of seismic events may have an impact on the formation of the landscape as the weak, fractured local rocks are vulnerable to failure under seismic shaking. An example occurred on 17 December 1941 with the Chia-Yi earthquake, which had a magnitude of $M_L = 7.1$ and a focus depth of about 10 km. The event is believed to have triggered many landslides, with an estimated combined death toll of about 350 people. The largest slide occurred at Tsao-Ling, about 28 km NE of the epicentre, where more than $100 \times 10^6\,\mathrm{m^3}$ of rock underwent failure, forming a dam 140 m in height across the Ching-Sui River.

Thus, in summary, the landscape of Taiwan can be considered to have been formed as a consequence of the active tectonic processes. The data suggest that the landforms probably evolved over a maximum of 4 Ma. Taiwan, therefore, represents a good example of a young landscape evolving under neotectonic conditions.

Climatic environment

A critical factor in the evolution of the landscape is the climatic conditions to which the area is subject. The climate of Taiwan island is humid, sub-tropical with high annual rainfall and temperatures. Importantly, the area is currently affected by an average of three large tropical cyclones per annum. Tropical cyclones (typhoons), may be defined as 'non-frontal, synoptic scale, low pressure systems with organised convection and definite cyclonic surface wind circulation, formed over tropical or sub-tropical waters' (Simpson 1981). Generally they form over the Pacific Ocean in the vicinity of Guam before moving in a northwesterly direction (see Fig. 4 for example). Their internal dynamics allow the system to be sustained or intensified, and generally they do not dissipate until they enter higher latitudes where the ocean is colder or until they strike a land mass. Late in their evolution they may enter the mid-latitude westerly wind belt and recurve to the northeast.

Tropical cyclones initiate strong winds, large waves, high tidal levels (storm surges) and intense rates of precipitation. The strong winds and storm surges may have a significant effect on the evolution of coastal landforms; sustained wind speeds of $85\,\mathrm{m\,s^{-1}}$ and storm surges of 13 m have been measured (Coch 1994). Of interest here are the very high precipitation intensities that accompany the passage of a tropical cyclone, resulting from the strong convection that is associated with the development of the weather system. Intensities of 100 mm per hour and 1000 mm per day are not particularly uncommon, although precipitation totals of 600 mm per tropical cyclone are probably more normal (Coch 1994). The majority of the precipitation occurs in a period of approximately five to eight hours as the centre of the tropical cyclone

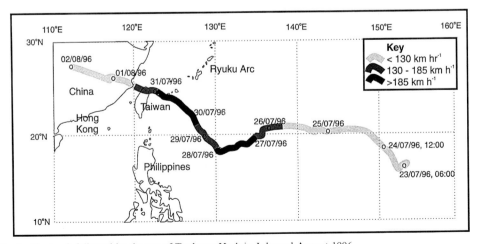

Fig. 4. The track followed by the eye of Typhoon Herb in July and August 1996.

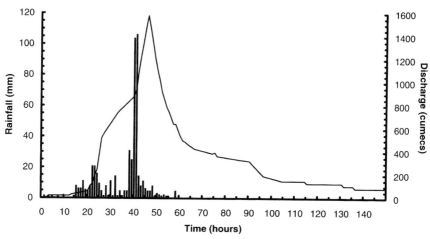

Fig. 5. The precipitation associated with Typhoon Ofelia, 1991, as measured at Taroko (line graph and left-hand axis) and the discharge of the Li-Wu River (bar graph and right-hand axis) (after Longley *et al.* 1992).

passes. An example of the precipitation generated in the east of Taiwan during the passage of a typical tropical cyclone, Typhoon Ofelia (1991), is provided in Fig. 5. It can be seen that high intensities occurred over a period of about 40 hours, with precipitation intensities of over 20 mm per hour occurring for six hours. For two hours (hours 40 and 41) the instensity exceeded 100 mm per hour.

The very high precipitation intensities lead to the development of intense surface activity. In typhoons, the very intense rainfall is often preceded by 20 or more hours of heavy, sustained precipitation that is sufficient to induce saturation of the soil and regolith and allow the initiation of landslides and rockfalls. The very heavy rainfall subsequently triggers soil erosion and major failures. The volumes of material involved in these failures may be very significant. Lee (1977) examined the catchments of 22 streams and rivers throughout Taiwan between 1965 and 1977 and catalogued a total of 7810 failures with a total surface area of $104\,\mathrm{km}^2$ and a total volume of $151 \times 10^6\,\mathrm{m}^3$. There are also a number of well documented major slope failures that have been induced by torrential rainfall, including the Wuchia Landslide of 1944, which destroyed a hydro-electric power station, a $1 \times 10^6\,\mathrm{m}^3$ slide on the Central Cross Island Highway in 1972, various slides totalling an area of 33 hectares in the catchment of Tuchian River in 1963, the debris from which buried a village (Moh 1977), and many more (see Hung (1987) for further examples).

In addition to the initiation of landslides, prolonged and intense rainfall induces catastrophic floods which may lead to the movement of very large volumes of sediment through both Hortonian and saturated overland flow which forms rills and gullies and strips the soil. The effects of these processes may be somewhat attenuated by the density of the tropical forest that covers the island, but the effects are very pronounced where there has been uncontrolled clearance of vegetation.

The extreme precipitation rates and totals induce high rates of denudation. Rates of denudation for the last 3 Ma have been in the order of 5 to $9\,\mathrm{m\,ka}^{-1}$, with contemporary measured physical and chemical rates being in the range of $4\,\mathrm{m\,ka}^{-1}$ for the south of the island, 3.9 to $8.1\,\mathrm{m\,ka}^{-1}$ for the Coastal Range, 0.6 to $3.0\,\mathrm{m\,ka}^{-1}$ for the northern end of the island and $1.6\,\mathrm{m\,ka}^{-1}$ for the west coast (Table 1) (Brunsden & Lin 1991). As the rate of denudation currently balances approximately the rate of uplift in the main part of the island, Brunsden (1990) and Brunsden & Lin (1991) proposed that a state of topographic equilibrium may exist, explaining the remarkable degree of summit conformity in the Central Mountains. However, this simple idea explaining summit conformity needs reappraisal to allow for the influence of drainage density, the effects of active tectonic processes, including orogenic (mountain) collapse which might help balance uplift by lowering the mountain peaks without denudation (evidence for the processes of orogenic collapse on the eastern margin of the Central Mountains has been cited by Lee *et al.* (1996)). Further, it is likely that there would be some isostatic rebound as a result of the unloading (e.g. Gilchrist & Summerfield 1991).

The remainder of this paper examines the topography and features of one part of the eastern margin of the Central Mountains, that of Taroko Gorge, and investigates whether the general principles for the evolution of the landscape in this area allow modelling of the landforms that are now seen.

Taroko Gorge

Taroko Gorge comprises a marble canyon up to 800 m deep, located between Taroko on the coast and Tien Hsiang inland, within Taroko National Park on the eastern side of the Central Mountains (Fig. 6). It has formed as a result of rapid fluvial down-cutting during Pleistocene uplift (Petley *et al.* 1997) and thus represents a neotectonic landform. It has most of the diagnostic characteristics for such a terrain (after Brunsden 1987; Petley *et al.* 1997), as follows:

1. rapid uplift matched by the rates of river incision and denudation;
2. accentuated base-level or sea-level changes, with associated geomorphological changes;
3. exposure to extreme climatic conditions;
4. high probability of formative seismic activity;
5. widespread structural weaknesses in the rocks, including discontinuities and juxtaposed rock units;
6. considerable variability in ground and slope water conditions;
7. mature vegetation;
8. sensitivity of the landscape to human activity.

Geologic province

Regional setting

The rocks of Taroko Gorge are located in the pre-Tertiary metamorphic complex, part of the Central Range geologic province (Ho 1987) (Fig. 3). The metamorphic belt is oriented roughly N–S for a distance of about 250 km, with a maximum width of about 30 km and a surface area of about 4600 km^2. All of the rocks in this area are grouped together as the Tananao Schist (Ho 1987). The thickness of the complex is not known with any precision, but is likely to be in the order of several thousand metres. The metamorphic complex has been subdivided into two belts, the western Tailuko belt and the eastern Yuli belt; Taroko Gorge lies wholly within the Tailuko belt. The contact between the belts

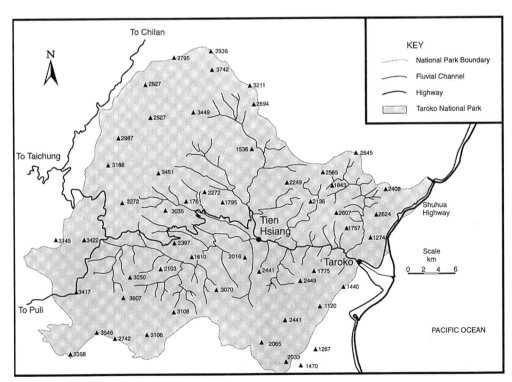

Fig. 6. The area covered by Taroko National Park. Taroko Gorge is located between Tien Hsiang and Taroko.

is poorly defined, but may represent a major fault, known as the Shoufeng Fault (Stanley et al. 1981).

Rock types

Within Taroko Gorge the geology primarily consists of Permian metamorphic basement, representing Eurasian crust, including marble, green schists, black (pelitic) and siliceous schists, and isolated bodies of gneiss (Longley et al. 1992; Yen et al. 1951). The general structure is that of a complex set of repeated thrust sheets formed as a result of east–west compression (M. E. Jones pers. comm. 1997). The marble, which is the predominant rock type, is massive and thickly bedded, although there is extensive jointing and small-scale faulting as a result of intense local tectonic deformation, high rates of crustal uplift and the relaxation of stress as a result of the formation of the gorge. It is, however, of sufficiently high quality to be quarried, although not within the study area. The black schists and the siliceous schists, both of which have pronounced cleavage, weather rapidly to form small, weak, angular fragments. The green schist is very soft and is found in bands up to 15 m thick within the marble. Its most notable aspect is that, due to its vulnerability to intense weathering, it has a strong control on the development of the local drainage. Within the other rocks are bodies of gneiss which were intruded as granite during the late Mesozoic, and which have been subsequently metamorphosed. There are also some localized Quaternary deposits which include colluvium, formed as a result of regolith falling from mountain slopes, and fluvial (river terrace) deposits, some of which are found on slopes in excess of 200 m above the valley floor, illustrating the intensity of the local uplift.

Formation of Taroko Gorge

Uplift and denudation rates

As discussed earlier, Taroko Gorge shows many of the characteristics of an active tectonic regime. Four factors are significant in the formation of such a narrow, deeply incised landform:

1. high rates of uplift resulting from the active tectonic setting;
2. intense down-cutting on the floor of the gorge;
3. properties of the local rock types, including high strength but a vulnerability to denudation;
4. high-magnitude climatic events that induce high rates of surface processes.

Whilst data on uplift rates at Taroko Gorge itself are not yet available, suspended sediment data, collected from a hydrological monitoring station within the gorge, have been obtained. Daily measurements are made of the discharge of the river, and weekly measurements are made of the suspended sediment content, with a greater frequency of measurement during periods of higher discharge. Records of discharge and suspended sediment content have been analysed to give a sediment budget, allowing the calculation of a net denudation rate for the period between January 1986 and December 1995. It is calculated that the total mass of sediment transported through the measuring point for the period of study is 3.51×10^7 t for an area of 434.6 km^2. This gives an average sediment yield of 8076 t km^{-2} per annum, suggesting an average contemporary rate of denudation of the land-surface of approximately 4.9 m ka^{-1}, allowing for porosity and organic content. This rate is close to that estimated by Brunsden & Lin (1991) (Table 1).

Sediment yield can also be estimated using the equations of Jansen & Painter (1974) who suggested that in a tropical climate with a high precipitation rate, sediment yield can be predicted from climate and topography using the following equation:

$$\log S = 0.1 \log D - 0.314 \log A + 0.75 \log H$$
$$+ 1.104 \log P + 0.368 \log T$$
$$- 2.324 \log V + 0.786 \log G - 2.032$$

where S = sediment yield (t km^{-2}); D = annual discharge = 2710×10^3 m^3 km^{-2} for Taroko Gorge; A = basin area = 434.60 km^2; H = greatest height of the terrain = 3742 m; P = average precipitation rate = 3200 mm; T = Average annual temperature = 18°C; V = vegetation index = 3 (forest and bare-rock); G = geology index = 5 (strong Cenozoic rocks). Therefore the estimated sediment yield is 8608 t km^2, giving an annual denudation rate of 5.2 m ka^{-1}, allowing for porosity and organic content. Thus, the measured and calculated rates of denudation show a high level of consistency. Sediment yield throughout Taiwan has also been described by Jansson (1988) who measured rates in the range 4800 to 21 000 t km^{-2}, such rates being unusually high for this climatic regime, probably as a consequence of the dynamic nature of the environment.

The formation of the gorge must have occurred as a result of the local uplift (Petley

1995). The geologic setting of Taiwan (Fig. 2) suggests that it is probably now on the boundary between the two tectonic zones of Brunsden & Lin (1991), but formation of the gorge would have occurred in the southern (Luzon Arc) regime. This implies that over the last 500 000 years the average rate of uplift has been approximately 5.5 m ka^{-1} (Brunsden & Lin 1991). It is reasonable to assume that, for a river draining a large mountainous area such as that of the Li-Wu drainage basin, rates of incision at the river bed would have been equivalent to the rate of uplift. If the depth of the main part of the gorge is taken to be 600 m, the time period required for the cutting of the gorge is approximately 109 000 years. This must be treated with some caution as pronounced climate change would greatly alter this figure, and global climate change has been significant over this period. Unfortunately, there is little information on the palaeoenvironment of Taiwan over the last 200 000 years. Liew & Tseng (1996) suggest, on the basis of Quaternary sediments from the Taipei Basin, that the last glacial was marked by colder, drier environments than the present, suggesting that denudation rates in the vicinity of the gorge would have been lower than those currently recorded. The sediments also suggest that there was an increase in precipitation totals and/or intensities at the start of the Holocene, with two moist events at about 6.5 ka BP and <2 ka BP, during which it is likely that denudation rates would have increased. Thomas & Thorp (1995) suggest that erosion rates for the humid tropics as a whole have been close to the current rate for the last 14 000 years. Thus, it appears that it is reasonable to assume that the denudation of the base of the gorge would have been equal to the uplift rate.

The calculation above gives an estimation of the time required to excavate 600 m of rock but does not provide an insight into the time required to create the landform itself. If it is assumed that local topography was eroded at the current rate of denudation (c. 5.0 m ka^{-1}) whilst the gorge was being excavated at an uplift rate of 5.5 m ka^{-1} (Brunsden & Lin 1991), the net rate at which the gorge would be created is 0.5 m ka^{-1}, allowing the cutting of a gorge landform of 250 m in depth during the last 500 000 years. In the period between 1 Ma and 500 ka BP uplift is estimated at 7.0 m ka^{-1} and denudation at 5.0 m ka^{-1} (Brunsden & Lin 1991), giving a net rate of landform (gorge) creation of 2.0 m ka^{-1}. The total time required for the creation of the gorge landform is thus 675 000 years, assuming a simple set of environmental conditions and steady rates of uplift and denudation. During this time the total amount of downcutting by the river would be 3.98 km whilst the total amount of denudation of the slopes immediately above the gorge would be 3.38 km. Of course all of these figures assume that the estimated rates apply for the whole period and rely on simplifications about the rates of denudation and tectonic processes, but they do serve to illustrate the approximate rates of landform evolution.

The location of the gorge at the northern end of the Coastal Range suggests that the gorge and surrounding mountains are now slowly entering the Ryuku Arc tectonic regime. Brunsden & Lin (1991) suggest that within this regime the uplift rate (and hence the rate of down-cutting of the floor of the gorge) is c. 2.0 m ka^{-1} whilst the denudation rate of the local topography is 2.6 m ka^{-1}. Assuming this to be correct, the depth of the gorge would be expected to decrease at approximately 0.6 m ka^{-1}. In addition the slopes will retreat, leading to a widening of the gorge, and may reduce in gradient if they are currently out of equilibrium with environmental conditions.

Computer modelling of the gorge evolution

Taroko Gorge presents a rare opportunity to understand the evolution of a tectonic landscape through time. This is principally because the landscape is still actively evolving due to the high rates of geomorphological processes. As estimates of current rates of processes are available, modelling in a simple form may be undertaken relatively easily. It is notable, however, that along its length the gorge has evolved a stable form, with steep or near-vertical lower walls and less steep higher slopes (Fig. 7).

Modelling is underway using sets of simplifying but realistic mathematical equations. This research is at an early stage and requires continued refinement, but it has allowed the development of a model that simulates the development of the morphology of the gorge reasonably well.

The model described here has been designed on a time-step basis utilizing five parameters.

1. Constant uplift rate throughout the area of 5.5 m ka^{-1}.
2. Constant denudation rate at base-level of 0.5 m ka^{-1}.
3. An additional denudation rate that is proportional to height above base-level. The increase in denudation rate with altitude is very difficult to estimate, but for simplicity it has been assumed to be a linear relationship. The rate has been set such that it is

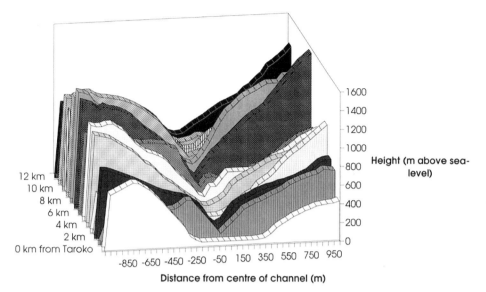

Fig. 7. Schematic cross-sections of Taroko Gorge, taken at 1 km intervals along the length of the centre of the Li-Wu River channel. Note that in the main (central) part of the gorge, the slopes have a characteristic steep lower section and a less steep higher portion.

zero at base-level and 5.5 m ka^{-1} at 4000 m. This has been calculated on the basis that the current highest peaks in Taiwan show summit conformity at about 4000 m (the highest peak in Taiwan is Yushan at 3997 m), which Brunsden & Lin (1991) suggested indicates a balance between denudation and uplift at this height. As mentioned previously, this assumption is probably an over-simplification (see below) but is a reasonable starting point.

4. A term to simulate the denudational activity of shallow landslides and slope wash at a gradient of less than 45°. This term has a linear form, and assumes that this value is zero for a horizontal surface and 2 m ka^{-1} for a slope of 45°.

5. A term to simulate the activity of rockfalls and rock avalanches for slopes with a gradient of greater than 45°. It is assumed that this term is a rate of 2 m ka^{-1} at 45° and 100 m ka^{-1} at 80° or above. This term was given a non-linear form, and was designed to encourage the rapid evolution of slopes from very steep gradients. Testing of the model suggested that such a term produced the closest fit to natural conditions, and the model at low levels tended towards a stable slope angle of about 70°, which is very close to that observed in the gorge.

The model allowed all of the terms to be varied individually or together, and it has been shown that the above terms produce the simulation most closely related to the observed landscape.

Since the main rock type forming Taroko Gorge is a metamorphosed reefoid limestone that is being uplifted as a result of intra-plate collision, the starting point was taken to be a planar surface onto which a consequent drainage channel system had been imposed, consisting of a depression with a depth of 10 m, a planar bed with a width of 10 m, and side walls with a gradient of 45° (Fig. 8). As the landscape has evolved under a climate that includes frequent high discharge events, caused by the precipitation associated with typhoons, it was assumed that the channel bed underwent denudation at a rate equal to the uplift, allowing the form of the long profile to remain constant. This assumption was made using the example of the Li-Wu river channel within Taroko Gorge, which has a low gradient despite high uplift rates, presumably because of the high discharges associated with typhoons.

Each iteration of the model was set to simulate a five year period, using the output from the previous step as an input. The model was, in the first instance, allowed to run for a period that simulated 100 000 years. The results of the simulation are shown in Fig. 9, with the evolution of the gorge depicted at 10 000 year steps. Initially, the rate of increase in the relief of the area is high, but it is clear from the model that this declines as the rate of denudation increases with greater altitude. The base of the gorge is

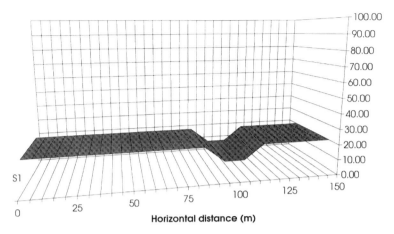

Fig. 8. Initial conditions for modelling the evolution of Taroko Gorge. Starting conditions consisted of an uplifted plate with a depression located within it. The base of the depression was flat with a width of 10 m and the walls of the depression had a gradient of 45°.

seen to remain at the same elevation, as specified in the model, whilst the gorge develops. Development of the gorge is characterized by two facets. First, the gorge increases in depth as the elevation increases, and second the width of the landform increases. Whilst these two facets are inevitable given the type of model that is being used, the nature of the processes is interesting.

As initial gorge development occurs (i.e. in the first 20 000 years) the gradient of the gorge walls increases from the initial 45° to a gradient of about 75°. As the time sequence continues the height of this lower slope increases whilst the gradient remains constant.

Above the very steep lower slope, a second slope with a lower gradient (c. 49°) develops.

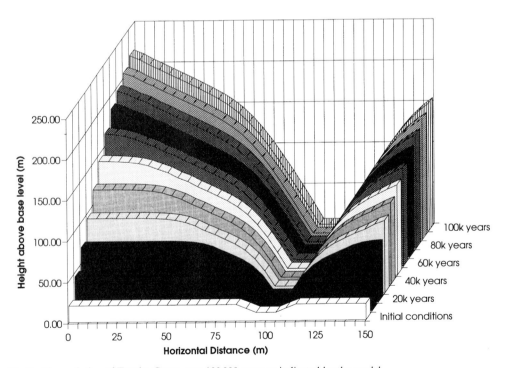

Fig. 9. The evolution of Taroko Gorge over 100 000 years as indicated by the model.

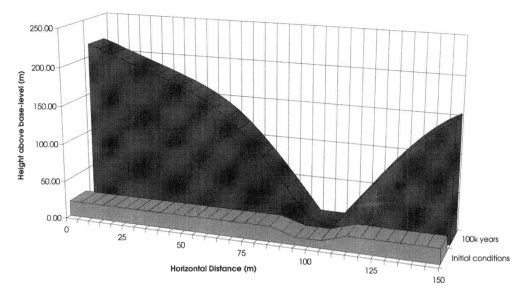

Fig. 10. The form of the model of the Gorge after 100 000 years compared with the starting conditions.

Again, as the model develops the length of this slope increases, and the width of the whole gorge system increases rapidly, including the whole width of the model area after about 60 000 years. The gradient of this higher slope remains at a constant 49°.

The final form of the model is remarkably similar to that measured in the field (compare Fig. 10 with Fig. 6), with steep lower slopes and more gentle higher slopes. Two important points can be made based on the output of the model. First, the decrease in gradient higher up the slope is not related to changes in the rate of uplift (uplift was constant with time) but is a consequence of the changes in denudation rate with altitude and slope angle. Second, the model suggests that summit conformity would not be indicative of a former land surface but would be a consequence of the processes acting on the landscape.

It is accepted that this model is simplistic in both the manner in which it models the processes acting and in the way it assumes simple geological conditions in two dimensions. The model requires further refinement, in particular in the reliance on denudation rates for a relatively large area, which consists of a complex range of landforms including the gorge and the regional uplands. However, despite these problems, the model still provides an important insight into the development of a landscape. It is intended that further development of the model will be undertaken to reflect more accurately the active processes.

Conclusions

In conclusion it is clear that Taroko Gorge has formed as a result of the high rates of uplift and denudation occurring in this area. Modelling of the development of such a landscape is possible using sets of simplifying equations accounting for the net processes occurring over time. Such models demonstrate that the gorge was created as uplift continued and that the morphology of the gorge walls is a consequence of the tectonic and geomorphological processes occurring in combination with the geological conditions. Finally, the model suggests that changes in slope gradient may be a consequence of the variation in process rates and types with height rather than changes in the rate of uplift.

The authors would like to thank most sincerely the Superintendent and staff of Taroko National Park, eastern Taiwan, for their support and assistance for this project. Especial thanks are due to Mr Morgan Chen and Dr Jason Liu. The authors are also grateful for the inputs to the project provided by Prof. Mervyn Jones of the University of Portsmouth, Prof. Hongey Chen of National Taiwan University, and Dr Chen-Hui Fan of Chinese Petroleum Corporation.

References

BONILLA, M. G. 1975. *A review of recently active faults in Taiwan*. United States Geological Survey Open File Report, 41–75.

BRUNSDEN, D. 1987. Principles of hazard assessment in neotectonic terrains. *Memoir of the Geological Society of China*, **9**, 305–334.
—— 1990. Tablets of Stone: towards the Ten Commandments of Geomorphology. *Zeitschrift für Geomorphologie Supplement*, **79**, 1–37.
—— & LIN, J-C. 1991. The concept of topographic equilibrium in neotectonic terrains. *In*: COSGROVE, J. & JONES, M. E. (eds) *Neotectonics and Resources*. Bellhaven, London, 120–143.
COCH, N. K. 1994. Geologic effects of hurricanes. *Geomorphology*, **10**, 37–63.
GILCHRIST, A. R. & SUMMERFIELD, M. A. 1991. Denudation, isostasy and landscape evolution. *Earth Surface Processes and Landforms*, **16**, 555–562.
HO, C. S. 1982. *Tectonic Evolution of Taiwan*. Ministry of Economic Affairs, Taipei, Taiwan.
—— 1986. A synthesis of the geological evolution of Taiwan. *Memoir of the Geological Society of China*, **7**, 15–29.
—— 1987. A synthesis of the geologic evolution of Taiwan. *Memoir of the Geological Society of China*, **9**, 1–18.
HSU, M. T. 1975. On the degree of earthquake risk in Taiwan. *Meteorological Bulletin*, **21**, 33–40.
HUNG, J-J. 1987. Landslides and related researches in Taiwan. *Memoir of the Geological Society of China*, **9**, 23–44.
JANSEN, J. M. L. & PAINTER, R. B. 1974. Predicting sediment yield from climate and topography. *Journal of Hydrology*, **21**, 371–380.
JANSSON, M. B. 1988. A global study of sediment yield. *Geografiska Annaler*, **70**, 81–98.
LEE, P. J. 1977. Rate of early Pleistocene uplift in Taiwan. *Memoir of the Geological Society of China*, **2**, 71–76.
LEE, Y-H., WOO, A. & YANG, C-N. 1996. Mountain building collapse in Taiwan mountain belt. *Proceedings of the 3rd Sino-British Geological Conference*, Taipei, 33.
LIEW, P-M. & TSENG, M-H. 1996. Late Quaternary climatic environment of the Taipei Basin. *Proceedings of the 3rd Sino-British Geological Conference*, Taipei, 37.
LIN, J-C. 1991. The structural landforms of the Coastal Range of eastern Taiwan. *In*: COSGROVE, J. & JONES, M. E. (eds) *Neotectonics and Resources*. Bellhaven, London, 65–74.
LONGLEY, E., FAN, C-H., IBSEN, M-L., NORTH-LEWIS, D., PENDRY, M., & PETLEY, D. N. 1992. *The Environmental Impact of Tropical Cyclones*. Final report of the University of London Expedition to Taiwan, 1991.
MA, K-F. & LIU, N-J. 1996. Lateral velocity variation and subduction slabs in Taiwan region. *Proceedings of the 3rd Sino-British Geological Conference*, Taipei.
MOH, Z-C. 1977, Landslides in Taiwan; some case reports. *In*: PENZIEN, J., SHENG, T. M. & YANG, Z. (eds) *Advisory Meeting on Earthquake Engineering and Landslides*. National Science Foundation, Washington, 199–218.
PENG, T. H., LI, Y-H. & WU, F. T. 1977. Tectonic uplift rates of the Taiwan island since the early Holocene. *Memoir of the Geological Society of China*, **2**, 57–69.
PETLEY, D. N. 1995. Engineering hazards in Taroko Gorge, eastern Taiwan. *Proceedings of the International Conference on Geohazards and Engineering Geology*, Coventry. 261–270.
——, LIU, C-N. & LIOU, Y-S. 1997. Geohazards in a neotectonic terrain, Taroko Gorge, eastern Taiwan. *Journal of the Geological Society of China*, **40**, 135–154.
SIMPSON, R. H. 1981. *The Hurricane and its Impact*. Blackwell, Oxford.
STANLEY, R. S., HILL, L. B., CHANG, H. C. & HU, H. N. 1981. A transect through the metamorphic core of the Central Mountains, southern Taiwan. *Memoir of the Geological Society of China*, **4**, 443–473.
THOMAS, M. F. & THORP, M. B. 1995. Geomorphic response to rapid climatic and hydrologic change during the late Pleistocene and early Holocene in the humid and sub-humid tropics. *Quaternary Science Reviews*, **14**, 193–207.
WANG, C-H. & BURNETT, W. C. 1991. Holocene mean uplift rates across an active plate-collision boundary in Taiwan. *Science*, **248**, 204–206.
YEN, T. P., SHANG, C. G. & KENG, W. P. 1951. The discovery of Fusuline Limestone in the metamorphic complex of Taiwan. *Bulletin of the Geological Survey of Taiwan*, **3**, 23–26.

Preliminary observations on the geomorphic evolution of the Guide Basin, Qinghai Province, China: implications for the uplift of the northeast margin of the Tibetan Plateau

P. A. FOTHERGILL[1] & H. MA

Centre for Quaternary Research, Department of Geography, Royal Holloway, University of London, TW20 0EX, UK
[1] *Present address: Department of Geology, University of Leicester, University Road, Leicester LE1 7RH, UK*

Abstract: Landform development in the Guide Basin has been influenced by the late Tertiary and Quaternary tectonic evolution of northeast Tibet. In the south of the Guide Basin, the remnants of three palaeosurfaces are preserved on top of flat-topped landforms. The two highest palaeosurfaces are separated by faults, which are thought to be extensional. Normal faulting may have begun in the middle–late Pleistocene, possibly indicating that the Guide Basin had reached its maximum elevation by this time. Yellow River drainage development within the Guide Basin may also have occurred in middle–late Pleistocene times, following the onset of extensional faulting.

This paper uses new geomorphic evidence from the Guide Basin, Qinghai Province, northwest China, to assess the history of uplift along the northeast margin of the Tibetan Plateau. A model is presented to explain the late Tertiary and Quaternary development of the Guide Basin, with respect to the tectonic evolution of the region.

The northeast margin of Tibet (Fig. 1) is an area of key tectonic and palaeoclimatic significance. It marks the transition between the high Tibetan Plateau in the west and the Loess Plateau further east. It has been uplifted to an average altitude of approximately 4000 m as a result of the late Palaeocene collision between India and Eurasia more than a 1000 km away (Molnar & Tapponnier 1975). Consequently, understanding the mode of formation of northeast Tibet is important for evaluating the effect of continental plate collisions.

The raising of Tibet has had a dramatic influence on both regional and global climate, by triggering the southeast Asian monsoon (Prell & Kutzbach 1992). Uplift information is required, therefore, from all areas of the Plateau, to help refine global climate models (e.g. Ruddiman & Kutzbach 1989). Investigations into the uplift of northeast Tibet are also important for understanding the evolution of the Yellow River, whose present-day drainage pattern has largely been fashioned by tectonics in this region (Barbour 1933; Lehmkuhl & Spönemann 1994).

Li (1991) and Li et al. (1995, 1996) have suggested that the northeast margin of Tibet has been subjected to a series of intense, very recent (Plio-Quaternary) uplift events, and that prior to around 3.4 Ma the Tibetan Plateau was at a relatively low altitude (<1000 m), with subdued relief and a stable tectonic regime. According to this viewpoint, uplift rates accelerated during the Quaternary, with present-day altitudes along the northeast margin of the Tibetan Plateau not being attained until Holocene times. This model does not, however, appear to be supported by the results of independent tectonic investigations, which suggest that Cenozoic tectonism in this region began in the late Miocene (Gaudemer et al. 1995). The data presented in this paper will help to elucidate the chronology of uplift along the northeast margin of the Tibetan Plateau.

A major complicating factor when interpreting geomorphological data is that landscapes often respond in a very similar fashion to episodes of tectonism as they do to climate change events (Molnar & England 1990). With this in mind, attention has been given in this study to identifying fault scarps and fault traces, both in the field and on satellite imagery. Faults are, by definition, irrefutable evidence of earth movement.

In the following section, current opinions on the tectonic evolution of northeast Tibet are reviewed, thus helping the reader to understand fully the regional significance of the results presented in this paper. One of the objectives of this study is to consider the geomorphological results in conjunction with structural geological evidence. Only in this way can conclusions be drawn with any confidence about the chronology of uplift in northeast Tibet.

From: SMITH, B. J., WHALLEY, W. B. & WARKE, P. A. (eds) 1999. *Uplift, Erosion and Stability: Perspectives on Long-term Landscape Development.* Geological Society, London, Special Publications, **162**, 183–200. 1-86239-030-4/99/ $15.00 © The Geological Society of London 1999.

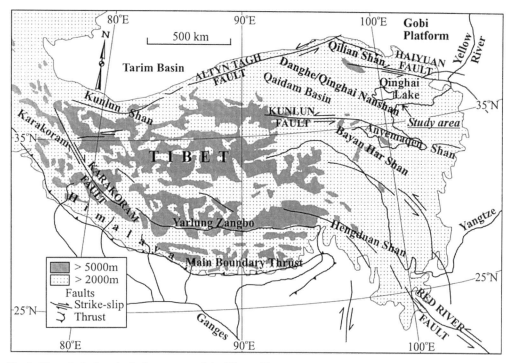

Fig. 1. The principal geological structures of the Tibetan Plateau.

The tectonic evolution of northeast Tibet

Evidence suggests that the south of the Tibetan Plateau has been uplifted by the wholesale underthrusting of Indian crust (Zhao et al. 1993; Ni et al. 1996), whilst north of the Lhasa Terrane Plateau formation is a result of distributed shortening (Chang et al. 1986; England & Searle 1986) and convective instability (England & Houseman 1988; Molnar et al. 1993; Owens & Zandt 1997). The northeastern margin of the Tibetan Plateau, however, is underlain by Asian lithosphere and has uplifted due to thrusting and crustal thickening, associated with movement along large, left-lateral strike-slip fault complexes (Fig. 1) (Molnar & Tapponnier 1975).

The mountains (Chinese: 'shan') to the northeast of the Qaidam Basin (Fig. 1) can be interpreted as crustal ramp anticlines, developed above thrust faults, which originate as splays from the left-lateral Altyn Tagh Fault (Peltzer et al. 1989; Gaudemer et al. 1995; Tapponnier et al. 1990, 1996). These splay faults can be traced for hundreds of kilometres as a series of southeast trending, elongate mountain ranges, which include the Qilian Shan, the Danghe Nan Shan and the Qinghai Nan Shan (Fig. 1). In the area around Qinghai Lake (Figs 1 and 2), NE–SW directioned compression, associated with movement along the Altyn Tagh Fault, has thickened the continental crust to 58–60 km (Pan et al. 1990). This has resulted in crustal loading and the formation of large compressional

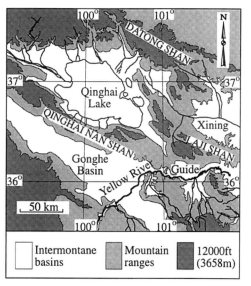

Fig. 2. The major mountain ranges and intermontane basins near to the study area, at the northeast margin of the Tibetan Plateau.

intermontane basins, such as the Gonghe, Xining and Qinghai Lake Basins (Fig. 2) (Pan et al. 1990).

To the east of the Qilian Shan (Fig. 1), activity along the Haiyuan Fault facilitates movement between northeast Tibet and the southern edge of the Gobi Platform (Gaudemer et al. 1995). Changes in the trend of the left-lateral Haiyuan Fault have resulted in the development of strike-slip basins at left-stepping releasing bends and push-up mountain ranges at right-stepping restraining bends (Burchfiel et al. 1991; Gaudemer et al. 1995). Thrust faulting in the Tianjing Shan and the Mibo Shan, Ningxia Province, seems to indicate that the Tibetan Plateau is extending to the northeast with time (Burchfiel et al. 1991).

Movement along the Altyn Tagh and Haiyuan Fault systems is thought to have begun in the late Miocene (Gaudemer et al. 1995; Tapponnier et al. 1996) and has continued to the present day. There is a high incidence of recent seismic activity in northeastern Tibet (Peltzer et al. 1988; Gaudemer et al. 1995). The last major earthquake (M = 6.9) in the Qinghai Lake area occurred in April 1990 and had its epicentre beneath the Gonghe Basin (Xiang & Zhao 1991). Movement along the earthquake fault, which strikes 102° and dips 42° to the SSW, was composed of 79 cm of reverse dip-slip and 5 cm of left-lateral strike-slip (Chen et al. 1992). Whilst the 1990 Gonghe earthquake may well have been initiated by the rapid filling of the Longyangxia Reservoir (see Rothé 1970; Gupta & Rastogi 1974), its slip direction is consistent with the regional stress regime.

The Guide Basin

The 1135 km^2 Guide Basin ranges in altitude from 2200 m adjacent to the Yellow River at Guide, to 3500 m at the foot of the Zhihai Shan. It is bounded to the northeast by the Laji Shan and the Guide Dong Shan and is separated from the Gonghe Basin to the west by the Wali Gong Shan. The Guide Basin is currently drained by the Yellow River and its tributaries (Fig. 2). The Yellow River enters and exits the Guide Basin through rectilinear bedrock gorges at Longyangxia (c. 1000 m) and Songba (c. 650 m) respectively.

The high mountain peaks surrounding the basin, especially the Zhihai Shan in the south (c. 5000 m) and the Riyue Shan in the north, are likely to have supported small-scale ice cover during the last two glaciations (Shi et al. 1991), but there is no firm evidence for widespread Quaternary glaciation in this region (Derbyshire et al. 1991). The present-day climate is cold, with an average annual temperature of 0.6°C (Xiang 1990). Semi-arid conditions prevail, with almost all the precipitation falling during the summer months, due to the controlling influence of the southeast Asian monsoon.

Faulting around the Guide Basin consists predominantly of southeast trending, high angle, northeast and southwest dipping thrust faults (Qinghai Geology Bureau 1964, 1972). Granite plutons have been offset and juxtaposed with highly contorted Palaeozoic rocks and an unconformable sequence of folded Mesozoic sediments (Qinghai Geology Bureau 1964, 1972). Ophiolite suites in the Laji Shan indicate that there was a northward subduction zone in this region in the lower Palaeozoic (Zhang et al. 1984). During the Mesozoic, thick flysch sediments deposited in the Guide Basin area (Zhang et al. 1984) were thrusted and folded by major tectonic events associated with the successive suturing of microcontinental blocks on to the southern margin of Eurasia (Chang et al. 1986; Wu et al. 1993). The Cenozoic reactivation of faults in the Guide Basin area, due to the intercontinental collision between India and Eurasia, has resulted in thrusting, crustal thickening and basin formation. The Guide Basin has, therefore, developed as an intermontane, syntectonic ramp basin in a compressional environment (see Cobbold et al. 1993).

Molasse deposition in the Guide Basin began in the Miocene (Qinghai Geology Bureau 1972; Pan et al. 1990). The main thickness of sediments (c. 1000 m) belongs to the Neogene Guide Formation, which is composed of intercalated parallel-bedded lacustrine siltstones and trough cross-bedded gravelly sands. The lacustrine deposits dominate in the depocentre of the basin, but are absent in the south and east of the field area. The trough cross-bedded gravelly sands were deposited in perennial braided-river environments (Fothergill, unpublished data). Mammalian fossil fauna have been found in the Guide Formation to the east of Guide, near the Yellow River (Pan 1994). The oldest fossil is *Hipparion fassatum*, which is characteristic of late Miocene deposits (Pan 1994). The main thickness of the Guide Formation contains *Hipparion Cf. tchicoicum* and *Hipparion platyodus*, known to be of Pliocene age (Pan 1994). Fluvial incision of the Guide Formation, together with the deposition and reworking of extensive Quaternary deposits, has produced a detailed morphostratigraphic record of basin evolution.

Fig. 3 (A) SPOT satellite image, showing the south of the Guide Basin.

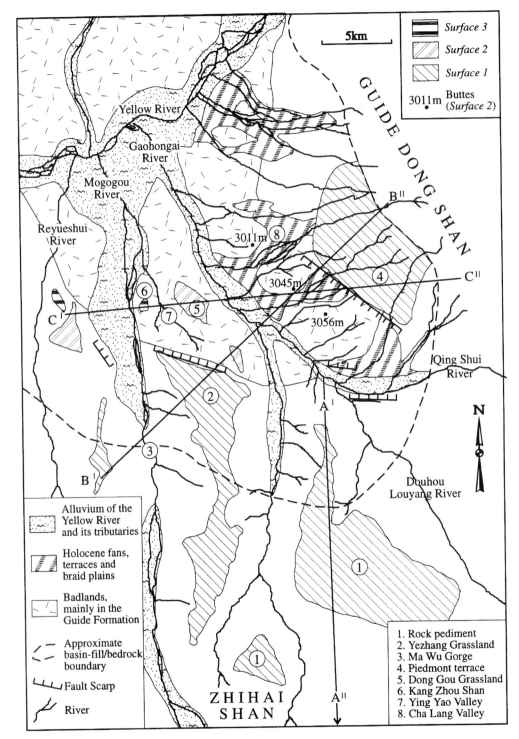

Fig. 3 (**B**) Geomorphological and sedimentological interpretation of the area shown in 3(A). The locations of key sites, referred to in the text, are shown. The positions of the three geological and topographic cross-sections, shown in Fig. 4, are also marked.

Methods

Investigations were concentrated in the area south of the Yellow River (Fig. 3A,B), mainly in the interfluve between the Mogogou and the Gaohongai rivers and in the piedmont of the Guide Dong Shan. Analysis of the landforms and their geomorphic relationship to one another was carried out using field sketches backed up with photographic data. Scarp and terrace gradients were calculated with the aid of a hand-held clinometer, and sedimentary lithologies, textures and structures were logged using the methods of Tucker (1982). Locations and altitudes above sea-level were recorded using a Global Positioning System receiver and a barometric altimeter. Chinese topographic maps (1:50 000 and 1:200 000) were used to draw accurate cross-sections, and SPOT satellite data were processed using ER Mapper software.

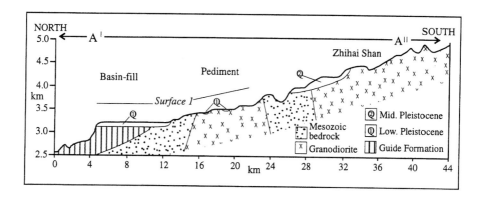

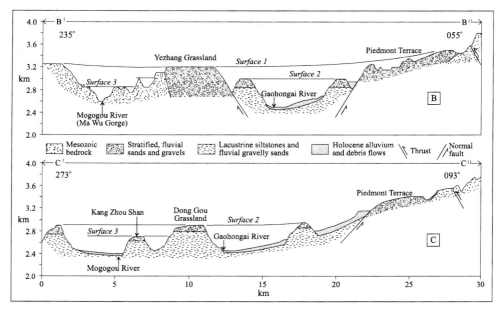

Fig. 4. (**A**) Cross-section from the southeast of the Guide Basin, showing the composite nature of Surface 1. Adapted from 1:200 000 geological and topographic maps (80 m contours). Sub-surface geological contacts are theoretical. Vertical scale ×5. (**B**) Cross-section located approximately SW–NE across the field area, showing the variable composition of Surface 1 and the downfaulting of Surface 2. (**C**) Cross-section from the centre of the field area, located approximately W–E, showing the development of mesas and buttes above the flood plains of the Gaohongai and Mogogou rivers. Both (**B**) and (**C**) have a vertical exaggeration of ×3.76 and were adapted from 1:200 000 geological and 1:50 000 topographical maps (10 m contours). Sub-surface geological contacts and continuity of the faults are theoretical.

Geomorphology south of the Yellow River

The geomorphology in the south of the Guide Basin is characterized by meso- to macroscale, flat, elevated palaeosurfaces, which have been incised around their edges to form areas of gullies and badlands (Fig. 3A,B). Evident on the SPOT satellite imagery are large, linear escarpments, which are clearly the product of fault activity (Fig. 3A,B). Much smaller faults, which offset the Guide Formation, can be identified in the field. Alluvial fans and terraces exist above the present-day flood plains of the Mogogou and Gaohongai rivers (Fig. 3A,B).

The Mogogou, Gaohongai and Reyueshui rivers, which develop *via* a trellis drainage network of tributary streams at the foot of the Zhihai Shan, are intensely braided, have a high-angle contact with the Yellow River and are discordant with the main geological structure. Drainage of the eastern margin of the Guide Basin occurs via predominantly straight, ephemeral river channels, which are arranged radially around the Guide Dong Shan (Fig. 3A,B).

Field observations and topographical cross-sections (Fig. 4A,B,C) show that distinct surfaces have developed within the Guide Basin. The highest and most extensive surface (Surface 1) is preserved on top of different landforms and can be traced across the majority of the field area.

Surface 1

In the southeast of the Guide Basin, Surface 1 is at an altitude of 3500 m, on a northward sloping rock pediment which has developed mainly on granodioritic bedrock (Fig. 4A). The pediment surface connects to the north with a much flatter depositional surface, composed of Pleistocene sediments overlying the Guide Formation. This depositional surface is conformable in altitude and slope with a large (60 km^2) plateau in the centre of the field area, covered by the Yezhang Grassland (Fig. 4B). Surface 1 shows its best preservation on top of this plateau.

Despite the uniformly flat nature and continuity of the Yezhang Grassland, it is underlain by a variable geology. Towards the south, only the sediments comprising the upper part of the plateau are exposed. Sections here consist of 3 m of loess, overlying thick (>50 m), mainly matrix-supported massive pebbles and cobbles. In the centre of the Yezhang Grassland thick sections of Jurassic bedrock (Qinghai Geological Bureau 1972), with a relatively steep dip (20–36°) to the northwest, can be seen. This rock type is exposed over a height of 640 m from the base of the Ma Wu Gorge (Figs 3B and 4B) up to a thin (*c.* 3 m) veneer of loess below the Yezhang Grassland surface. The northern part of the Yezhang Grassland, at an altitude of 3200 m, is underlain by thick (>200 m) exposures of unconsolidated trough cross-bedded sandy gravels, interbedded with planar units of sand. The gravels are predominantly rounded and clast-supported granitic and arenaceous pebbles and cobbles, becoming more indurated with depth.

West of the Yezhang Grassland, Surface 1 can be traced across the Ma Wu Gorge (Fig. 4B), where it forms an extensive plain surrounded by bedrock gullies. In the east of the Guide Basin, Surface 1 is preserved as a piedmont terrace (Fig. 3B), which is conformable with the Yezhang Grassland (Fig. 4B). The piedmont terrace is at an altitude of 3300–3500 m, has a gentle slope (2–3°) towards the southwest and forms a high-angle contact with the Guide Dong Shan (Fig. 5). Incision of the piedmont terrace has occurred along ephemeral, WSW flowing rivers, exposing the sediments that comprise this surface (Surface 1). Beneath a thin (2.7 m) covering of sandy loess, there are at least 300 m of angular to sub-angular poorly indurated, mainly clast-supported, coarse sandy gravels, with secondary lithological units of thinly to thickly laminated and trough cross-laminated medium sand. The gravel clasts are predominantly Triassic sandstone and mudstone.

Surfaces 2 and 3

To the north of the Yezhang Grassland, Surface 2 is preserved as the Dong Gou Grassland, at an altitude of approximately 3000 m on top of a 5 km^2 mesa (Fig. 6). Surface 2 is also preserved to the east of the Gaohongai river, on top of three buttes, each with an altitude of around 3000 m (Figs 3B and 5). To the west of the basin, Surface 2 forms a flat-topped 4.5 km^2 summit, with a height of 2950 m (Figs 3B and 4C). Surface 2 is fringed by extensive badlands (Fig. 6), which have formed in the fissile siltstone and gravelly sand deposits that characterize the Guide Formation in the centre of the basin. These badlands characteristically have narrow interfluves and linear slope profiles, very similar to those reported from the classic badland areas of the mid-western USA (Howard 1994). Fluvial sands and gravels cap the less resistant Guide Formation deposits in the badland areas.

Surface 3 lies some 200–300 m below Surface 2. It is preserved on top of a mesa, locally referred to as Kang Zhou Shan (Figs 3B and

Fig. 5. The northeast margin of the field area, showing the Guide Dong Shan (1) and its piedmont region. The high piedmont terrace (2) (Surface 1) and the fault scarp (3) that borders it can be seen beneath the mountain front. In the middle distance, a 3045 m high butte (4) is a remnant of Surface 2. In the near distance, a large sediment fan (5), with limited fan foot incision, can be seen above the flood plain of the Mogogou river.

Fig. 6. The Dong Gou Grassland (Surface 2), which is situated on top of a 5 km^2 mesa. Widespread incision and extensive badland development can be seen.

4C), as well as on top of an elevated grassland in the west of the basin (Fig. 3B) and possibly as an erosional terrace in the Ma Wu Gorge (Fig. 4B). The Kang Zhou Shan has a similar composition to the Dong Gou Grassland, with fluvial deposits overlying basin fill sediments of the Guide Formation. It is the least extensive surface developed within the Guide Basin and can only be observed above the present-day drainage systems of the Mogogou and the Reyueshui rivers. Surfaces 2 and 3 are, in places, offset from Surface 1 by fault escarpments.

Faulting

The northern edge of the Yezhang Grassland is delimited by a steep, linear, 600 m high scarp (Fig. 7), which is characterized by sharply defined triangular facets (Fig. 8), usually associated with extensional faulting (e.g. Armijo et al. 1986). The trend of the scarp is to the ESE. Scarps with similar trends, also attributed to faulting, can be observed close to the junction of the Qing Shui and the Douhou Louyang rivers (Fig. 3B), beneath the piedmont of the Guide Dong Shan (Fig. 9).

Southeast trending faults are also evident along the margin of the high piedmont terrace, which terminates abruptly at an extremely straight, 250 m high scarp (Fig. 9). In addition, small faults with southeast strikes and southwest dips are exposed in the Guide Formation badlands, at the foot of the Guide Dong Shan piedmont in the Cha Lang Valley (Figs 9 and 10). Occasionally, these small faults offset colour banding within the lacustrine siltstones and their movement can be identified as normal (Fig. 10). A southeast trending scarp is seen in the west of the field area, above a well formed terrace of the Mogogou river (Figs 7 and 11). This scarp, which dips to the northeast and has a pronounced linearity over approximately 2 km, is also the product of fault movement. No evidence for faulting was observed within the recent fluvial sediments, which unconformably overlie the Guide Formation deposits along the foot of the Guide Dong Shan piedmont.

Alluvial fans and terraces

The development of V-shaped river valleys in the piedmont terrace below the Guide Dong Shan,

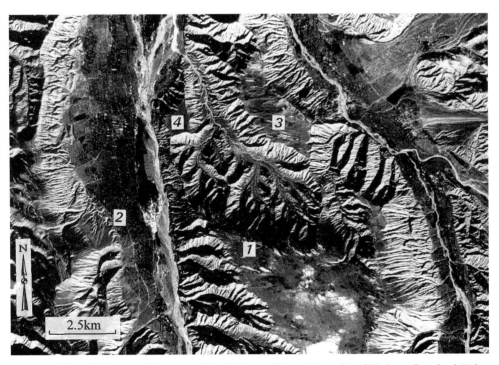

Fig. 7. SPOT satellite image of the centre of the field area. The northern edge of Yezhang Grassland (1) is marked by a steep fault scarp. A fault trace can also be seen above the Mogogou river (to the SW of 2). The Dong Gou Grassland (3) and the Kang Zhou Shan (4) are mesas, formed by intense fluvial downcutting.

Fig. 8. The northern edge of the Yezhang Grassland (looking southwest). The 600 m high scarp is characterized by well formed triangular facets (1).

which connect with large braid plains and sediment fans away from the mountain front (Figs 5 and 12), represents the latest phase in the geomorphological evolution of the Guide Basin. Incision of the fans is restricted to the foot area (Fig. 5). The braid plain at Cha Lang Valley (Figs 3B, 9 and 12) is composed of poorly bedded, clast-supported, imbricated gravels. It is separated from a fluvial terrace above it by a steep (35°), 37 m high scarp, with angular breaks of slope (Fig. 12). The terrace is composed of massive sandy silt, overlying a basal exposure (5 m) of rounded and clast-supported pebbles, cobbles and boulders interbedded with lenses of thinly to thickly laminated sand.

Terraces have been developed above the Mogogou river (Fig. 11). Terrace remnants, approximately 80 m above the river, can be seen on both sides of the valley, between the Ma Wu Gorge and the Kang Zhou Shan. In the Ying Yao Valley (Fig. 3B), recent terrace development onlaps the badlands at the base of the Dong Gou Grassland (Fig. 6). This terrace, which has been heavily gullied down to the present-day level of the river, consists of thinly bedded, rounded, matrix-supported, unconsolidated granules and pebbles interbedded with thinly laminated silt and massive coarse sand.

Discussion

In the south of the field area, Surface 1 is preserved on granodioritic bedrock (Fig. 4A). In this area, Surface 1 is considered to be the remnant of an erosion surface which antedates basin formation. Prior to the formation of the Guide Basin in the late Miocene, an erosion surface existed across this area (Fig. 13/I). Relief was probably relatively subdued, and long-term geomorphological and base-level stability had led to the development of a pediment on the granodioritic bedrock at the foot of the mountains in the south of this area (the present-day Zhihai Shan).

Since the onset of late Miocene tectonism, this erosion surface has been undergoing destruction. However, it has also acted as a zone of transition, across which material eroded from the uplifting Zhihai Shan has been transported northwards towards the subsiding basin centre (Fig. 13/II). In the south of the basin, therefore, Surface 1 can be considered as a relict of a Miocene rock pediment (see Oberlander 1989; Dohrenwend 1994), in its final stages of disintegration.

In the centre and to the east of the basin, Surface 1 is depositional, lying on top of the molasse deposits that infill the basin. Surface 1, therefore,

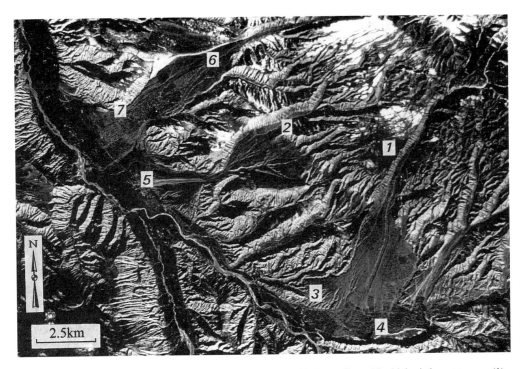

Fig. 9. SPOT satellite image from the piedmont zone of the Guide Dong Shan. The high piedmont terrace (1) and the NW-SE oriented fault scarp that delimits it (2) are clearly visible. Faults with roughly E-W trends and downthrows to the N/NE (3 and 4) can also be identified. The foot of the large alluvial fan (5) is at a slightly higher level than the flood plain of the Gaohongai river, suggesting recent incision. Terrace formation can be seen above the Cha Lang Valley (6). Location 7 shows the site of small extensional faults observed in the Guide Formation (e.g. Fig. 10).

is time-transgressive, being formed on top of an erosional Miocene pediment in the south of the field area and on a Pleistocene depositional surface further north. Prior to dissection, Surface 1 formed the uppermost drainage level within the Guide Basin (Fig. 13/III).

The thick, coarse gravels that comprise Surface 1 at the foot of the Guide Dong Shan may have been deposited in large prograding fan complexes (Fig. 13/II). In the centre of the basin, deposition took place in alternating lacustrine and distal perennial braided-river environments (Fig. 13/II). The fact that such thick molasse deposits have accumulated in the Guide Basin suggests that their deposition took place in closed drainage conditions, in which case the Yellow River was not present in the basin during molasse deposition. The fact that a remnant of a Miocene pediment surface has been preserved in the south of the Guide Basin also suggests that base-level lowering, due to Yellow River drainage development, is a recent phenomenon. Once base-level lowering has occurred, then pediments are susceptible to very rapid dissection (Oberlander 1989; Dohrenwend 1994).

Two large sub-parallel fault scarps, with opposite dip directions, delimit the extent of Surface 1: one to the north of the Yezhang Grassland and the other to the southwest of the piedmont terrace (Fig. 3B). The triangular facets on the fault scarp to the north of the Yezhang Grassland suggest that it is extensional. The occurrence of small normal faults in the Guide Formation deposits beneath the piedmont terrace scarp (e.g. Fig. 10) indicate that this area has also undergone extension. The two faults which delimit the extent of Surface 1 may, therefore, form a graben in the centre of the basin, which has resulted in the downfaulting of Surface 2 with respect to Surface 1 (Figs 4B and 13/IV). Following this line of argument, Surfaces 1 and 2 were once conformable and they are comparable in age. They have been classified as two separate surfaces in this paper on the basis of their discontinuity.

The age of the onset of extensional faulting in the Guide Basin is not clear. According to

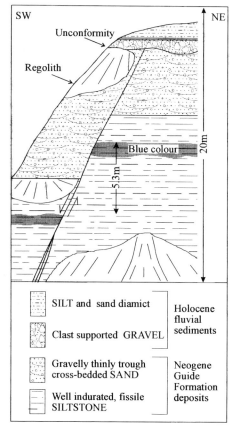

Fig. 10. Sketch of a small normal fault within the Guide Formation, at the foot of the Cha Lang Valley.

geological maps, early Pleistocene fluvial deposits are present just below the surface of the Yezhang Grassland (Surface 1). This suggests that in early Pleistocene times, Surface 1 still formed the main drainage level within the basin (Fig. 13/III) and that extensional faulting had yet to occur. A maximum age of the lower Pleistocene can, therefore, be inferred for extensional fault movement.

Field investigations, however, show the deposits beneath the Yezhang Grassland surface to be unconsolidated and fresh, possibly indicating that they have been deposited more recently than lower Pleistocene times. If this is the case, then a much younger maximum age for fault activity must be assumed. Work currently being carried out, using optically stimulated luminescence (OSL) dating to investigate the age of loess deposition on top of Surface 2, will provide a minimum age for the onset of extensional faulting in the Guide Basin. Loess only began to accumulate on Surface 2 after it had been downfaulted with respect to Surface 1, and base-level lowering and fluvial incision had begun (Fig. 13/IV).

Normal faulting, with a maximum middle Pleistocene age, can be seen in the loess/palaeosol profile within the neighbouring Xining Basin. Assuming that the tectonic histories of the Guide and Xining Basins are interrelated, the first occurrence of normal faulting in this area may have been in the middle–late Pleistocene.

The occurrence of extensional faulting in this area may indicate that uplift has ceased. If this is the case, peak elevations in this part of northeast Tibet may already have been attained. A similar argument has been used from the south of Tibet, where the formation of N–S oriented grabens, transverse to the direction of thrusting, marks the onset of crustal collapse (Armijo *et al.* 1986; England & Houseman 1989; Molnar *et al.* 1993; Coleman & Hodges 1995; Fort 1996). However, the data presented in this paper are unlikely to be indicative of large-scale gravitational collapse in northeast Tibet, for the reasons outlined below.

The normal faults in the Guide Basin have a strike roughly parallel with the mapped thrusts (Qinghai Geological Bureau 1964, 1972). Extensional faulting in the Guide Basin is perhaps more analogous with that observed in the Peruvian Andes than in southern Tibet. In Peru, recent normal faulting occurs on planes approximately parallel to the mountain range and is thought to be a manifestation of buoyancy forces caused by the high mountain chains and deep crustal roots (Dalmayrac & Molnar 1981). Next to the Peruvian Andes, both horizontal compression and crustal shortening still occur at low altitudes, whereas in some of the high altitude regions extension is taking place, with the least compressive stress being perpendicular to the mountain ranges (Dalmayrac & Molnar 1981). A similar situation may have occurred in the Guide Basin. If this is the case, extension in the Guide Basin may be a local phenomenon and is probably not indicative of extensive stretching of the crust in northeast Tibet. Nevertheless, the occurrence of extensional tectonism in the Guide Basin may suggest that this area is no longer undergoing uplift.

Following the formation of Surface 2, the Mogogou and Gaohongai rivers began to incise into the basin fill deposits (Fig. 13/IV). The Ma Wu Gorge began to form at this stage, due to drainage superimposition (Fig. 13/IV). Transverse drainage in the centre of the basin may also have started to develop at this time resulting in the first stages of downcutting at Longyang

Fig. 11. Northeast facing escarpment (1) developed above a terrace of the Mogogou river (2). The linearity of the scarp (over c. 2 km) is suggestive of fault movement. The terrace is approximately 80 m above the present-day course of the Mogogou river.

Gorge and Songba Gorge and the initial interconnection of drainage between the series of intermontane basins in this area. Surface 3 is limited in extent and is only evident close to the present-day course of the Mogogou and Reyueshui rivers. Surface 3 can be interpreted as a pediment terrace (Barsch & Royse 1972), formed when the drainage in the Guide Basin was higher than today.

Terrace development above the present-day alluvium of the Yellow, Mogogou and Gaohongai rivers may be related to climate change, with episodes of downcutting occurring during the wetter interglacial periods. Climate change has had a profound effect on landform evolution within the Guide Basin, thus complicating the interpretation of the geomorphological and sedimentological evidence. The discussion above, however, suggests that the formation of Surface 1, and its initial downfaulting to form Surface 2, was controlled by prolonged thrusting followed by extensional faulting.

Surfaces 2 and 3 are surrounded by extensive badlands. Badland formation may be due to a variety of reasons (Campbell 1989; Howard 1994), but in the Guide Basin it is primarily a reflection of lithological controls and high erosion rates. Factors such as differential permeability and regularly spaced (5 cm) joints within the Guide Formation favour the development of badlands, by leading to the extensive development of piping (Campbell 1989; Jones 1981) and an increase in the erodibility of the rock (Nicholas & Dixon 1986). The rate of badland scarp erosion dictates the speed of dissection of the mesas and buttes upon which Surface 2 is positioned (Oberlander 1989). Gullying, pipe-induced collapse, rockfalls, debris flows and mudflows all act to degrade the Guide Basin badlands.

All three surfaces within the Guide Basin are undergoing dissection, leading to the development of a landscape dominated by an embayed plateau, mesas and buttes. Stream capture may have occurred where the Mogogou and Gaohongai rivers meet the Yellow River (see Fig. 3A,B). Young (1984) suggests that landscapes dominated by scarp dissection and river capture are indicative of a lowering of the regional base-level and the incision of major rivers. Bryan et al. (1987) suggest that badland formation may be indicative of base-level changes. In the Guide Basin, the first appearance of the Yellow River during Pleistocene times caused a drop in the base-level at the basin outlet. This event was responsible for intense incision and badland formation (Fig. 13/V).

Fig. 12. The Cha Lang Valley. The peaks of the Guide Dong Shan (1) have a high-angle contact with the upper piedmont terrace (2) (Surface 1). The upper piedmont terrace has been incised by a V-shaped valley which connects with the braid plain of the Cha Lang Valley. A terrace (3) is separated from the braid plain by a steep, 37 m high scarp.

The most recent geomorphological evolution of the Guide Basin may be related to small-scale glaciation and deglaciation in the surrounding mountain ranges, together with a change to a more humid regime in the early Holocene. Equilibrium Line Altitude (ELA) depressions of approximately 600 m have been suggested for both the Qinghai Nan Shan (Porter et al. 1994) and the Laji Shan (Yugo Ono pers. comm.) during the Last Glacial Maximum (LGM). Assuming similar ELA depressions in the Guide Dong Shan, which has a present-day snow line altitude of approximately 4800 m (Wang & Derbyshire 1987) and summit altitudes of around 4600 m, then small glaciers would have occupied the upper valleys of this range during the LGM.

The limited extent of entrenchment at the foot of the large alluvial fan beneath the Guide Dong Shan (Fig. 5) implies that incision has occurred here only very recently. The sharp morphology of the scarp separating the terraces in the Cha Lang Valley also suggests that this is a very young feature (Pierce & Colman 1986). The large alluvial fan and the terrace in the Cha Lang Gou were probably formed in early Holocene times. Deglaciation in the Guide Dong Shan would have produced abundant sediment, which would have been dumped at the foot of the range. Subsequent enhanced fluvial activity during the Holocene is responsible for incision. Similarly, the terrace in the Ying Yao Valley is formed of reworked deposits, probably eroded from the surrounding badlands during an early Holocene shift to more humid conditions.

Conclusion

Three separate surfaces can be identified on top of the landforms in the south of the Guide Basin. Surface 1, which is preserved as both a remnant pediment and a depositional surface, was once continuous across the field area, forming the uppermost level of drainage within a closed basin system. It has formed, in places, on top of a thick sequence of molasse, which began to be deposited in the late Miocene in response to crustal thickening in northeast Tibet. Large fault scarps delimit the present-day extent of Surface 1 to the north of the

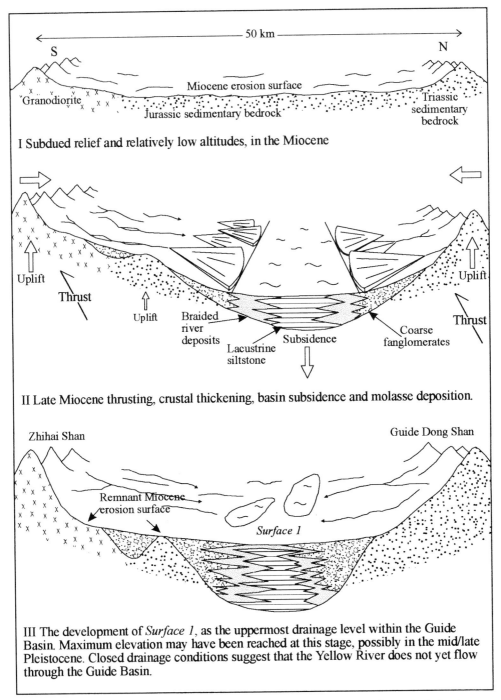

Fig. 13. Schematic diagram, presenting a model for the late Tertiary/Quaternary sedimentological and geomorphological development of the Guide Basin, with respect to tectonics. Stages I–III: compression, basin formation and the development of Surface 1. Stages IV and V: extension, incision and the establishment of the Yellow River drainage system through the Guide Basin.

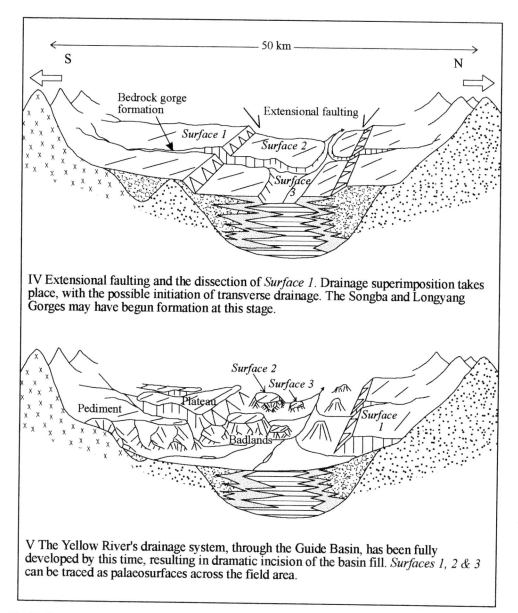

Fig. 13. (*continued*).

Yezhang Grassland and to the west of the Guide Dong Shan piedmont terrace. These faults, which are thought to be extensional, may have become active in the middle–late Pleistocene, possibly indicating that the Guide Basin had reached its maximum elevation by this time.

Surfaces 1 and 2 may have been offset from one another by extensional block faulting. Surface 3 is a pediment terrace, formed above the present-day drainage of the Mogogou and Reyueshui rivers. Surfaces 2 and 3 have been incised to such an extent that they are presently only preserved on top of mesas and buttes, surrounded by extensive badlands. Fluvial downcutting has occurred in response to the first appearance of the Yellow River in the Guide Basin, causing a drop in the local base-level. Yellow River drainage may have first developed in the Guide Basin in middle–late Pleistocene times, following the

onset of extensional faulting. Further work is required to place more accurate constraints on the kinematics and age of normal faulting within the Guide Basin, as well as to determine the relationship between extensional tectonism in this area and the evolution of the Yellow River.

P. Fothergill is sponsored by NERC (GT4/94/366/G). He also received generous donations from the BGRG Research and Publications sub-committee and the QRA Young Researchers Award, to help fund the field work. Thanks go to Angela Morrison at NERC for obtaining the SPOT satellite data. Thanks also go to Dr Lewis Owen, Dr David Petley and an anonymous reviewer for their very helpful comments.

References

ARMIJO, R., TAPPONNIER, P., MERCIER, J. L. & HAN, T. L. 1986. Quaternary extension in southern Tibet: Field observations and tectonic implications. *Journal of Geophysical Research*, **91**(B14), 13803–13872.

BARBOUR, G. B. 1933. Pleistocene history of the Huangho. *Geological Society of America Bulletin*, **44**, 1143–1160.

BARSCH, D. & ROYSE, C. F. 1972. A model for development of Quaternary terraces and pediment-terraces in the southwestern United States of America. *Zeitschrift für Geomorphologie N.F.*, **16**(1), 54–75.

BURCHFIEL, B. C., ZHANG, P., WANG, Y. et al. 1991. Geology of the Haiyuan fault zone, Ningxia-Hui autonomous region, China, and its relation to the evolution of the northeastern margin of the Tibetan Plateau. *Tectonics*, **10**(6), 1091–1110.

BRYAN, R. B., CAMPBELL, I. A. & YAIR, A. 1987. Postglacial development of the Dinosaur Provincial Park badlands, Alberta. *Canadian Journal of Earth Science*, **24**, 135–146.

CAMPBELL, I. A. 1989. Badlands and badland gullies. *In*: THOMAS, D. S. G. (ed.) *Arid Zone Geomorphology*. Pinter, London, 159–183.

CHANG, C., CHEN, N., COWARD, M. et al. 1986. Preliminary conclusions of the Royal Society and Academia Sinica 1985 geotraverse of Tibet. *Nature*, **323**, 501–507.

CHEN, Y. T., ZHAO, M. & WU, F. T. 1992. Focal processes of the Gonghe, Qinghai, China earthquake of April 26, 1990 from levelling and seismic data. *EOS Transactions of the American Geophysical Union*, **73**(25 Supplement), 65.

COBBOLD, P. R., DAVY, P., GAPIAS, E. A. et al. 1993. Sedimentary basins and crustal thickening. *Sedimentary Geology*, **86**, 77–89.

COLEMAN, M. & HODGES, K. 1995. Evidence for Tibetan Plateau uplift before 14 Myr ago from a new minimum age for east–west extension. *Nature*, **374**, 49–52.

DALMAYRAC, B. & MOLNAR, P. 1981. Parallel thrust and normal faulting in Peru and constraints on the state of stress. *Earth Planetary Science Letters*, **55**, 473–481.

DERBYSHIRE, E., SHI, Y., LI, J., ZHENG, B., LI, S. & WANG, J. 1991. Quaternary glaciation of Tibet: the geological evidence. *Quaternary Science Reviews*, **10**, 485–510.

DOHRENWEND, J. C. 1994. Pediments in arid environments. *In*: ABRAHAMS, A. D. & PARSONS, A. J. (eds) *Geomorphology of Desert Environments*. Chapman & Hall, London, 321–353.

ENGLAND, P. & HOUSEMAN, G. 1988. The mechanics of the Tibetan Plateau. *Philosophical Transactions of the Royal Society of London*, **A326**, 301–320.

—— & —— 1989. Extension during continental convergence, with application to the Tibetan Plateau. *Journal of Geophysical Research*, **9**(B12), 17561–17579.

—— & SEARLE, M. 1986. The Cretaceous–Tertiary deformation of the Lhasa block and its implications for crustal thickening in Tibet. *Tectonics*, **5**(1), 1–14.

FORT, M. 1996. Late Cenozoic environmental changes and uplift on the northern side of the central Himalaya: a reappraisal from field data. *Palaeogeography, Palaeoclimatology, Palaeoecology*, **120**, 123–145.

GAUDEMER, Y., TAPPONNIER, P., MEYER, B. et al. 1995. Partitioning of crustal slip between active faults in the eastern Qilian Shan, and evidence for a major seismic gap, the 'Tianzhu Gap', on the western Haiyuan fault, Gansu (China). *Geophysical Journal*, **120**, 599–645.

GUPTA, H. K. & RASTOGI, B. K. 1974. Investigations of the behaviour of reservoir-associated earthquakes. *Engineering Geology*, **8**, 29–38.

HOWARD, A. D. 1994. Badlands. *In*: ABRAHAMS, A. D. & PARSONS, A. J. (eds) *Geomorphology of Desert Environments*. Chapman & Hall, London, 213–242.

JONES, J. A. A. 1981. *The Nature of Soil Piping – a review of research*. British Geomorphological Research Group, Research Monograph **3**, GeoBooks, Norwich.

LEHMKUHL, F. & SPÖNEMANN, J. 1994. Morphogenetic problems of the upper Huang He drainage basin. *GeoJournal*, **34**(1), 31–40.

LI, J. 1991. The environmental effects of the uplift of the Qinghai-Xizang Plateau. *Quaternary Science Reviews*, **10**, 479–483.

—— et al. 1995. Uplift of Qinghai-Xizang (Tibet) Plateau and global changes. *XIV INQUA Congress, 1995, Berlin*. Lanzhou University, Lanzhou.

——, FANG, X., MA, H., ZHU, J., PAN, B. & CHEN, H. 1996. Geomorphological and environmental evolution in the upper reaches of the Yellow River during the late Cenozoic. *Science in China D*, **39**(4), 380–390.

MOLNAR, P. & ENGLAND, P. 1990. Late Cenozoic uplift of mountain ranges and global climate change: chicken or egg? *Nature*, **346**, 29–34.

—— & TAPPONNIER, P. 1975. Cenozoic tectonics of Asia: effects of a continental collision. *Science*, **189**, 419–426.

——, ENGLAND, P. & MARTINOD, J. 1993. Mantle dynamics, uplift of the Tibetan Plateau, and the Indian monsoon. *Reviews of Geophysics*, **31**(4), 357–396.

NI, J., SANDVOL, E., ZHAO, W., KIND, R., NABELEK, J. ZHAO, L. & INDEPTH-II BROADBAND TEAM. 1996. Seismic images from the INDEPTH-II broadband experiment: Implications for underthrusting models. *30th International Geological Congress, Beijing 4–14th August*. Special Symposia on the tectonic evolution and the mechanism of uplift of the Qinghai-Xizang (Tibet) Plateau.

NICHOLAS, R. M. & DIXON, J. C. 1986. Sandstone scarp form and retreat in the Land of Standing Rocks, Canyonlands National Park, Utah. *Zeitschrift für Geomorphologie N.F.*, **30**(2), 167–187.

OBERLANDER, T. M. 1989. Slope and pediment systems. *In*: THOMAS, D. S. G. (ed.) *Arid Zone Geomorphology*. Pinter, London, 55–84.

OWENS, T. J. & ZANDT, G. 1997. Implications of crustal property variations for models of Tibetan Plateau evolution. *Nature*, **387**, 37–43.

PAN, B. 1994. A study on the geomorphic evolution and development of the upper reaches of Yellow River in Guide Basin. *Arid Land Geography*, **17**(3), 43–50 (in Chinese).

PAN, G., WANG, P., XU, Y., JIAO, S. & XIANG, T. 1990. *Cenozoic tectonic evolution of Qinghai-Xizang Plateau*, P.R. China Ministry of Geological and Mineral Resources, Geological Memoirs Series 5, No. 9, Geological Publishing House, Beijing (in Chinese).

PELTZER, G., TAPPONNIER, P. & ARMIJO, R. 1989. Magnitude of Late Quaternary left-lateral displacements along the north edge of Tibet. *Science*, **246**, 1285–1289.

——, ——, GAUDEMER, Y. et al. 1988. Offsets of late Quaternary morphology, rate of slip, and recurrence of large earthquakes on the Chang Ma fault (Gansu, China). *Journal of Geophysical Research*, **93**(B7), 7793–7812.

PIERCE, K. L. & COLMAN, S. M. 1986. Effect of height and orientation (microclimate) on geomorphic degradation rates and processes, late-glacial terrace scarps in central Idaho. *Geological Society of America Bulletin*, **97**, 869–885.

PORTER, S., AN, Z. & WU, X. 1994. Variation in summer monsoon strength along the north front of the Qinghai Nan Shan since the Last Glacial Maximum. *International Symposium and Workshop on Palaeoenvironmental Records of Desert Margins and Monsoon Variation during the last 20 ka*. Xi'an, China, August, 14–23.

PRELL, W. L. & KUTZBACH, J. E. 1992. Sensitivity of the Indian monsoon to forcing parameters and implications for its evolution. *Nature*, **360**, 647–652.

QINGHAI GEOLOGY BUREAU. 1964. *Geological Map of Xining, 1/200 000*. Xining, China.

—— 1972. *Geological Map of Gou Ma Ying, 1/200 000*. Xining, China.

ROTHÉ, J. P. 1970. Seismes Artificiels. *Tectonophysics*, **9**, 215–238.

RUDDIMAN, W. F. & KUTZBACH, J. E. 1989. Forcing of late Cenozoic northern Hemisphere climate by plateau uplift in southern Asia and the American west. *Journal of Geophysical Research*, **94**(D15), 18409–18427.

SHI, Y., LI, B., LI, J. et al. 1991. *Quaternary glacial distribution map of the Qinghai-Xizang (Tibet) Plateau*. Science Press, Beijing.

TAPPONNIER, P., MEYER, B., AVOUAC, J. P. et al. 1990. Active thrusting and folding in the Qilian Shan, and decoupling between the upper crust and mantle in northeastern Tibet. *Earth and Planetary Science Letters*, **97**, 382–403.

——, ——, GAUDEMER, Y. & VAN DER WOERD, J. 1996. Strike-slip driven growth of the Tibetan Plateau and other SE Asian reliefs. *30th International Geological Congress, Beijing 4–14th August*. Special Symposia on the tectonic evolution and the mechanism of uplift of the Qinghai-Xizang (Tibet) Plateau.

TUCKER, M. 1982. *The Field Description of Sedimentary Rocks*. Geological Society of London Handbook, Wiley, Chichester.

WANG, J. & DERBYSHIRE, E. 1987. Climatic geomorphology of the northeastern part of the Qinghai-Xizang Plateau, Peoples Republic of China. *Geographical Journal*, **153**(1), 59–71.

WU, G., XIAO, X., LI, T. et al. 1993. Lithospheric structure and evolution of the Tibetan Plateau: the Yadong-Golmud geoscience transect. *Tectonophysics*, **219**, 213–221.

XIANG, G. & ZHAO, G. 1991. General account of the earthquake (M6.9) occurring in Gonghe and Xinghai counties of Qinghai Province. *Recent Developments in World Seismology*, **1**, 7–13 (in Chinese).

XIANG, L. 1990. *Basic Manual of Qinghai*. Qinghai People's Press, Xining, China (in Chinese).

YOUNG, R. A. 1984. Geomorphic evolution of the Colorado Plateau margin in west-central Arizona: A tectonic model to distinguish between the causes of rapid, symmetrical scarp retreat and scarp dissection. *In*: MORISAWA, M. & HACK, J. T. (eds) *Tectonic Geomorphology, Proceedings of the 15th Annual Binghamton Geomorphology Symposium, September 1984*. Allen & Unwin, London, 261–278.

ZHANG, Z. M., LIOU, J. G. & COLEMAN, R. G. 1984. An outline of the plate tectonics of China. *Geological Society of America Bulletin*, **95**, 295–312.

ZHAO, W., NELSON, K. D. & PROJECT INDEPTH TEAM. 1993. Deep seismic reflection evidence for continental underthrusting beneath Tibet. *Nature*, **366**, 557–559.

The landscape evolution of Nemegt Uul: a late Cenozoic transpressional uplift in the Gobi Altai, southern Mongolia

LEWIS A. OWEN[1], W. DICKSON CUNNINGHAM[2], BRIAN F. WINDLEY[2], J. BADAMGAROV[3] & D. DORJNAMJAA[3]

[1] *Department of Earth Sciences, University of California, Riverside, California 92521, USA*
[2] *Department of Geology, University of Leicester, Leicester LE1 7RH, UK*
[3] *Geological Institute, Mongolian Academy of Sciences, Peace Avenue, Ulaanbaatar 20351, Mongolia*

Abstract: The geomorphology and structural geology of Nemegt Uul, Southern Mongolia, is examined as an example of a mountain range that has formed within a restraining bend along a major intracontinental strike-slip fault system, the Gobi–Tien Shan fault system. Structural and geomorphological analysis demonstrates that the mountain belt is young and has been differentially tilted and eroded. A geomorphological model is developed showing that uplift and erosion have resulted in the formation of deeply incised mountains, alluvial fans, badlands, desert pavements and dunes.

The Gobi Altai Mountains of southern Mongolia comprise a series of discontinuous ranges that trend approximately east–west into northwest China (Fig. 1). The Gobi Altai rise from an extensive desert surface, which has elevations of between 1000 and 2000 m, to a maximum of 3957 m in Ih Bodg. Cunningham *et al.* (1996, 1997) and Cunningham & Windley (1995) have shown that these mountains are transpressional uplifts that formed during the Cenozoic along a series of strike-slip faults that define a corridor of left-lateral transpressional deformation from southern Mongolia to NW China. The longest fault system within this corridor is the Gobi–Tien Shan fault system which can be traced for over 800 km from the easternmost Tien Shan to the Gurvan Sayhan Ranges of the SE Gobi Altai. Cunningham *et al.* (1996) proposed that this corridor of left-lateral transpressional deformation represents a distant response to the continued northeastward indentation of India into Asia since approximately 50 Ma. Understanding the evolution of the Gobi–Tien Shan fault system, as well as other strike-slip fault systems, is important for understanding the overall Cenozoic deformation field of Central Asia (cf. Baljinnyam *et al.* 1993; Davy & Cobbold 1988; Tapponnier & Molnar 1979). The geomorphology of the Gobi Altai directly reflects the late Cenozoic active tectonics of southern Mongolia and provides important constraints on the geometry, style and timing of uplift in the region.

This paper describes the tectonic and geomorphological characteristics of one of the main mountain ranges, Nemegt Uul, situated in southernmost Mongolia (Fig. 1). In it, a model for the landscape evolution of the region is presented that may be applied to the other transpressional mountain ranges in Mongolia and in similar mountain regions elsewhere in the world.

Methodology

Geomorphological and geological field mapping was undertaken at a variety of scales ranging from 1:5000 to 1:20 000 following established procedures using field survey techniques aided by Global Positioning Systems, barometric altimetry and levelling in selected regions of Nemegt Uul (cf. Owen *et al.* 1997). Mapping was assisted by the use of Landsat MSS images (Fig. 2) and aerial photographs. Transects into the mountain range were made to study the structural geology, geomorphology and sedimentology.

Topographic maps (1:100 000 scale) were used to examine the regional geomorphology of Nemegt Uul. A geographical information system (GIS, using ARC/INFO) was used for digital terrain analysis and to calculate geomorphological indices for part of central Nemegt Uul (Fig. 2B). Using the GIS, hypsometric curves were constructed using the methods of Strahler (1952) for the whole of the Nemegt Uul study area, north and south of the main drainage divide, and for individual catchment areas. An index was then derived to assess the skewness of elevations (catchment area skewness) within individual catchment areas as an aid to assessing their relative geomorphological maturity. In addition, geomorphological indices

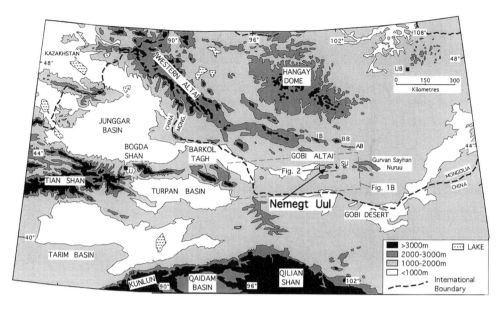

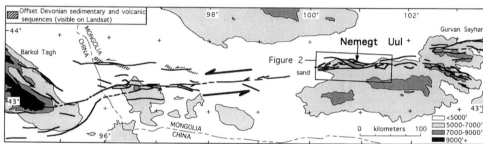

Fig. 1. (**A**) The geographical setting of Nemegt Uul in southern Mongolia within the Gobi Altai Mountains. IB, Ih Bogd; BB, Baga Bogd; AB, Arsta Bogd; SU, Severy Uul; UB, Ulaan Baatar; U, Urumchi. (**B**) Structural setting of Nemegt Uul along the Gobi–Tien Shan fault system (after Cunningham *et al.* 1996).

were used to compare regions throughout Nemegt Uul. These included mountain front sinuosity, using the methods of Bull & McFadden (1977), and a transverse topographic symmetry factor using Cox's (1994) method.

Tectonic setting

Two dominant left-lateral strike-slip fault systems are present in southern Mongolia and northwest China: the North Gobi Altai and the Gobi–Tien Shan fault systems (Cunningham *et al.* 1996). These fault systems occupy the central part of a wide corridor of left-lateral crustal movement between the Hangay Dome to the north and the Qilian Shan on the north side of the Tibetan Plateau. The North Gobi Altai fault system continues for over 300 km from the eastern Gobi Altai to the western Altai mountains. To the south, the Gobi–Tien Shan fault system can be traced for over 800 km and terminates in the east in a series of ranges (Gurvan Sayhan Nuruu) that have the geometry of a horse-tail splay (Fig. 1B), while in the west it passes into the easternmost Tien Shan terminating in the sigmoidal-shaped restraining bends of Barkol Tagh and Bogda Shan. The study area, Nemegt Uul, is near the eastern end of the Gobi–Tien Shan fault system and constitutes a broad sigmoidal-shaped restraining bend (Figs 1B and 2B).

Rock types and structural geology

Unpublished mapping by the Mongolian Geological Survey coupled with field observations

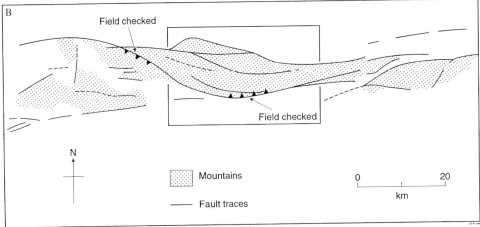

Fig. 2. (A) Landsat image of Nemegt Uul; (B) structural interpretation based on Landsat image and field checking. The area highlighted by the box is covered by the GIS.

from this study indicate that Nemegt Uul is dominantly composed of Devonian–Carboniferous greenschist-grade metasedimentary and metavolcanic rocks and unmetamorphosed volcanics and clastic sedimentary rocks. Subordinate lithologies include small granitic intrusions, serpentinite and ophiolitic rocks, and Mesozoic terrigenous deposits. The range lies along an E–W strike-belt dominated by low-grade metavolcanic, volcaniclastic and sedimentary sequences believed to have been deposited in an arc/backarc environment (Lamb & Badarch 1995).

The structural geology of the range is dominated by an array of approximately E–W trending faults that either cut through the range or help define the mountain fronts (Figs 2 and 3). The sigmoidal curvature of some of the longer faults has resulted in the sigmoidal shape of the range. This strongly suggests that the distribution of uplift is structurally controlled. Two structural transects were completed to gain an understanding of the kinematic nature of the major faults and the range's cross-strike structure (Fig. 3). Section A–B transects the northern half of the range, and is dominated by several large thrusts including a 0.5 km wide thrust zone that defines the northern mountain front and deforms recent alluvial deposits. Another major thrust zone has uplifted ophiolitic rocks in the centre of the range and forms a prominent topographic escarpment. The structural vergence

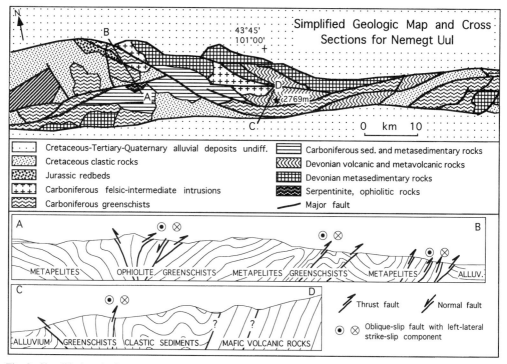

Fig. 3. Simplified geologic map and cross-sections of Nemegt Uul. Map largely taken from unpublished mapping by the Mongolian Geological Survey. Cross-sections completed by authors.

for most of the faults along section A–B is northward, whereas the geometry of large fold structures in compartments between major faults suggests that they are kinematically separate, and that they may be reactivated Palaeozoic structures.

The southern half of the range is dominated by a major S-directed thrust fault that has emplaced Palaeozoic phyllites and slates over undated foreland alluvial deposits (Fig. 3; Cunningham et al. 1996, fig. 12). Cross-section C–D indicates that the frontal ridge is a triangular wedge bounded by a steep left-lateral thrust on its northern side. North of this fault are unmetamorphosed sedimentary and volcanic rocks. Thus, along section C–D, the greatest amount of thrust-related uplift and exhumation of metamorphic basement rocks has occurred at the southern front and not in the centre of the range.

Our observations indicate that Nemegt Uul has bilateral thrust vergence with mountain-bounding thrusts dipping into the centre of the range. Extrapolation of these faults to depth suggests that the range has a complex flower structure geometry. Left-lateral components of motion have occurred on several major faults as indicated by oblique-slip slickenlines and stretching lineations suggesting a transpressional origin for the mountain along the trend of the Gobi–Tien Shan fault system (Cunningham et al. 1996).

Geomorphology

The discontinuous mountain ranges of the Gobi Altai rise from an extensive desert surface, which has elevations of between 1000 and 2000 m, to a maximum of 3957 m on Ih Bogd, 3590 m on Baga Bogd Uul, 2477 m on Artsa Bogd Uul, 2825 m on Gurvan Sayhan Uul, 2632 m on Sevrey Uul and 2769 m on Nemegt Uul (Fig. 1). The climate in the Gobi is of semi-arid, continental type with summer temperatures that exceed 40°C and winter temperatures that frequently drop below −40°C. In the winter, the region is influenced by the Mongolian High Pressure System (MHPS) which drives strong westerly wind systems and produces snow. Vegetation is relatively sparse, being usually restricted to xerophytes and grasses, but lush grassy pastures are present where the water table is very high, or along spring lines, and at higher altitudes within the mountains.

Throughout the Gobi, the desert is dominantly of reg type, comprising alluvial fans and deeply weathered Mesozoic and early Tertiary bedrock with badlands topography. Owen *et al.* (1997) described the characteristics of the alluvial fans and showed their formation as dominated by climatically driven processes. Areas of active sand dunes are restricted in extent but, where present, are of barchan and seif type.

Most streams are ephemeral and have their source in the mountains. These fill during heavy rainstorms and/or as snow melts at higher locations within the mountains during springtime, particularly in alluvial fan areas, where the water table is locally as close as 2 m below the surface. Rare perennial streams are located at spring lines and generally can only be traced for several hundred metres. North of Baga Bogd Uul and Ih Bogd Uul are large lakes which are internally drained and have brackish to saline waters.

Figures 4 and 5 illustrate the main geomorphological characteristics of Nemegt Uul and the adjacent foreland. Figure 5 also shows the main lithofacies associations within each environment. Geomorphologically, Nemegt Uul and its foreland can be broadly divided into four zones: the mountains; the alluvial fan environments; badlands; and dunes and desert pavements (Fig. 5).

The mountains

Nemegt Uul is extremely elongate with a length–width ratio of approximately 1:8. Figure 6 shows the outline of Nemegt Uul with the calculated mountain front sinuosity for selected reaches. The mountain front is extremely straight where it is bounded by faults, particularly on the northern side of the range (cf. Figs 2B and 6). The mountain front sinuosity increases considerably within a number of embayments on the southern side of the range that mark stepover zones between the range-bounding faults.

Figures 7 and 8 illustrate the topography of the GIS study area. A clear asymmetry in the topography is picked out by the comparatively larger catchment areas north of the main east–west drainage divide compared with the smaller catchments to the south (Fig. 4). Furthermore, Fig. 8 shows that the distribution of elevations south of the main east–west drainage divide is more skewed towards the higher elevations compared with areas north of it. Towards the eastern and western ends of the main range, northward-flowing streams cross the whole range with much of their catchment on the southern foreland of the mountain range. This suggests that the northern margin of the mountain range has undergone a longer period of denudation than the southern margin, and it implies that uplift in the southern region is younger, or that the mountain belt as a whole has a northward tilt.

Figure 9 shows the area–altitude distributions and hypsometric curves of three different types of catchments within the study area, and Fig. 10 the catchment area skewness (CAS) for each catchment area in the GIS. Catchment areas with high positive CAS values are flat-iron-shaped catchments that correspond to fault facets, while catchment areas with negative CAS values approaching zero are well drained, with deep box-shaped valleys in their lower reaches.

Figure 11 illustrates the transverse topographic symmetry factor for each of the catchment areas. This shows that the streams are symmetrical within each catchment, and that they trend oblique to perpendicular to the main structural grain.

The high peaks are deeply weathered (LFA6 in Fig. 5) and, although they have steep sides ($>60°$), they are surrounded by extensive regolith and the valley heads open up into broad high pastures (Fig. 12A). Mass movement is mainly by rock fall and creep (Fig. 12B), and talus deposits up to 3 m thick with slopes of between 25° and 35° are common along most of the valleys. These talus deposits are commonly stratified (Fig. 12C) and in limestone-rich areas they are loosely cemented with calcite (LFA5 in Fig. 5); most are deeply incised by gulleys and rills, and the talus slopes are often truncated at their toes.

Low-angled ($<10°$) debris flow fans are present within broad valley stretches where small tributary valleys converge. These fans comprise angular, polymictic matrix-supported pebbles and cobbles that are crudely stratified parallel to the surface of the fan. Most fans are truncated at their toes (Fig. 13A) and incised towards their heads. River terraces are also present showing various phases of construction and incision (Figs 13B and 14).

Small pebbles and sands are dominant within the recent dry stream channels. In some valleys, however, large imbricated clusters of boulders (up to 2 m in diameter) are present representing large flood events (Fig. 13C).

Alluvial fans

Alluvial fans surround the mountain range and radiate from the main valleys to form extensive bajadas (Figs 2A and 4); they stretch for as much

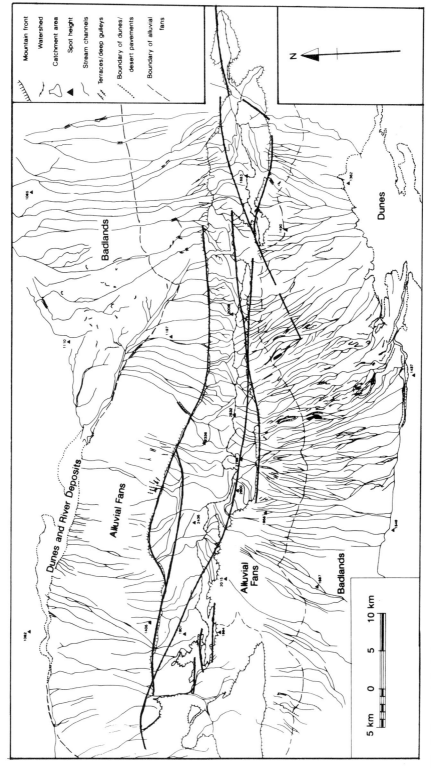

Fig. 4. Geomorphological map of Nemegt Uul and its forelands based on topographic maps and Landsat imagery. The bold dashed lines indicate major fault traces.

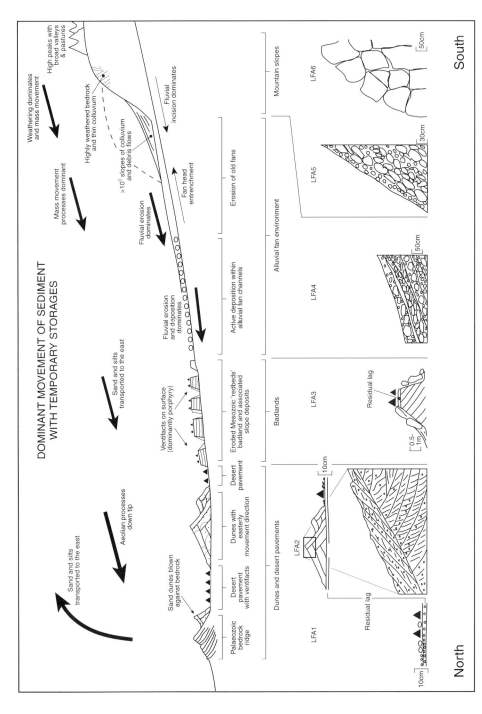

Fig. 5. Schematic section across the northern margin of Nemegt Uul and its pediment showing the main geomorphological characteristics, lithofacies associations (LFA) and the paths of sediment transfer.

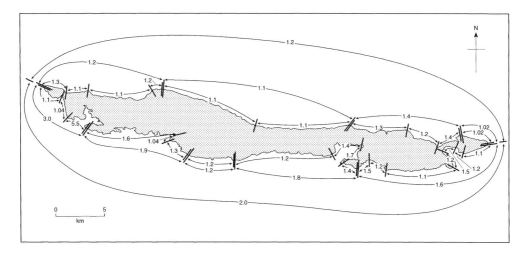

Fig. 6. Map showing the main Nemegt Uul range with the mountain front sinuosity (total length of the mountain front along its foot/straight-line length of mountain front) calculated for selected lengths of the range. The thick black lines mark major fault traces.

as 10 km away from the mountain front where they are replaced by desert pavements. Desert pavements also exist on some of the older alluvial fan surfaces. The alluvial fans are most extensive along the northern side of Nemegt Uul, whereas to the south they are steeper and stretch for only a few kilometres. Most of the major fans are entrenched at their heads to a depth of a few metres, frequently down to bedrock. The channels are several tens to hundreds of metres wide and comprise sand, pebbles and occasional small boulders. In some alluvial fans (e.g. NE corner of Fig. 4), entrenchment is concentrated in the mid-fan stretches. These probably represent areas of localized uplift associated with foreland propagating blind thrusts. The alluvial fan sediments comprise angular to sub-rounded cobbles and pebbles which are clast-supported or supported in a matrix of medium to coarse sand. Many of the clasts are imbricated and the bedding is gradational with crude decimetre-thick units which dip at 1° to 4° sub-parallel to the surface of the alluvial fan (LFA4 in Fig. 5 and Fig. 15). Small debris flows are present in the upper reaches of the alluvial fan, adjacent and near to the mountain front. Debris flows are rare in the mid-fan areas and the surfaces of the fans are depleted of fine sands and silts, and armoured with pebbles and cobbles. Basalt boulders are present on many of the fan surfaces, but have no obvious local source in Nemegt Uul. This suggests that they may be a residual deposit of former extensive basalt flows. In the distal fan areas the surfaces are depleted of fine sand and silt, and pebble-sized ventifacts are common. Calcretes have developed on some alluvial fan surfaces. Some alluvial fan sediments have been deformed by thrust faults (Fig. 16) that diverge away from the mountain range.

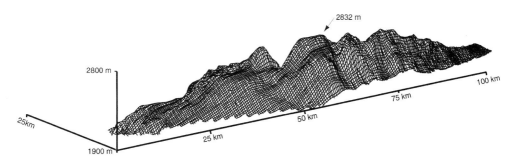

Fig. 7. Digital model of part of Nemegt Uul viewed from the southwest. The total length of the model is 40 km with a relative relief of 1300 m (vertical exaggeration is 10 times the horizontal scale).

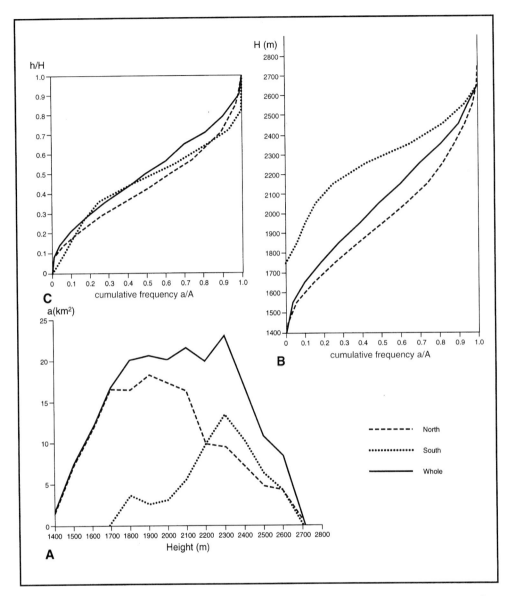

Fig. 8. (**A**) Area–altitude distribution and (**B** and **C**) hypsometric curves for the whole of the Nemegt Uul study area and the hypsometric curves for the areas north and south of the main watershed (A = total area of the catchment; a = area of catchment between 100 m contours; H = maximum elevation of the catchment area − minimum elevation of the catchment area; h = elevation of each 100 m contour).

Badlands

Badlands are best developed along the southern side of Nemegt Uul, where horizontal or gently dipping Cretaceous marlstones and sandstones have been deeply eroded. In some areas, the valleys reach depths of several tens of metres, with valley width exceeding hundreds of metres. Screes are thin (generally <1 m thick) and there is little sediment on the valley floors (LFA3 in Fig. 5).

Dunes and desert pavements

Dunes are present along the lowest elevations between the pediments of Nemegt Uul and the adjacent mountain ranges. The presence of

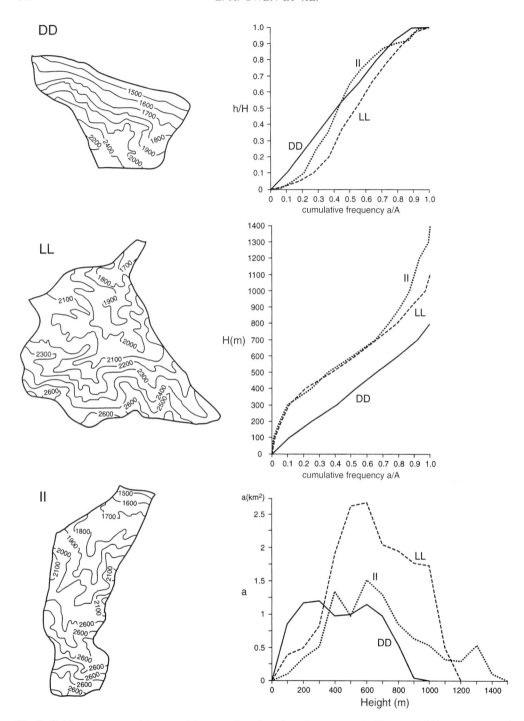

Fig. 9. Catchment areas and hypsometric curves for selected catchment areas in Nemegt Uul (A = total area of the catchment; a = area of catchment between 100 m contours; H = maximum elevation of the catchment area − minimum elevation of the catchment area; and h = elevation of each 100 m contour).

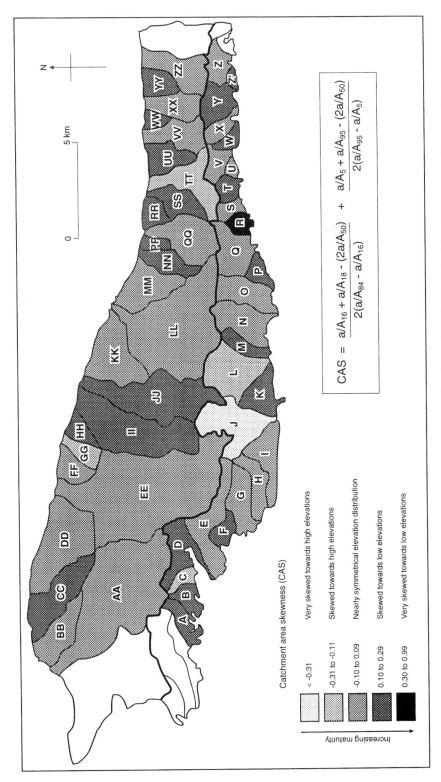

Fig. 10. Map showing the catchment area skewness (CAS) for catchment areas within Nemegt Uul. CAS is derived from the hypsometric curves of each catchment where a/A_x is altitude of the x percentile. The greater the skewness towards lower elevations, the greater the denudation of the catchment. (NB: CAS was not calculated for the unshaded catchment areas).

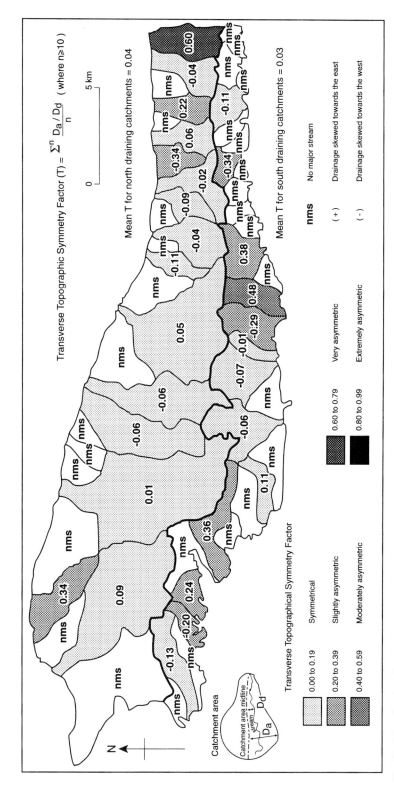

Fig. 11. Map showing the transverse topographic symmetry factor (T) for each of the catchment areas within Nemegt Uul.

Fig. 12. (**A**) View looking south down one of the main tributary valleys from the top of Nemegt Uul (43°39.2′N 100°53.8′E, 2450 m). Note the craggy peaks surmounted by talus slopes. (**B**) Talus slope from near the top of Nemegt Uul. Note the lobate forms on the surface that indicate creep (43°39.2′N 100°53.7′E, 2450 m). (**C**) Stratified scree deposit from within one of the main valleys in Nemegt Uul (43°43.3′N 100°38.7′E, 1690 m).

both barchan and seif dunes indicates that westerly winds dominate. In some areas the dunes are partially vegetated and have become temporarily stabilized, but the majority are active and over-ride vegetation (LFA2 in Fig. 5).

Desert pavements are common on the distal fan surfaces, pediments and in the badlands. The pavements comprise armoured surfaces with abundant pebble- to boulder-sized ventifacts, but little silt-sized sediment (LFA1 in

Fig. 13. (**A**) Dissected debris fan within one of the main valleys in Nemegt Uul (43°43.4'N 100°38.7'E, 1690 m). (**B**) High river terraces and dissected fans along one of the main valleys in Nemegt Uul (see Fig. 14 for detailed map) (43°42.9'N 100°39.1'E, 1690 m). (**C**) Flood deposit on one of the main valleys of Nemegt Uul (43°42.9'N 100°39.2'E).

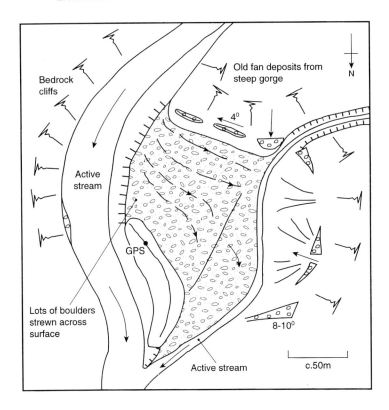

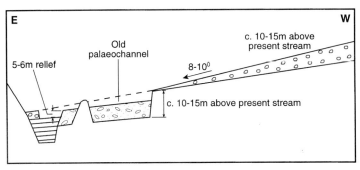

Fig. 14. Map and section showing river terraces and dissected fans in one of the main valleys of Nemegt Uul (GPS position: 43°42.9′N 100°39.1′E, 1670 m).

Fig. 5). Some of the pavements have poorly developed calcretes.

Discussion

Structural analysis shows that Nemegt Uul is an asymmetric flower structure that formed within a broad restraining bend. The straight stretches of the mountain front clearly support the assumption that the mountain range is very young and still active. Faulted Quaternary and Neogene alluvial fan sediments on the north and south sides indicate Quaternary faulting along the mountain belt. The straightness of the mountain front is attributed to thrust and oblique-slip faulting that has placed Palaeozoic rocks over Mesozoic and Cenozoic rocks. Mountain front sinuosity is higher, and embayments are present where there are stepover zones between the main faults. Deformation within the foreland is supported not only by faulted alluvial fan sediments, but by subtle changes in fan gradient picked out by entrenchment within the mid-fan

Fig. 15. View of typical alluvial fan deposits, showing centimetre- to decimetre-thick crude low-angled stratification (43°44.7'N 100°38.5'E, 1440 m).

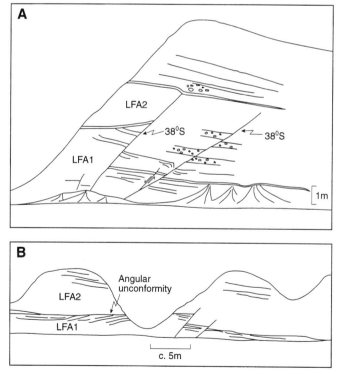

Fig. 16. Schematic section looking west at faulted alluvial fan deposits on the north side of Nemegt Uul (43°44.7'N 100°38.5'E, 1440 m): (**A**) details of the faulting; (**B**) the overall section. LFA1 – Brown coloured, poorly stratified, centimetre-thick beds of matrix-supported (sandy silt) sub-angular pebbles. Some of the clasts are imbricated. Occasional centimetre-thick lenses of silty sand are present. LFA2 – Grey coloured, crudely stratified, matrix-supported (sandy silt) polymictic medium to large pebbles (> 3 cm diameter) with occasional cobbles and small boulders.

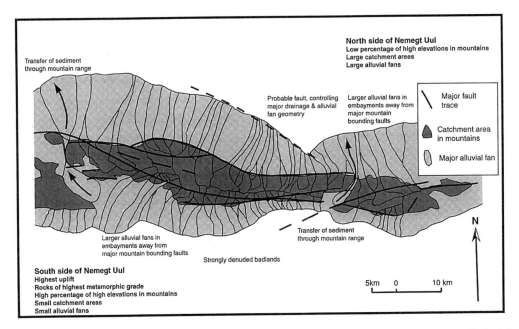

Fig. 17. Map showing the zones of sediment transfer and the relationship between catchment areas and alluvial fans with respect to the differential uplift and denudation across Nemegt Uul.

reaches and small terraces in the distal reaches of the foreland. In addition, a NW trending drainage channel north of Nemegt Uul (Figs 4 and 17), oblique to the northward regional slope, is probably controlled by faulting that parallels the northern edge of Nemegt Uul.

The geomorphological indices and digital terrain analysis support the view that the northern margin of the mountain range has been more deeply eroded and possibly reflects differential tilting of the mountain range. This is supported by the highest elevations concentrated in the southern half of the range. In addition, the highest grade metamorphic rocks along the cross-section crop out at the southern end (Fig. 3) suggesting greater uplift and exhumation in the south in this region. Consequently, the alluvial fans are larger on the northern side of the range, whereas alluvial fan development on the southern side is tightly constrained to areas near the mountain front and erosion in the distal reaches of the foreland is greater than the sediment supply from the mountains. The heterogeneity of rock types in the range (Fig. 3) precludes the possibility that the topographic asymmetry between north and south sides is due to differences in lithology-controlled resistance to erosion. Moreover, Owen et al. (1997) showed that the alluvial fans developed as a consequence of the tectonic setting, being adjacent to rapidly uplifting mountains, and that the main sedimentological and geomorphological characteristics of the alluvial fans were controlled by climate change and the associated changing hydrological conditions.

There is a clear geographical control of sedimentary facies associated with Nemegt Uul; this is shown in Fig. 5. Sediment produced by weathering and mass movement processes within the mountains is quickly transported to the foreland to form alluvial fans and there is little sediment storage within the valleys. The alluvial fans are themselves eroded by stream incision and their surfaces are deflated of fine sediment. Where alluvial fan sediments are very thin or absent, the Mesozoic bedrock has been deeply eroded by fluvial and aeolian processes. Eroded alluvial fan and pediment sediments are carried to the distal parts of the foreland where they form sand dunes. The dunes migrate eastward and the silt fraction is progressively deflated by aeolian processes.

Figure 17 shows the spatial relationships between sizes of catchment areas and alluvial fans on both the northern and southern sides of Nemegt Uul in relation to the major faults. The rocks with the highest grade metamorphism are present along the southern side of Nemegt Uul as a consequence of greater uplift. This is reflected in the higher percentage of land area at high altitudes, the smaller catchment areas and alluvial fans, and the more deeply eroded

badlands on the southern side of the mountain range as compared with the northern side. The largest alluvial fans on both sides of the range, however, coincide with embayments away from major mountain bounding faults and towards the eastern and western end of the range in sediment transfer zones.

Conclusion

Nemegt Uul is an example of a desert mountain range that has formed within a restraining bend along a major left-lateral strike-slip fault system. Topography, valley development, drainage patterns, alluvial fans and badlands development are controlled by the rapid uplift, denudation and sediment transfer. Denudation by deep weathering, mass movement processes and the rapid transfer of sediment out of the mountain have produced deep valleys perpendicular or very oblique to the main structural grain of the mountain range. The greater catchment area and alluvial fan development on the northern side of the range, and the geomorphological indices and altitudinal distribution suggest that the mountain range has been differentially tilted to the north, possibly due to larger magnitudes of thrust displacement and greater uplift along the southern margin. Figure 17 shows the spatial relationships between the size of catchment areas and alluvial fans in relation to the major faults on both sides of Nemegt Uul. There is a clear relationship between the high percentage of land at high elevations, the smaller catchment areas and alluvial fans, and the more deeply eroded badlands on the southern side of the mountain range with the higher uplift that has exposed higher grade metamorphic rocks as compared with the higher percentage of lower elevations, larger catchment areas and larger alluvial fans on the northern side of Nemegt Uul. The largest alluvial fans on both sides of the range, however, coincide with embayments away from major mountain bounding faults and occur towards the eastern and western end of the range in stepover zones where there is transport of sediment from the southern foreland through the mountain range to the northern foreland.

This research was undertaken as part of NERC grant GR9/01881 awarded to B.F.W. and L.A.O. Thanks to Justin Jacyno for drafting the diagrams, British Petroleum for supplying the Landsat MSS images and two anonymous referees for their careful editting and constructive comments.

References

BALJINNYAM et al. 1993. Ruptures of major earthquakes and active deformation in Mongolia and its surroundings. *Geological Society of America Memoir*, **181**.

BULL, W. B. & MCFADDEN, L. D. 1977. Tectonic geomorphology north and south of the Garlock fault, California. *In*: DOEHRING, D. O. (ed.) *Geomorphology in Arid Regions. Proceedings of the Eighth Annual Geomorphology Symposium*. State University of New York at Binghamton, Binghamton, NY, 115–138.

COX, R. T. 1994. Analysis of drainage basin symmetry as a rapid technique to identify areas of possible Quaternary tilt-block tectonics: an example from the Mississippi Embayment. *Geological Society of America Bulletin*, **106**, 571–581.

CUNNINGHAM, W. D. & WINDLEY, B. F. 1995. Cenozoic transpression in southwestern Mongolia and the Gobi–Altai–Tien Shan connection. *Terra Nova*, Suppl. **1**, 7, 40.

——, ——, DORJNAMJAA, D, BADAMGAROV, G. & SAANDER, M. 1996. Late Cenozoic transpression in southwestern Mongolia and the Gobi Altai–Tien Shan connection. *Earth and Planetary Sciences*, **140**(1–4), 67–82.

——, ——, OWEN, L. A., BARRY, T., DORJNAMJAA, D. & BADAMGAROV, G. 1997. Geometry and style of partitioned deformation within a Late Cenozoic transpressional zone in the eastern Gobi Altai Mountains, Mongolia. *Tectonophysics*, **277**(4), 285–306.

DAVY, P. & COBBOLD, P. 1988. Indentation tectonics in nature and experiment. 1, Experiments scaled for gravity. *Bulletin of Geological Institute, Uppsala*, **14**, 129–141.

LAMB, M. A. & BADARCH, G. 1995. The arc mess monster: sedimentation, volcanism and tectonics of southern Mongolia during the Palaeozoic. *GSA Abstracts with Programs*, A-1391.

OWEN, L. A., WINDLEY, B. F., CUNNINGHAM, W. D., BADAMGAROV, J. & DORJNAMJAA, D. 1997. Quaternary alluvial fans in the Gobi of southern Mongolia: evidence for neotectonics and climate change. *Journal of Quaternary Science*, **12**(3), 239–252.

STRAHLER, A. N. 1952. Hypsometric (area–altitude) analysis of erosional topography. *Geological Society of America Bulletin*, **63**, 1117–1142.

TAPPONNIER, P. & MOLNAR, P. 1979. Active faulting and Cenozoic tectonics of the Tien Shan, Mongolia, and Baikal regions. *Journal of Geophysical Research*, **84**, 3425–3459.

Temporal and spatial variation in suspended sediment yields from eastern North America

CATHERINE T. CONRAD & HOUSTON C. SAUNDERSON

Department of Geography and Environmental Studies, Wilfrid Laurier University, Waterloo, Ontario, Canada N2L 3C5

Abstract: Sediment transport data from 189 gauging stations on rivers in eastern USA and eastern Canada were used to estimate sediment yields. The length of record ranged from one to 42 years. Most stations had marked fluctuations in tonnage from one year to the next. Yields ranged spatially from $1.5 \, t \, km^{-2} \, a^{-1}$ to $3157 \, t \, km^{-2} \, a^{-1}$ when the full length of record was used. Multiquadric surfaces were generated to show time-averaged yields using five-year, ten-year, and total intervals of record. The patterns of suspended sediment yield changed when different lengths of record were used, suggesting caution when generating results from a limited record of suspended sediment data.

Suspended sediment data have been used to assess regional and global patterns of erosion (Walling & Webb 1996a). Researchers have also estimated denudation using suspended sediment yield as their indicator of surface lowering (Dole & Stabler 1909; Fournier 1960; Schumm 1963; Judson & Ritter 1964; Ahnert 1970; Yoshikawa 1985; Bartolini et al. 1996). The reliability of studies which use sediment yield data is critical, since this information can ultimately form a basis for theories of landscape evolution, the effects of human activities on the land surface and perturbations of the global geochemical cycle. Walling & Webb (1996b) report that important uncertainties regarding the reliability of suspended sediment data exist because of differing periods of record, non-stationary river behaviour, and data reliability. This paper reports on the implications of using different record lengths when generating maps of variations in suspended sediment yield. Although long-term records are probably required to establish any trends in sediment yields (Walling 1997), the length of sampling interval required to identify the presence or absence of trends is still an open question. Short intervals (years) produced different patterns of yield from longer intervals (decades) from the data set covering eastern North America.

Geographical setting

The study area extends from the southern border of the state of Georgia, USA, to the northern tip of the province of Newfoundland, Canada, and from the Atlantic coast in the east to the Appalachian Plateau in the west. The Appalachian Mountains form the dominant topographic feature of the region, lying between the stable interior of North America, comprising flat-lying Palaeozoic rocks, and the Coastal Plain, where Mesozoic (Cretaceous) and Cenozoic sediments partially or completely bury older rocks (Williams 1995). The southern segment of the Appalachian Mountains can be divided into four geological provinces. The Appalachian Plateau in the west consists predominantly of Palaeozoic sedimentary rocks. The Ridge and Valley province and the Blue Ridge province, to the east of the Appalachian Plateau, are predominantly igneous Palaeozoic rocks, and the Piedmont province is made up of metamorphosed Palaeozoic and Precambrian rocks. The northern segment of the Appalachians cannot be readily divided into sub-divisions (Fig. 1).

The Pleistocene continental ice sheet covered the entire northern segment of the Appalachians, and extended approximately as far as latitude $38°N$. Beyond the glacial limit, the cover of glacial drift is replaced by material weathered from the underlying bedrock, talus, colluvium and alluvium (Rodgers 1970).

Description of data sets

Daily suspended sediment load data (Fig. 2) were obtained from the United States Geological Survey (USGS WATSTORE database) and from Environment Canada. Intervals of record ranged from one to 42 years. Table 1 gives a breakdown of the number of sites and years of record (by state and province). Most stations have less than ten years of record, but some have more than 20 years. Some states have many more years of record than others (e.g. Pennsylvania vs. Maine).

From: SMITH, B. J., WHALLEY, W. B. & WARKE, P. A. (eds) 1999. *Uplift, Erosion and Stability: Perspectives on Long-term Landscape Development*. Geological Society, London, Special Publications, **162**, 219–228. 1-86239-030-4/99/ $15.00 © The Geological Society of London 1999.

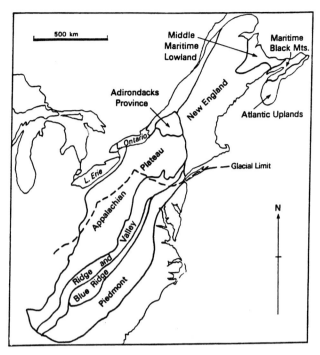

Fig. 1. The Appalachian geological provinces (after Bridges 1990).

Multiquadric spatial interpolation

Maps depicting spatial variation of suspended sediment yield were generated using the multiquadric (MQ) method of interpolation. Hardy's (1971) multiquadric method can be used to generate surfaces where $z = f(x, y)$, z being the scalar dependent variable and x, y the locational coordinates of z. The MQ method is an exact method because the surface goes through the initial sample points, as opposed to statistical methods that involve smoothing of the original point data (Stone and Saunderson 1996). Software (Saunderson 1994), used initially for other fluvial applications (Saunderson 1992; Saunderson & Brooks 1994), was implemented to map the pattern of sediment yields derived from the gauge data.

Mapping the regional pattern of time-averaged yields (t km^{-2} a^{-1}) required, as a first step, solution of the following multiquadric equation:

$$\sum_{j=1}^{n} c_j [(x_i - x_j)^2 + (y_i - y_j)^2 + R^2]^{0.5} = z_i \quad (1)$$

where x_{ij} and y_{ij} are the Cartesian coordinates of gauging stations, R is a shape parameter affecting the smoothness of the surface and z_i are the time-averaged sediment yields estimated from the gauging data. For $R = 0$, the basis function becomes that of a right circular cone, the model used here. The unknown coefficients, c_j, were obtained from singular value decomposition of the system of linear simultaneous equations. Knowledge of the column vector c_j then made it possible, as a second step, to interpolate to any number of new locations with coordinates x_p and y_p and sediment yields z_p using the following equation:

$$\sum_{j=1}^{n} c_j [(x_p - x_j)^2 + (y_p - y_j)^2 + R^2]^{0.5} = z_p \quad (2)$$

Results and discussion

Data variability

There are large ranges and variability in suspended sediment yields within and between data sets from the rivers of eastern North America. The graph of suspended sediment (Fig. 3) for the Potomac River in Maryland, USA, shows an example of variability in the suspended load over time. Had data collection at this station been concluded prior to the dramatic event indicated by the peak on the graph (November 1985), the yield calculated at this station would have been very different.

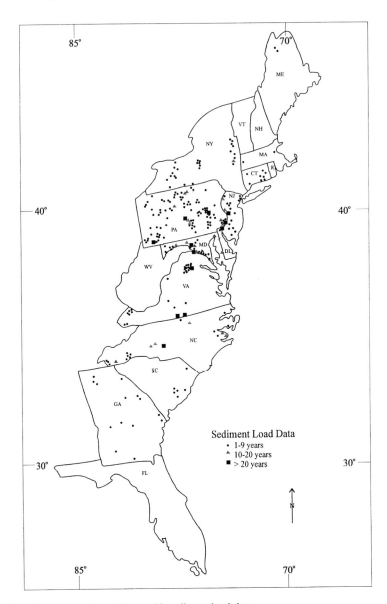

Fig. 2. Location of stream gauging stations with sediment load data.

Table 2 shows the sediment yield for each state and province and the maximum and minimum yields within that state/province. Sediment yields for the 14 states/provinces range from 12.4 to 372.9 t km^{-2} a^{-1}. Maximum yields (from the 189 individual sites) ranged from 18.4 to 3156.7 t km^{-2} a^{-1}. Minimum yields (at a station) ranged from 1.5 to 66.7 t km^{-2} a^{-1}. There is a marked difference (Table 2) between the mean sediment yield for the eastern states (96.7 t km^{-2} a^{-1}) and the mean sediment yield for the eastern provinces (16.8 t km^{-2} a^{-1}).

These values indicate that the sediment yield of the eastern USA and Canada may be much greater than previously reported (Table 3). There are a number of reasons which can explain the larger estimates determined by the present authors. The first difference is one of research area. All of the authors (who are referenced in Table 3) derived their sediment yield estimates on a global area, whereas in the present paper, sediment yield was determined over a smaller, regional area. Sediment yields were calculated using the standard estimate at the sample

Table 1. *Available suspended sediment data for eastern North America*

State/Province	Number of sites	Years of data			Max. period of record
		<10	10–20	>20	
Connecticut	7	7	0	0	1981–1990
Delaware	2	1	0	1	1948–1980
Georgia	13	13	0	0	1959–1963
Maine	2	2	0	0	1966–1972
Maryland	13	6	5	2	1959–1993
Massachusetts	2	2	0	0	1966–1972
New Jersey	15	12	2	1	1949–1982
New York	20	20	0	0	1954–1979
North Carolina	10	7	2	1	1951–1993
Pennsylvania	60	54	5	1	1951–1993
South Carolina	4	4	0	0	1966–1972
Virginia	19	16	0	3	1951–1993
New Brunswick	9	6	2	1	1966–1988
Newfoundland	5	5	0	0	1979–1990
Nova Scotia	8	4	4	0	1967–1988
Total	189	159	20	10	

Data used in this study can be broken down into number of years of record. Although most sites have less than ten years of continuous, daily data, there are some which have a significantly long period of continuous record (1951–1993)

stations. What differs in the present approach is the method of spatial interpolation used to map the regional estimates. This means that overgeneralization and smoothing of the data have been a problem in previous studies.

A second difference is one of data quantity. Sediment transport data from 189 gauging stations on rivers across the eastern United States and Canada were used to estimate sediment yields for this paper. At the global scale, however, researchers (such as those in Table 3) generally use only a few sites in each region. Milliman & Meade (1983), for example, used only two data points when determining the sediment yield for the same portion of the globe as the present study. Sites on major rivers of the world are generally used, with smaller rivers and tributaries excluded. The result has been a

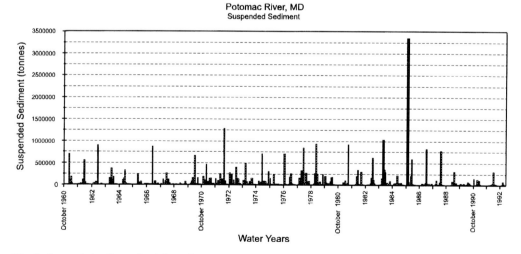

Fig. 3. Suspended sediment load from the Potomac River at Point of Rocks, Maryland, demonstrating the temporal variations between October 1960 and February 1993. The large peak occurred on 6 and 7 November 1985.

Table 2. *Sediment yield for the individual eastern states and provinces*

State/Province	Sediment yield (t km^{-2} a^{-1})	Max. yield (t km^{-2} a^{-1})	Min. yield (t km^{-2} a^{-1})
Connecticut	20	30.6	9.3
Delaware	119.9	173	66.7
Georgia	62.3	152.9	1.5
Maryland	62.9	155.4	5.7
Massachusetts	13.8	18.4	9.2
New Jersey	18.8	58.3	2.2
New York	108.5	425.2	4.6
North Carolina	157.3	1017	17.5
Pennsylvania	114.4	1436.2	12
South Carolina	12.6	21.8	3.4
Virginia	372.9	3156.7	3.2
New Brunswick	25.2	69.4	2
Newfoundland	12.9	30.7	3.5
Nova Scotia	12.4	34.8	5.6
Eastern States:	96.7	3156.7	1.5
Atlantic Canada:	16.8	69.4	2

Overall sediment yield, maximum and minimum yield averaged for the entire USA and Canadian sections of the study area are given at the bottom of the table for comparison

regional, and probably world-wide underestimation of the world's sediment yields. Obviously, the more data sites, the closer the values will be to the true sediment yields of any region. Small rivers and tributaries (including some from eastern North America), with smaller floodplain storage than major rivers, can have larger sediment yields. In fact, a number of small rivers may contribute as much or more sediment than large ones. Stave Run near Reston, Virginia, with a basin area of only 0.2 km^2, has a sediment yield of 3157 t km^{-2} a^{-1}, whereas the Hudson River in New York State, with a basin area of 12 000 km^2, has a yield of only 20 t km^{-2} a^{-1}.

The final difference is temporal. Not only are more data points over a smaller area included in this study, but a longer period of time is also covered. A problem with estimates of global suspended sediment yield can therefore be seen when comparing the results to a detailed regional analysis. Differences in values of these magnitudes remain a problem when researchers are using different rivers spanning different time periods at different scales.

Multiquadric maps

The number of years used to determine sediment yield affects the yield results, implying that maps generated from different time intervals may show changes in the pattern of yield. In an effort to demonstrate how patterns change and shift, maps of yields averaged over a five-year interval, ten-year interval, and total interval of record, were produced. The pattern of suspended sediment yields based on five years of record (Fig. 4) shows largest yields associated with tributaries of the Potomac ($c.$ 400 t km^{-2} a^{-1}) and in the western corner of Virginia, associated with Ohio River tributaries ($c.$ 580 t km^{-2} a^{-1}). The area around Washington, DC, also has high five-year sediment yield values.

Figure 5 shows the results of mapping suspended sediment yields which have been averaged over a ten-year interval of record. This map depicts a smaller region since there are now fewer stations which have a continuous ten-year record. The areas around Washington, DC and Reston, Virginia, still emerge as high yield regions on the ten-year record, with largest yields of up to $c.$ 250 t km^{-2} a^{-1}.

Figure 6 is based on 79 of the total 189 stations (every second record in the entire data set).

Table 3. *Estimates of suspended sediment yield in eastern North America, according to various authors*

Reference	Suspended sediment yield (t km^{-2} a^{-1})
Fournier (1960)	10–60
Strakhov (1967)	10–50
Milliman & Meade (1983)	17
Dedkov & Mozzherin (1984)	50–100
Walling & Webb (1987)	10–500
Lvovitch et al. (1991)	20–200
This study	0.2–3157

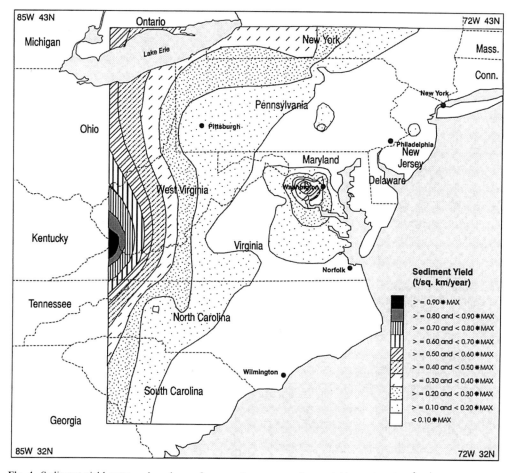

Fig. 4. Sediment yield patterns based on a five-year time-average (max. yield = $c.\,580\,\text{t}\,\text{km}^{-2}\,\text{a}^{-1}$).

Lengths of record ranged from one to 42 years. Yields are now as high as $c.\,900\,\text{t}\,\text{km}^{-2}\,\text{a}^{-1}$; such values are found in Pennsylvania, where tributaries to the Susquehanna yield large amounts of sediment. There is a marked difference between Fig. 4, the map of five-year, time-averaged yields, and Fig. 6, the map of entire length of record. For example, there are large yields in North Carolina, whereas the former high yield centred around Washington is no longer evident. These results indicate that the spatial patterns of suspended sediment variability may change depending on the number of years of record used.

Interpretation of results

To account for global or regional variations in sediment yield, the complexity of the controls involved must be considered. Any explanation of generalized pattern must, for example, take account of the influence of rock type, relief, tectonic stability, land use and human activity as well as that of climate (Walling & Webb 1987). Although not the express purpose of the present research, some preliminary explanations may be presented for the spatial variation in suspended sediment yields.

In an investigation of the possible relationship between discharge and suspended sediment load, plots were compared for those stations with at least five years of continuous data. Results from this study show that the peak discharge and peak sediment load correlate in approximately half of the sites (out of 48 stations, maximum peak discharge and maximum peak sediment load matched at 21 of the sites). This indicates that although meteorological conditions explain high discharges of water and

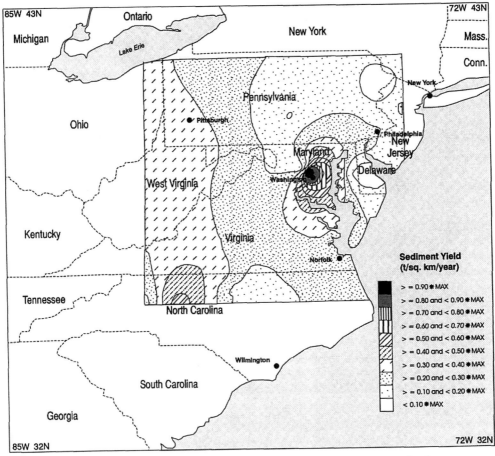

Fig. 5. Sediment yield patterns based on a ten-year time-average (max. yield = $c.\ 250\ \text{t km}^{-2}\ \text{a}^{-1}$).

sediment in approximately half of the cases, some alternative factors must account for the other high peaks.

The influence of rock type is evident in the difference between yields in the eastern USA and eastern Canada (Table 2). The lower yields in the Canadian provinces are likely to be a result of the exposure of less erodible Palaeozoic granites and metamorphic rocks, especially in the rivers of Newfoundland, where channels traverse bedrock. Many of the channels in the eastern regions of the United States, on the other hand, traverse thick alluvium and coastal deposits. Generally, lower sediment yields are typical of rivers in areas that were stripped of sediment during the last glaciation. The influence of human activity in the high yield area surrounding Washington has been noted by Costa (1975, p. 1285), who suggested that construction activity, which was rapidly increasing in the 1960s and 1970s, '...may help to explain some of the amazingly large sediment loads...'. Expanding urban land uses in this heavily populated region, as well as the availability of soft, erodible alluvium are probably factors contributing to these high yields.

Perhaps a critical minimum number of years of data collection is necessary in order to account for temporal fluctuations in suspended sediment load. Eight of 14 stations with a minimum record of 15 years (Table 4) have higher yields for the longest interval available. Six of 14 stations have equal or lower values. These differences over time may be a result of land use changes, including reservoir development. Further analysis of station data is on-going to address the statistical variation in the time-series and to deduce whether these variations relate to land use change.

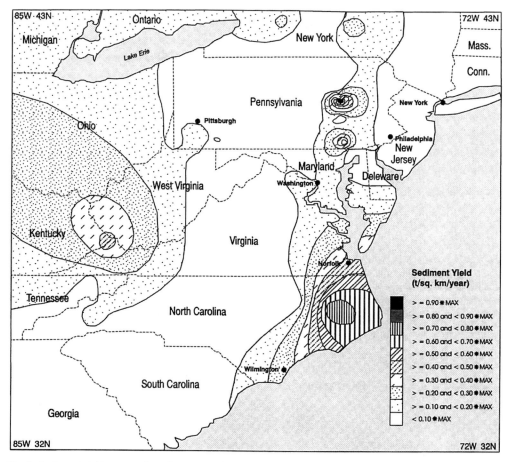

Fig. 6. Sediment yield patterns based on a sample of 79 records (max. 42-year record) (max. yield $= c.\,900\,\mathrm{t\,km^{-2}\,a^{-1}}$).

Conclusion

Sediment yield data for the world's rivers can provide a valuable means of studying the global denudation system, but a number of questions still need to be addressed.

1. The eastern region of North America has been modified significantly by humans, making projection from post-colonial to geological time uncertain.
2. The present study has considered only *suspended* sediment and needs to be compared with estimates derived from bed and solute loads.
3. This study has included both small and large river channels. Both should be included in global-scale studies of denudation in order to include the total contribution of fluvial sources of sediment. It is often the case that large rivers of the world have the longest records. Using only large rivers to generate maps may lead to underestimation of yields. Maps including several sites, many with only a few years of record (Fig. 6) may, in fact, be more indicative of the regional-scale changes.
4. Results from this study reaffirm Fournier's (1960) caution regarding extrapolation from modern sediment yields to longer-term landscape development. The relationship between length of record and spatial pattern remains a question. This study points to the need for a more thorough investigation of the non-stationarity and non-linearity of observations in the time-series containing sediment data. Migrating maxima in yields may well be the norm rather than the exception for long intervals of record, and even more so for geological time intervals.

Table 4. *Values for sediment yield, calculated over five, ten and 15 years, and total record yield*

Station	Length of record (years)	Sediment yield ($t\,km^{-2}\,a^{-1}$) calculated over			
		5 years	10 years	15 years	Entire record
Annapolis R., NS	18	7.41	8.92	8.93	9.19
Kennebecasis R., NB	21	26.65	30.09	33.42	30.73
Peticodiac R., NB	16	10.18	13.01	14.29	14.25
Brandywine C., DE	32	63.9	54.5	62.6	66.7
Potomac R., MD	32	33.7	30.4	39.6	43.7
Delaware R., NJ	32	46.9	48.8	42.7	37.6
Little R., NC	15	36.3	27	37.4	37.4
Yadkin R., NC	42	116.7	149.8	139.6	142.7
Brandywine C., PA	15	39.8	59.3	63.4	63.4
Juniata R., PA	40	36.6	31.7	27.2	26.3
Bixler R., PA	16	28.8	24.8	22.4	23.3
Rappahannock R., VA	42	46.4	50.1	46.5	55.8
Roanoke R., VA	26	62.2	58.6	47.2	52.5
Dan R., VA	26	91.3	104.6	85.7	87

Sediment yield was calculated for several time-averages for those sites with at least 15 years of record in an effort to demonstrate the variability of the results which are derived depending on the number of years which are used

The authors would like to thank Hydrosphere Data Products, Inc. for providing the compiled USGS data at a reduced cost, and WLU Cold Regions Research Centre for purchasing the Hydat cd-rom from Greenland Engineering Group (Environment Canada data). Thanks to Pam Schaus for producing Figures 4, 5 and 6 from our original colour slides. Thanks also to Dr Friesen of WLU for translating the data published in Table 3 from Dedkov & Mozzherin.

References

AHNERT, F. 1970. Functional relationships between denudation, relief, and uplift in large mid-latitude drainage basins. *American Journal of Science*, **268**, 243–263.

BARTOLINI, C., CAPUTO, R. & PIERI, M. 1996. Pliocene–Quaternary sedimentation in the Northern Apennine Foredeep and related denudation. *Geological Magazine*, **133**(3), 255–273.

BRIDGES, E. M. 1990. *World Geomorphology*. Cambridge University, Cambridge.

COSTA, J. 1975. Effects of agriculture on erosion and sedimentation in the Piedmont Province, Maryland. *Geological Society of America Bulletin*, **86**, 1281–1286.

DEDKOV, A. P. & MOZZHERIN, V. T. 1984. *Eroziya I Stock Nanosov na Zemle*. Izdatelstvo Kazanskogo Universiteta.

DOLE, R. & STABLER, H. 1909. Denudation. *United States Geological Survey Water-Supply Paper* **234**, 78–93.

FOURNIER, F. 1960. *Climat et Erosion*. Presses Universitaires de France, Paris.

HARDY, R. L. 1971. Multiquadric equations of topography and other irregular surfaces. *Journal of Geophysical Research*, **76**(8), 1905–1915.

JUDSON, S. & RITTER, D. F. 1964. Rates of regional denudation in the United States. *Journal of Geophysical Research*, **69**(16), 3395–3401.

LVOVICH, M. I., KARASIK, G. YA., BRATSEVA, N. L., MEDVEDEVA, G. P. & MALESHKO, A.V. 1991. *Contemporary Intensity of the World Land Intracontinental Erosion*. USSR Academy of Sciences, Moscow.

MILLIMAN, J. D. & MEADE, R. H. 1983. World-wide delivery of river sediment to the oceans. *Journal of Geology*, **91**, 1–21.

RODGERS, J. 1970. *The Tectonics of the Appalachians*. Wiley-Interscience, New York.

SAUNDERSON, H. 1992. Multiquadric interpolation of fluid speeds in a natural river channel. *In:* KANSA, E. (ed.) *Advances in the Theory and Applications of Radial Basis Functions. Computers & Mathematics with Applications*, **24**(12), 187–193.

—— 1994. Multiquadric surfaces in C. *Computers & Geosciences*, **20**(7/8), 1103–1122.

—— & BROOKS, G. 1994. Multiquadric sections from a fluid vector field. *Journal of Applied Science & Computations*, **1**(2), 208–226.

SCHUMM, S. 1963. The disparity between present rates of denudation and orogeny. *United States Geological Survey Professional Paper* **454-H**.

STONE, M. & SAUNDERSON, H. 1996. Regional patterns of sediment yield in the Laurentian Great Lakes basin. *International Association of Hydrological Sciences Publication* No. **236**, 125–131.

STRAKHOV, N. M. 1967. *Principles of Lithogenesis*. Oliver & Boyd, Edinburgh.

WALLING, D. E. 1997. The response of sediment yields to environmental change. *International Association of Hydrological Sciences Publication* No. **245**, 77–89.

WALLING, D. E. & WEBB, B. W. 1987. Material transport by the world's rivers: evolving perspectives. *International Association of Hydrological Sciences Publication* No. **164**, 313–329.

—— & —— (eds) 1996a. *Erosion and Sediment Yield: Global and Regional Perspectives. International Association of Hydrological Sciences Publication* No. **236**.

—— & —— 1996b. Erosion and sediment yield: a global overview. *International Association of Hydrological Sciences Publication* No. **236**, 3–19.

WILLIAMS, H. 1995. Introduction. *In*: WILLIAMS, H. (ed.) *Geology of the Appalachian–Caledonian Orogen in Canada and Greenland*. Geological Survey of Canada, Ottawa, 4–19.

YOSHIKAWA, T. 1985. Landform development by tectonics and denudation. *In*: PITTY, A. (ed.) *Geomorphology: Themes and Trends*. Barnes & Noble, Maryland, 194–210.

Palaeolandsurfaces and neotectonic analysis in the southern Sierras Pampeanas, Argentina

CARLOS H. COSTA[1], ALDO D. GIACCARDI[1] & EMILIO F. GONZÁLEZ DÍAZ[2]

[1]*Departamento de Geología, Universidad Nacional de San Luis, Chacabuco 917, 5700 San Luis, Argentina*
[2]*Departamento de Ciencias Geológicas, Universidad de Buenos Aires, Pabellón II, Ciudad Universitaria, 1428 Buenos Aires, Argentina*

Abstract: The Sierras Pampeanas of northwestern Argentina are ranges underlain by crystalline basement rocks bounded on their western margins by N–S trending reverse faults along which uplift took place during the Andean orogeny. A pre-uplift regional erosion surface is commonly preserved on the eastern slopes which have been tilted in most cases towards the east. No stratigraphical units of Neogene age have been preserved on this surface. The remnants of this palaeolandsurface have been used in the southernmost Sierra de San Luis for reconstructing the palaeotopography, dismembered during Neogene tectonism. This analysis facilitated the identification of minor blocks and the calculation of the vertical component of neotectonic faults as well as a rough estimate of the rates and amplitudes of uplift along the range front. Simple geometrical techniques were applied for a preliminary balancing of cross-sections and for estimating shortening and the detachment depth of faults.

The Sierras Pampeanas (Pampean Ranges) are a series of broad mountain ranges rising up from the plains of the Andean foreland of central western Argentina (27–33°S) (Fig. 1). Neogene uplift took place along faults generally located at the western margins of basement blocks during movements of the Andean orogenesis (González Bonorino 1950; Gordillo & Lencinas 1979; Criado Roque *et al.* 1981). These orogenic episodes have been related to an increase in the convergence rate along the active margin of the South American plate, due to the break-up of the Farellones plate *c.* 25 Ma (Pilger 1984; Pardo Casas & Molnar 1987; Ramos 1988). Some characteristics of the neotectonic deformation are still poorly understood, such as uplift magnitude, faulting geometry and relative movements among minor blocks. This is primarily due to (a) the paucity and quality of Neogene stratigraphic records and (b) the concealment of structural relationships along the uplifted front by Quaternary alluvial deposits. Furthermore, the uncertain age of the palaeolandsurface in the interior ranges constitutes a serious limitation to a better understanding of the Neogene evolution of these ranges. On the other hand, the mere presence of this palaeolandsurface as a dominant geomorphological feature does provide an important means of addressing some of the problems of neotectonics.

In the Sierra de San Luis (southern Sierras Pampeanas) (Fig. 1), this palaeolandsurface is widely preserved on the eastern slope but in places it has been disrupted by Neogene tectonism. Because of the wide extent of the palaeosurface and the reasonable degree of documentation for the range bounding faults, this range was selected as a test area for analysing the contribution of palaeosurfaces to understanding neotectonic activity. This paper presents examples of the results obtained.

Geological setting

The Sierras Pampeanas have been identified as a broken foreland of the Andean orogeny (Jordan & Allmendinger 1986), as well as one of the surface geological features which characterize the flat-lying segment of the subducting Nazca plate (Barazanghi & Isacks 1976; Jordan *et al.* 1983). The last significant uplift event took place during the Late Tertiary and Quaternary (González Bonorino 1950; Gordillo & Lencinas 1979; Criado Roque *et al.* 1981) along north-trending reverse faults which commonly bound the west margin of the blocks. As a consequence of the geometry of the faulting, these uplifted blocks have been slightly tilted to the east.

The Sierra de San Luis consists for the most part of a basement complex of metamorphic and igneous rocks. The low- to medium-grade metamorphic units include schists, migmatites, gneisses and phyllites. They are considered to

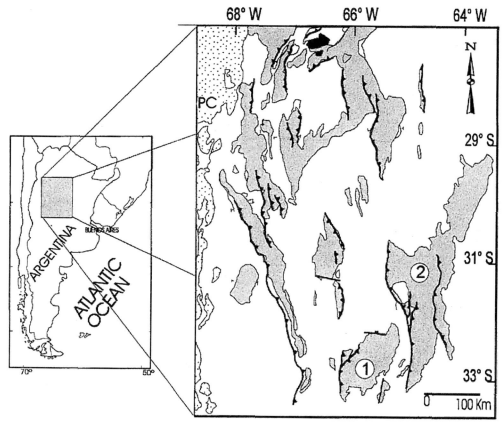

Fig. 1. Areal extent of the Sierras Pampeanas showing the main bounding faults and their relationships with Precordillera fold and thrust belt (PC). 1, Sierra de San Luis; 2, Sierra de Córdoba.

derive from Late Precambrian sequences and have been affected by at least three tectometamorphic and magmatic events before the Late Palaeozoic (Kilmurray & Dalla Salda 1977; Ortiz Suarez et al. 1992). These events are related to the emplacement of pre-, syn- and post-kinematic granitoids with ages ranging from Cambrian to Early Carboniferous.

The stratigraphic record continues with Gondwanan lacustrine deposits and local manifestations of Cretaceous and Neogene volcanic rocks. In addition, Neogene redbeds have been described from the southern half of the range (e.g. Santa Cruz 1979; Di Paola & Rivarola 1992; Sozzi et al. 1995) where scarce fossil records constrain the age of the upper part of the sequence to the Pliocene–Early Pleistocene. A clear angular unconformity between these sediments and fanglomerates of the alluvial fan environment suggests that the main episode of the last significant uplift took place during Late Pliocene–Pleistocene (Criado Roque et al. 1981; Costa 1992).

Currently, the San Luis range is located within a semi-arid region, but annual rainfall in the highlands normally reaches 800 mm.

The uplift geometry of the Sierra De San Luis

The range is roughly elongated along a NNE trend (Fig. 2) and is bounded in the west by the San Luis fault. In E–W cross-section the profile shows a short and steep western slope and a gentle and extensive eastern slope. The remnants of the palaeolandsurface are preserved on the latter.

Exposures of the fault surface along the 200 km of its trace are scarce. The documented outcrops indicate a reverse east-dipping fault or fault zone with average inclinations of 45°, but

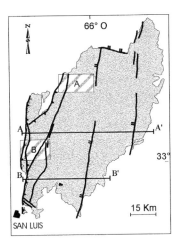

Fig. 2. Sketch of the Sierra de San Luis showing the main Neogene faults. Rectangles outline the areas described in the text: A, Mesilla del Cura; B, Río Nogolí. Locations of cross-sections correspond to the profiles displayed in Fig. 8.

ranging from 10° to 60° (Costa 1992, 1994; González Díaz et al. 1997).

Many authors have suggested that uplift took place along faults listric at depth with rotation along a horizontal axis (e.g. González Bonorino 1950; Jordan & Allmendinger 1986). According to Introcaso et al. (1987), gravimetric modelling in the nearby Sierra de Córdoba (Fig. 1) is in agreement with the shortening and uplift associated with listric faults. This type of geometry has also been imaged by seismic sections at the nearby San Luis Basin, westward of this range. However, deep seismic registers (Cominguez & Ramos 1990) and ongoing seismicity (Smalley et al. 1993) at the Sierras Pampeanas–Precordillera fold and thrust belt boundary, indicate a planar geometry for the block-bounding faults. Martino et al. (1995) suggested that such a geometry could also explain the Neogene structural relief achieved by the Sierra de Córdoba, similar to that observed in the Rocky Mountain foreland uplifts (Smithson et al. 1978).

Unfortunately there is neither enough subsurface information available nor reliable focal mechanisms in the Pampean uplift region for proper testing of these models.

The palaeolandsurface of the Sierra de San Luis

This palaeolandsurface can be easily recognized in aerial images by its smooth texture and homogeneous tone. It is characterized by a gentle, undulating landscape tilted to the east (Fig. 3) and exposing denuded basement rocks without sedimentary cover, except for a widespread Quaternary loessic cover. These deposits

Fig. 3. Panoramic view to the north from the range highlands. Note the continuity of the palaeolandsurface gently tilted to the east. Outcropping rocks are mostly granitoids and migmatites. Volcanic hills appear in the background (extreme right).

contribute to the homogeneity of the landscape and their maximum thickness is up to 20 m in the higher part of the range. These deposits are composed of silt from aeolian sources as well as sand and coarse colluvium of local origin and soil and palaeosol development is common. Sometimes a calcareous duricrust (10–50 cm thickness) is present.

This palaeosurface is widely preserved all along the range, except in the vicinity of major rivers and creeks where recent changes in the regional base-level have led to a rejuvenation of fluvial valleys. These phenomena are more significant near the west side of the range, where the uplift front is located.

The palaeosurface is disrupted by north-trending high-angle faults (most of them reverse) as outlined in Fig. 2. The resulting scarps reach their maximum amplitudes near the border of the range (up to 100 m), while vertical displacements vanish in the inner areas.

The surface monotony is also interrupted by Miocene–Pliocene volcanic domes and associated volcanic products (Fig. 3). This volcanism has been regarded as the ultimate volcanic activity along the flat-lying subduction segment, after the subducted Nazca plate became horizontal (Ramos et al. 1991). The volcanic hills reach maximum altitudes of 300 m above the palaeosurface and are grouped into three areas (La Carolina, Cerros Largos and Cerros del Rosario), aligned in a NW–SE direction. This alignment coincides with a smooth topographic bulge, which constitutes a drainage divide on the eastern slope of the range.

Some considerations on palaeolandsurface development

There are no thermochronometric data available which can illuminate the uplift history and palaeosurface development of the Sierra de San Luis. However, metamorphic conditions and plutonic emplacements in some Sierras Pampeanas suggest a burial depth ranging from 2.8 to 4.25 km for Triassic times (Jordan et al. 1989). In the nearby Sierra de Córdoba, Patiño & Patiño Douce (1987) indicated that Devonian intrusions were emplaced at 5 km depth.

The Sierra de San Luis was already exposed at the surface during Late Carboniferous–Early Permian times, as evidenced by the isolated presence of Gondwanan lacustrine deposits of Bajo de Veliz Fm in the northeastern sector. The relationship between the basement rocks and the sediments is not exposed and, consequently, it is difficult to know the palaeo-topography of this former environment. In other Pampean Ranges, González Bonorino (1972) described the palaeolandsurface as formed before the Carboniferous or Permian, even though according to thermochronometric information provided by Jordan et al. (1989) the Sierras Pampeanas was a mountainous area during the deposition of Gondwanan sediments. Jordan et al. (1989) also conclude that the development of such a regional palaeolandsurface was diachronous in the Sierras Pampeanas region, covering a timespan of 300–400 Ma. There is also general agreement that the palaeolandsurface had been partly exhumed and deformed as a consequence of Neogene tectonism.

In the Sierra de San Luis, this regional surface has been described as a peneplain (e.g. Pastore & Ruiz Huidobro 1952; Pastore & González 1954; Gordillo & Lencinas 1979; González Díaz 1981). The last author referred to it as the 'San Luis Peneplain'. It is also agreed that the landscape was basically shaped by fluvial systems.

Seismic cross-sections across the Cretaceous rift basin westward of the range (San Luis Basin) show that these continental sediments overly a smooth and continuous basement surface disrupted by Mesozoic and Cenozoic faulting (Criado Roque et al. 1981). Although the age of the basal stratigraphic units of this basin are not proven (Triassic?), it is understood that the basement surface could be correlated with the one exposed in the Sierra de San Luis. The volcanic products associated with the Neogene volcanism also testify to the presence of a smooth relief by that age. Tertiary red-beds thrusted at the western range front (Costa 1994), as well as tectonic inversion documented by Schmidt et al. (1995) in the nearby Cretaceous basins, indicate significant uplift of this area during the Neogene.

The information available, therefore, suggests that this palaeolandsurface seems to have developed between Carboniferous and Triassic–Cretaceous times.

Principles and methods

The palaeolandsurface widely preserved on the eastern slope of the range could be used to estimate the vertical component of neotectonic deformations by assuming that: (1) the development of such a surface took place prior to Neogene tectonism (González Díaz 1981; Jordan et al. 1989); and (2) the landscape of the Sierra on the eve of these tectonic episodes must have been characterized by very smooth relief. It seems likely that this smooth erosion

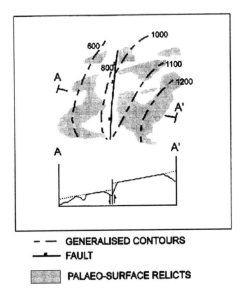

Fig. 4. Criteria used for identifying Cenozoic vertical deformations of the palaeosurface, according to the arrangement of tectoisohyps and for estimating the overall uplift magnitude due to Cenozoic tectonism.

Generalized contour lines traced from the palaeolandsurfaces area were superimposed onto the topographic contours and interpolated, with the aim of reconstructing the enveloping palaeotopographic surface (Fig. 4). Anomalies in the pattern of generalized contours lead to the identification of minor blocks standing above the surface of the eastern slope of the range. Field observations indicated that these blocks are indeed bounded by brittle shear zones or fracture belts. Cross-sections were also used for estimating the overall uplift of the range following deformation of the palaeolandsurface (Fig. 4).

Reconstruction of the palaeotopographic surface as well as the geometry of surface faulting provide the basic information required for interpreting fault geometry at depth. For instance, if a listric geometry of faults is assumed, the techniques developed by Erslev (1986) for the Rocky Mountains allow the construction of area-balanced cross-sections for probable geometries of faulting at depth (Fig. 5). These techniques have been applied to some Sierras Pampeanas by Jordan & Allmendinger (1986). In the Laramide region of North America, deformation characteristics of the basement-cored foreland uplifts are outlined by the overlying sedimentary strata. However, no sedimentary cover is present in most of the Sierras Pampeanas for imaging the Neogene structural style, but this absence could be substituted by reconstructing the present altitude of the regional palaeolandsurface.

surface was partly buried by Mesozoic/Cenozoic sediments.

Taking into account the low gradient of these landscapes, abrupt changes in elevation of the present surface could be indicative of the vertical component of neotectonic deformations or to the preservation of morphological relics (monadnocks), although these have not been identified in this range.

This geometrical approach also provides complementary criteria for testing models based on geophysical data.

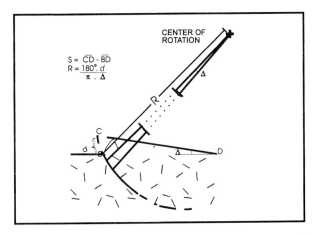

Fig. 5. Geometrical relationships involved in the reconstruction of geometry of faults at depth, adapted from Erslev (1986). S, Shortening; Δ, tilt angle of the palaeosurface. After Costa (1996).

Examples

El Realito–Mesilla del Cura area

In this sector located in the northwest part of the range, extensive palaeolandsurface remnants can be recognized at altitudes ranging from 1700 m to 1000 m and decreasing from west to east (Fig. 6). Minor remnants appear in a step-like pattern at different altitudes and their present configuration seems to be related to secondary branches of the Portezuelo Blanco fault (PB), one of the major faults of the range. The nature of these faults cannot be established on the basis of present data, because fault surfaces are almost vertical and no reliable information was found on these planes for reconstructing geometrical and kinematic features. However, in the Río Nogolí area (see Fig. 7) the Portezuelo Blanco fault does not disrupt the palaeolandsurface and the conspicuous slickenlines data indicate a strike-slip movement.

A major morphological feature in the area is the deeply incised canyon of the Luján River. The palaeosurface does not show any notable difference in altitude on opposite sides of the canyon. Any possible tectonic control on the canyon by a fault or shear zone, therefore, excludes its reactivation during Neogene diastrophism. East of this river, the Mesilla del Cura block (MCB) has been dismembered on its eastern side by several faults as indicated by the presence of small remnants of palaeosurfaces which remain between this block and the Quines river.

Shear zones and other fault-related phenomena were observed along zones of abrupt changes in altitude of the surface. The sense of movement of these faults could only be postulated on the basis of observed fragmentation of the palaeotopography. Other kinematic indicators or evidence for determining the sense of movement were hard to find in the granitoids and migmatites of the area.

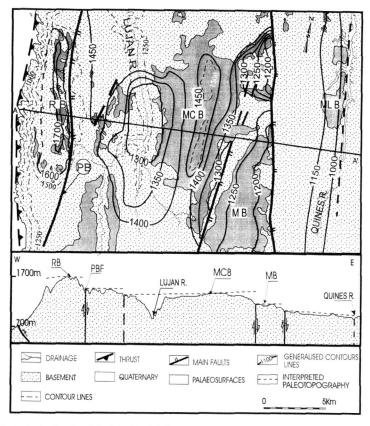

Fig. 6. Morphotectonic sketch of the Mesilla del Cura area. RB, Realito Block; PBF, Portezuelo Blanco Fault; MCB, Mesilla del Cura Block; MLB, Media Luna Block

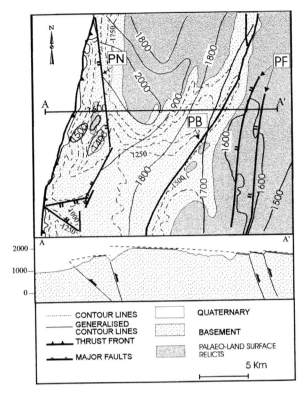

Fig. 7. Morphotectonic sketch of the Río Nogolí area. PN, Pantano Negro Fault; PB, Portezuelo Blanco Fault; PF, Pancanta Fault system.

Río Nogolí area

The generalized contour map shown in Fig. 7 suggests a secondary fault system (Pantano Negro fault – PN) which gave rise to the definition of a minor block interposed between the uplift front and the main range block. The geometry of the uplift front is here characterized by east-dipping reverse faults (45° average) which have thrusted Tertiary sediments (Costa 1992). The proposed boundary between these blocks is characterized by the pervasive Pantano Negro shear zone dipping 60° E on average and following the foliation and previous cataclastic/mylonitic planes of the basement rocks. Fault-slip data analysis has indicated the predominance of reverse movements (Costa 1992).

The Portezuelo Blanco fault (PB) is a major structure in the Sierra de San Luis (Fig. 2) with a sharp and continuous topographic expression. However, it does not seem to disrupt the erosion surface. In contrast, the Pancanta Fault system (PF), with a less pronounced topographic expression, is defined by secondary faults with a minor vertical displacement of Neogene age that clearly truncates the erosion surface with the generation of notable topographic scarps. Both fractures are defined by high-angle (75–85) east-dipping reverse faults (Costa 1992).

Uplift characteristics

Although discussion on possible uplift geometries is beyond the aim of this paper, some examples are presented in order to illustrate the contribution of palaeolandsurface analysis to this topic.

Figure 8 shows two different profiles obtained through reconstruction of the palaeolandsurface by means of generalized contours. Figure 8a shows a continuous slope gradient ranging between 2° and 4° to the east with faults and brittle shear zones at the thrust front dipping 45° E. Corresponding shortening is 4%. Meanwhile in the cross-section in Fig. 8b, which represents the southernmost part of the range (see Fig. 2), slope angles are up to 8° E and a topographic hinge in the palaeolandsurface can be distinguished. Dominant brittle shear zones present the same geometry as in Fig. 8a, even if the youngest faults affecting Quaternary

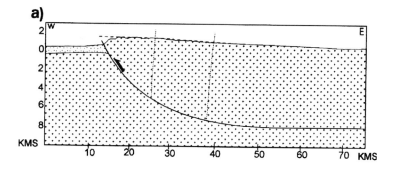

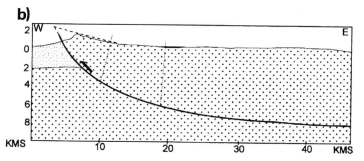

Fig. 8. Sections showing the fault geometry obtained after applying the criteria explained in Fig. 5. See Fig. 2 for locations.

(Pleistocene?) conglomerates at the thrust front are characterized by dip angles between 10° E and 35° E. An associated shortening of 6% was estimated.

Certainly, the structural implications of these changes in the enveloping surface have not been fully explored, but it constitutes an example of the contribution of this analysis towards understanding structural problems.

The results obtained relating to detachment depths of faults and associated shortening are in agreement with the data obtained by Jordan & Allmendinger (1986) in other Sierras Pampeanas, even if the detachment depths are lower than the ones predicted by gravimetry for the Sierra de Córdoba.

Maximum uplift of the Sierra de San Luis after disruption of the palaeolandsurface was calculated as being between 960 m and 1240 m. Putting constraints on the timing of the main uplift episodes between 5 and 0.5 Ma, would give estimates for rates of uplift of between $0.2 \, \text{mm a}^{-1}$ and $2.5 \, \text{mm a}^{-1}$. Morphometric attributes of the mountain front and morphotectonic comparisons with other intraplate uplifts like the Basin and Range, USA (Bull 1987), Wasatch Front, USA (Machette et al. 1991), and Ih Bogd, Mongolia (Ritz et al. 1995), strongly suggest that uplift rates are likely to be closer to the minimum value obtained.

Costa (1996) achieved comparable estimates of 0.2–$3.1 \, \text{mm a}^{-1}$ by integrating these criteria with the morphometric analysis of mountain fronts, according to the principles proposed by Bull (1987), using palaeotopography to estimate the uplift magnitude, and mountain front analysis for estimating the timing of uplift.

Concluding remarks

Utilization of the remnants of the palaeolandsurface preserved in the Sierra de San Luis as reference markers, has proved to be useful for addressing some neotectonic problems. This method, even with its inherently low degree of accuracy, nonetheless constitutes a valuable first approximation for areas with an almost total lack of geophysical and radiometric information. This simple procedure provides some basic criteria for constraining estimates of the vertical component of neotectonic deformation, uplift magnitudes, shortening rates and depths for fault detachment.

Considering the strong morphostructural similarities within the Sierras Pampeanas, it is

thought that the criteria of analysis developed for the Sierra de San Luis could be extended to the entire area of the Sierras.

Useful comments from two anonymous reviewers are gratefully acknowledged. An early version of this paper benefited from a review by H. Diederix. This work was supported by the Universidad Nacional de San Luis Project 348901 and Universidad de Buenos Aires UBACYT Ex-244.

References

BARAZANGHI, M. & ISACKS, B. 1976. Spatial distribution of earthquakes and subduction of the Nazca plate beneath South America. *Geology*, **4**, 686–692.

BULL, W. 1987. Relative rates of long term uplift of mountain fronts. *In*: CRONE, A. & OMDAHL, E. (eds), *Directions in Paleoseismology*. US Geological Survey Open-File Report **87–673**, Denver, 192–202.

COMINGUEZ, A. & RAMOS, V. 1990. Sísmica de reflexión profunda entre Precordillera y Sierras Pampeanas. *11° Congreso Geológico Argentino Actas*, **2**, 311–314.

COSTA, C. 1992. *Neotectónica del Sur de la Sierra de San Luis*. PhD thesis, Universidad Nacional de San Luis, San Luis.

—— 1994. The Neogene thrust front of the Sierra de San Luis, Pampean Ranges, Argentina. *Bulletin INQUA Neotectonics Commission*, **17**, 76.

—— 1996. Análisis neotectónico en las sierras de San Luis y Comechingones: Problemas y métodos. *13° Congreso Geológico Argentino Actas*, **2**, 285–300.

CRIADO ROQUE, P., MOMBRU, C. & RAMOS, V. 1981. Estructura e interpretación tectónica. *In*: YRIGOYEN, M. (ed.), *Geología de la Provincia de San Luis* 8° Congreso Geológico Argentino Relatorio, Buenos Aires, 155–192.

DI PAOLA, E. & RIVAROLA, D. 1992. Formación San Roque: Complejo fanglomerádico-fluvial terciario, sur de la sierra de San Luis. *Revista Asociación Geológica Argentina*, **47**, 23–29.

ERSLEV, E. 1986. Basement balancing Rocky Mountain foreland uplifts. *Geology*, **24**, 259–262.

GONZALEZ BONORINO, F. 1950. Algunos problemas geológicos de las Sierras Pampeanas. *Revista Asociación Geológica Argentina*, **5**, 81–110.

—— 1972. *Descripción geológica de la Hoja 13c, Fiambalá (Provincia de Catamarca)*. Dirección Nacional de Geología y Minería, Buenos Aires, Boletín **127**, 1–73.

GONZALEZ DIAZ, E. 1981. Geomorfología, *In*: YRIGOYEN, M. (ed.) *Geología de la provincia de San Luis*. 8° Congreso Geológico Argentino Relatorio, Buenos Aires, 193–236.

——, FAUQUE, L., COSTA, C., GIACCARDI, A. PALOMERA, P. & PEREYRA, F. 1997. La avalancha de rocas del "Potrero de Leyes", Sierras Pampeanas Australes, sierra Grande de San Luis, Argentina (32°30′ lat. S). *Revista Asociación Geológica Argentina*, **52**, 93–107.

GORDILLO, C. & LENCINAS, A. 1979. Sierras Pampeanas de Córdoba y San Luis. *In*: TURNER, J. (ed.) *Geología Regional Argentina*, Córdoba, **1**, 577–650.

INTROCASO, A., LION, A. & RAMOS, V. 1987. La estructura profunda de las sierras de Córdoba. *Revista Asociación Geológica Argentina*, **42**, 177–187.

JORDAN, T. & ALLMENDINGER, R. 1986. The Sierras Pampeanas of Argentina. A modern analogue of Rocky Mountains foreland deformation. *American Journal of Science*, **286**, 737–764.

——, ISACKS, B., ALLMENDINGER, R., BREWER, J., RAMOS, V. & ANDO, C. 1983. Andean tectonics related to geometry of subducted Nazca plate. *Geological Society of America Bulletin*, **94**, 341–361.

——, ZEITLER, P., RAMOS, V. & GLEADOW, A. 1989. Termochronometric data on the development of the basement peneplain in the Sierras Pampeanas, Argentina. *Journal of South American Earth Sciences*, **2**, 207–222.

KILMURRAY, J. & DALLA SALDA, L. 1977. Caracteres petrológicos y estructurales de la región central y sur de la sierra de San Luis. *Obra Centenario Museo La Plata*, **4**, 167–178.

MACHETTE, M., PERSONIUS, S., NELSON, A., SCHWARTZ, D. & LUND, W. 1991. The Wasatch fault zone, Utah-segmentation and history of Holocene earthquakes. *Journal of Structural Geology*, **13**, 137–149.

MARTINO, R., KRAEMER, P., ESCAYOLA, M., GIAMBASTIANI, M. & ARNOSIO, M. 1995. Transecta de las Sierras Pampeanas de Córdoba a los 32°S. *Revista Asociación Geológica Argentina*, **50**, 60–77.

ORTIZ SUAREZ, A., PROZZI, C. & LLAMBIAS, E. 1992. Geología de la parte sur de la sierra de San Luis y granitoides asociados. *Estudios Geológicos*, **48**, 269–277.

PARDO CASAS, F. & MOLNAR, P. 1987. Relative motion of the Nazca (Farallon) and South American Plates since Late Cretaceous time. *Tectonics*, **6**, 233–248.

PASTORE, F. & RUIZ HUIDOBRO, O. 1952. *Descripción geológica de la hoja 24 G (Saladillo), San Luis*. Dirección Nacional de Geología y Minería, Boletín **78**, Buenos Aires, 1–63.

—— & GONZALEZ, R. 1954. *Descripción geológica de la hoja 23 G (San Francisco), San Luis*. Dirección Nacional de Geología y Minería, Boletín **80**, Buenos Aires.

PATIÑO, M. & PATIÑO DOUCE, A. 1987. Petrología y petrogénesis del batolito de Achala, provincia de Córdoba a la luz de la evidencia de campo. *Revista Asociación Geológica Argentina*, **42**, 201–205.

PILGER, R. 1984. Cenozoic plate kinematics, subduction and magmatism: South American Andes. *Journal of the Geological Society, London*, **141**, 793–802.

RAMOS, V. 1988. The tectonics of the Central Andes: 30° to 33° S latitude. *In*: CLARK, S., & BURCHFIEL, C. (eds) *Processes in Continental Lithospheric Deformation*. Geological Society of America, Boulder, Special Paper, **218**, 31–54.

——, MUNIZAGA, F. & KAY, S. 1991. El magmatismo Cenozoico a los 33°S de latitud: Geocronología y relaciones tectónicas. *6° Congreso Geológico Chileno Resúmenes*, 892–896.

RITZ, J., BROWN, E., BOURLÈS, D. *et al.* 1995. Slip rates along active faults estimated with cosmic-ray exposures dates: Applications to the Bogd fault, Gobi-Altai, Mongolia. *Geology,* **23**, 1019–1022.

SANTA CRUZ, J. N., 1979. Geología de las unidades sedimentarias aflorantes en el área de las cuencas de los ríos Quinto y Conlara. Provincia de San Luis. República Argentina. *7° Congreso Geológico Argentino Actas,* **1**, 335–349.

SCHMIDT, C., ASTINI, R., COSTA, C., GARDINI, C. & KRAEMER, P. 1995. Cretaceous rifting, alluvial fan sedimentation and neogene inversion, southern Pampean Ranges, Argentina. *In*: TANKARD, A., SUÁREZ, R. & WELSINK, H. (eds) *Petroleum Basins in South America*. American Association of Petroleum Geologists, Tulsa, Memoir **62**, 341–358.

SMALLEY, R., PUJOL J., REGNIER, M., CHIU, J. ISACKS, B., ARAUJO, M. & PUEBLA, N. 1993. Basement seismicity beneath the Andean Precordillera thin-skinned thrust belt and implications for crustal and lithospheric behaviour. *Tectonics,* **12**, 63–76.

SMITHSON, S., BREWER, J., KAUFMAN, S., OLIVER, J. & HURICH, C. 1978. Nature of Wind River thrust, Wyoming, from COCORP deep reflection data and from gravity data. *Geology,* **6**, 648–652.

SOZZI, H., OJEDA, G. & DI PAOLA, E. 1995. Estratigrafía y sedimentología de abanicos aluviales en el área de Nogolí, San Luis, *Revista Asociación Geológica Argentina,* **50**, 165–174.

The significance of high planation surface in the Andes of Ecuador

MAURO COLTORTI[1] & CLIFF D. OLLIER[2]

[1] *Dipartimento di Scienze della Terra, Università di Siena, Italy*
[2] *Centre for Resource and Environmental Studies, Australian National University, Canberra 0200, Australia*

Abstract: A traverse of Ecuador crosses five planated units including two main horsts which constitute the Andean Cordillera. A single surface once extended across all units, and its formation was completed by Early Pliocene times. Vertical uplift occurred mainly in the Plio-Pleistocene. Most rock structures in the Andes result from pre-planation tectonics and have no connection with mountain building. Post-planation vertical uplift created the mountains, and uplift led to gravitational spreading. The causes of stillstand and uplift are debatable but subduction and compression fail to account for the nature, the symmetry and the timing of events in the Ecuadorian Andes.

The Ecuadorian Andes are part of the Andes mountain chain which runs along the western side of South America. In Ecuador the chain is 600 km long but is only 150–180 km wide. This is the narrowest part of the Andes, and very suitable for the study of a transect across the entire Andean chain which investigates the geomorphological evolution of the different structural units.

A striking characteristic of the higher part of the Cordillera is a planation surface, preserved as slightly dissected remnants extending for tens of kilometres. Planation surfaces are described on many mountain chains of the world and most authors agree that they represent erosion surfaces created close to past base-level (sea-level) and later uplifted. The rocks affected by planation were previously folded, faulted and sometimes overthrust for tens of kilometres. Such is the case in the Andes of Ecuador. The processes that folded and metamorphosed the rocks did not form the mountains, as mountain creation through uplift occurred much later.

This paper aims to illustrate the geomorphological evolution of the Ecuadorian Andes and especially the significance of the planation surface. The planation surface not only provides a starting point for examining the recent evolution of the Andean Cordillera, but puts constraints on the nature of tectonic processes involved in mountain building.

Morphotectonic units

The Costa

The Costa (Figs 1 and 2) is not a coastal plain as the name might suggest, but a dissected plateau at an elevation of about 500–600 m (Figs 2 and 3a). The basement consists of Lower Cretaceous basaltic rocks (Pinon Formation (Fm)) attributed to a volcanic arc (Faucher & Savoiard 1973; Kennerly 1980; Pilger 1981; Baldock 1982; Suarez et al. 1983). The planation surface cuts these rocks in the Jama Mache and Chogon Colonche Hills. It also cuts across Tertiary basins (Borbon, Progreso, Jambeli-Tumbes, Manabi, Esmeraldas) some of which reach 10 000 m in thickness. They are filled with deformed sandstones, siltstones, shales and clays with intercalated tuffs, limestones and calcareous shell layers deposited in a marine environment from Cretaceous to earliest Pliocene (Bristow & Hoffstetter 1977; Baldock 1982), indicating that the planation surface is no older than earliest Pliocene.

Recent uplift of this area is indicated by the presence of at least three marine terraces, the highest one at approximately 300 m (Tablazo Fm: Bristow & Hoffstetter 1977).

The Western Cordillera

The Western Cordillera consists of Mesozoic, mainly Cretaceous sediments, highly contorted and often vertical (Figs 1, 2 and 3a). The Mesozoic rocks were evidently deposited in a trough parallel to the present Andes, folded, and then bevelled by an erosion surface cut across both Mesozoic and older rocks. It is bounded to the west by the impressive fault escarpment which links the Jumbeli–Naranjal Fault to the Guajaquil–Babahojo–S. Domingo (GBSD in Fig. 2) and extends to the Colombian border, and to the east, by the extensional faults of the Inter-Andean Depression. The youngest clastic formations outcropping in the Western Cordillera are Late Miocene in age and represent continental

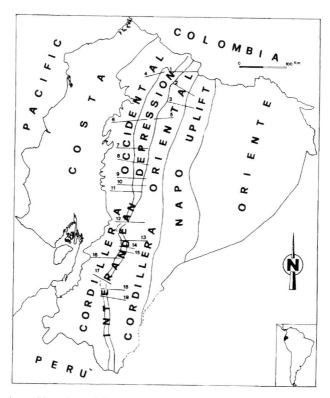

Fig. 1. The main physiographic regions of the Ecuadorian Andes showing the locations of the sections of Fig. 4.

basin-fill deposits which will be described in the following section. They show that the planation surface was eroded after this time. For example, the uppermost preserved sediments of the Mangan Fm, filling the Nabon basin of southern Ecuador, were deposited between 8.5 and 7.9 Ma (Hungerbuhler et al. 1995).

The Inter-Andean Depression

The Inter-Andean Depression is a graben elongated parallel to the Cordilleras and filled with a thick sequence of Plio-Pleistocene fluvial and lacustrine sediments to create an Accumulation Plateau (Figs 2 and 4). These sediments are interlayered with lava flows, agglomerates and pyroclastics coming from strato-volcanoes located along the fault escarpments bounding the Depression, indicating contemporaneous volcanic activity (Ficcarelli et al. 1992).

North of Quito, from Cayambe to Pifo, a thick sequence of lacustrine sediments has been recognized in the upper part of the conglomerates. A series of folds, faults and overthrusts affects these sediments which are underlain and overlain by undisturbed horizontal sediments, indicating that folding was the result of gravity sliding and not regional compression (Falsini 1995). This small-scale gravity feature provides a model for the larger-scale gravitational movements which, we suggest, have played a significant part in Andean structure.

In the Inter-Andean Depression in the Cuenca area, the Accumulation Plateau is formed by a thick sequence of conglomerates, fluvial sands, clays, tuffs and volcanic breccias (Turi Fm: Baldock 1982). In most parts of the Inter-Andean Depression, the Accumulation Plateau is deeply dissected by fluvial erosion, which creates gorges many hundreds of metres deep, such as the Apaqui River Gorge north of the Chota Valley, and the Guayalabamba Gorge east of Quito. The geomorphological setting is similar to the sediments of the Pre-Canon stage in northern Peru (Tosdal et al. 1984).

The present-day Inter-Andean Depression was created during the Plio-Pleistocene because the sediments are interlayered with lava flows which are younger than 1.8–1.5 Ma (Barberi et al. 1988). Moreover, in Colombia, on the northward extension of the Depression, the

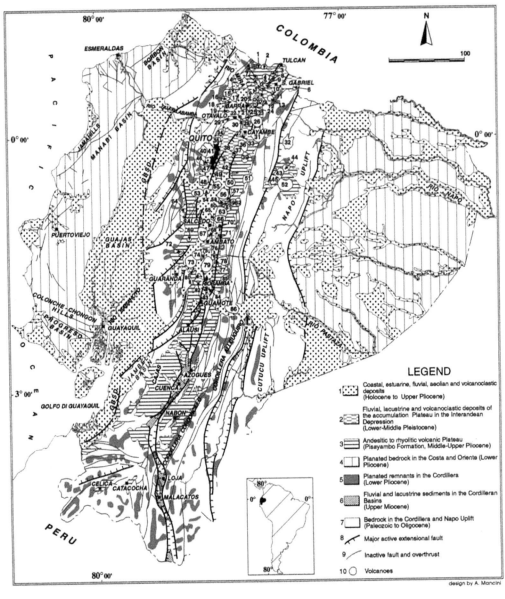

Fig. 2. Main geomorphological units of Ecuador showing places and features mentioned in the text. 'Volcano' stage of erosion: Intact (I); Planeze (P); Residual (R); Skeleton (S); Caldera (C); with recent cones (wc); active (a); collapsed (c). 1, Cerro Negro de Marasquer (Rwc); 2, Chiles (Pwc); 3, Chalpatan (C); 4, Potrerillos (C); 5, Chiltazon (R); 6, Soche (S); 7, Iguan (R); 8, Horqueta W (R); 9, Cerro Maiurco (R); 10, Azufral (R); 11, Boliche (R); 12, Cerotal (R); 13, Mangus (P); 14, Pilavo (S); 15, Negro Puno (R); 16, Yanaurco (S); 17, Chachimbiro (R); 18, Cotacachi (P); 19, Cuicocha (Ia); 20, Yahuarcocha (C); 21, Imbabura (P); 22, Cerro de Araque (I); 23, Cerro Cubilche (I); 24, Guangaba (R); 25, Redondo (R); 26, Rumiloma (R); 27, Laguna S.Pablo (C); 28, Cusin (P); 29, Cushnirrumi (R); 30, Moyanda (R); 31, Cayambe (P); 32, Reventador (I); 33, Cananvalle (I); 34, Pululagua (C); 35, Casitagua (P); 36, Pambamarca (P); 37, Las Puntas (Pwc); 38, Cotourcu (P); 39, Chacana (C), 40, Guagua Pichincia (Ia); 41, Ruhu Pichincia (P); 42, Ilalo (P); 43, Pan de Azucar (S); 44, Cerro Negro (R); 45, Loa Guacamayos (S); 46, Cerro Pailon (S); 47, Atacazo (P); 48, Corazon (P); 49, Pasochoa (P); 50, Sincholagua (R); 51, Antisana (Ia); 52, Sumaco (Ia); 53, Iliniza (P); 54, Loma Santa Cruz Chica (R); 55, Ruminahui (P); 56, Cotopaxi (Ia); 57, Morurco (C); 58, Agualongo Nord (R); 59, Quilindana (P); 60, Morro (R); 61, Chalupas (C); 62, El Descanso (R); 63, Titsulo (R); 64, Quilotoa (Ia); 65, Putzulagua (P); 66, Paturcu (R); 67, Sagoatoa (R); 68, Cerro Chuquirahua (S); 69, Unamancho (I); 70, not visited; 71, not visited; 72, Cerro Puevulo (S); 73, Chimborazo (Iwc); 74, Carihuairazo (P); 75, Punalica (I); 76, Llimpi (P); 77, Mulmul (P); 78, Tungurahua (Ia); 79, Igualata (P); 80, Cerro Shohulurcu (R); 81, Calpi (I), 82, Loma Luyuc (Iwc); 83, Tulabug (I); 84, Bellavista (I); 85, El Altar (Ic); 86, Sangay (Ia).

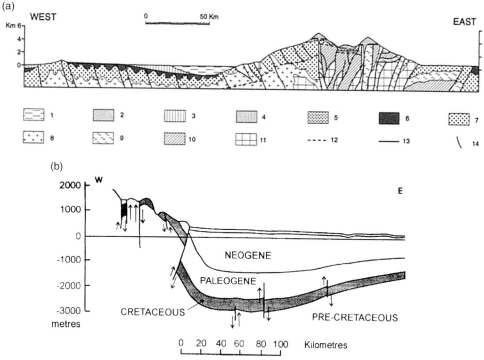

Fig. 3. (**a**) Geological cross-section across the Cordillera (modified after Litherland *et al.* 1993). Key: 1, fluviatile, estuarine and marine sediments (Pleistocene–Holocene); 2, volcanic sequences (Lower–Middle Pleistocene); 3, fluvial, estuarine and marine sediments (Middle–Upper Pliocene); 4, fluviatile and lacustrine (Accumulation Plateau, Upper Pliocene–Lower Pleistocene); 5, lavas and ignimbrites (Volcanic Plateau, Pisayambo Formation, Middle–Upper Pliocene); 6, alluvial fan, fluviatile, lacustrine deposits (Upper Miocene); 7, continental and marine sediments (Upper Miocene–Cretaceous); 8, volcanic rocks (Cretaceous); 9, marine sediments (Jurassic); 10, intrusive rocks; 11, basement (Palaeozoic); 12, hypothetical location of the main detachment surface; 13, planation surface; 14, major faults. This sketch shows the fan-like arrangement of faults in the Equadorian Andes, with high-angle faults as shown by such authors as Megard (1987) and Litherland *et al.* (1993). These authors presume subduction causes the faults in the west, but not in the east (**b**) Geological section of the Oriente (after Rosania 1989).

base of the continental sequences of the Bogotá area (Funza I and II) are 2.7 Ma old (Andriessen *et al.* 1994).

All along the Inter-Andean Depression these Plio-Pleistocene sediments lie unconformably over highly folded Upper Miocene fluvial and lacustrine deposits (i.e. Chota Fm, Mangan Fm: Kennerly 1980; Baldock 1982) which have been dated to the end of the Miocene (Barberi *et al.* 1988; Hungerbuhler *et al.* 1995). Lavenù *et al.* (1995) recognized progressive unconformities in these sediments suggesting a synsedimentary tectonic activity. However, as observed by Litherland *et al.* (1993), the bedrock frequently consists of metamorphic and plutonic rocks. In the Chota Valley and in the Punin area (south of Rio Bamba) (Fig. 4) severely folded and faulted Miocene and, in places, Upper Creta-

ceous and Palaeocene rocks were observed lying unconformably over horizontal Lower Cretaceous Red Beds (Silante Fm: Baldock 1982). This indicates that a major detachment, continuing at depth in the Western Cordillera, affected these rocks. These movements created gravitational trench-like features corresponding with the present-day Inter-Andean Depression which was contemporaneously filled with continental sediments. The Upper Miocene syndepositional deformations (Hungerbuhler *et al.* 1995; Lavenù *et al.* 1995) provide the age for this activity. Probably the superficial shortening indicated by the folds is connected with movements which denuded the Eastern Cordillera leading to the outcropping of the basement. However, these movements had already finished when the planation was modelled. It is suggested

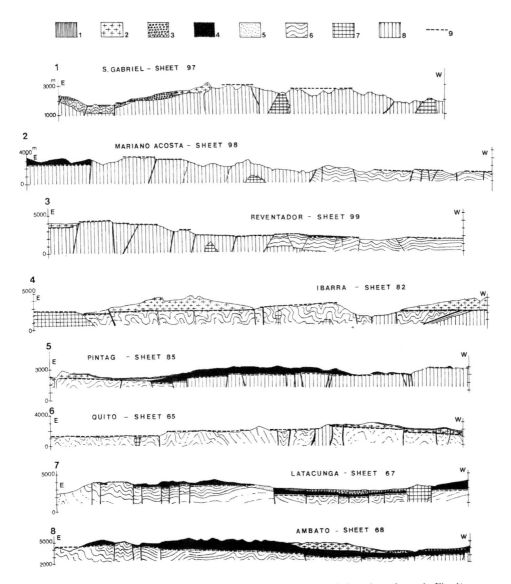

Fig. 4. Geological cross-sections across the Cordillera from north to south (locations shown in Fig. 1): 1, fluvial, lacustrine, aeolian, landslides and volcanoclastic (Holocene–Middle to Lower Pleistocene); 2, intrusive, lava flows and pyroclastics (Holocene–Middle to Lower Pleistocene); 3, conglomerates with subordinate pyroclastics and lavas (Accumulation Plateau, Upper Pliocene–Lower Pleistocene); 4, lavas and ignimbrites (Volcanic Plateau, Pisayambo Formation, Middle–Upper Pliocene); 5, alluvial fan, fluviatile and lacustrine deposits in the tectonic depressions of the Cordillera (Upper Miocene); 6, continental and marine sediments (Upper Miocene–Cretaceous); 7, intrusive rocks; 8, basement (Palaeozoic); 9, planation surface (modified from *Geological Map of Ecuador, scale 1:100 000*). In correspondence to the Chota valley (Section 4) and Punin (Section 11) the basement is unconformably covered by horizontal or gently dipping Cretaceous Red Beds. These are covered by strongly folded and overthrusted rocks from Cretaceous (Yunguilla Fm) to Upper Miocene (Chota Group) which allow us to recognize the presence of a detachment surface which affected the sedimentary cover. The planation surface, modelled during the Early Pliocene, has been recognized on top of the Napo Uplift as well as on the westernmost part of Oriente.

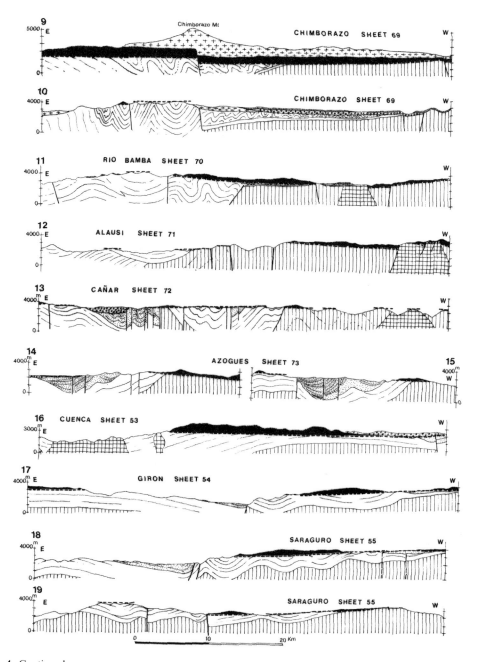

Fig. 4. Continued.

here that the planation surface possibly corresponds to the unconformity cutting the Upper Miocene folded rocks that were later buried under Plio-Pleistocene sediments. It was lowered by the high-angle extensional faults which created the present-day Inter-Andean Depression.

The Eastern Cordillera (Cordillera Real)

The Eastern Cordillera consists largely of highly folded and partly metamorphosed rocks, intruded by granites and severely planated (Figs 2, 3a and 4). Two late Precambrian orogenies have been postulated, and the greenschist

metamorphism is described as Palaeozoic. Aspen & Litherland (1992) describe the metamorphic geology in terms of five lithotectonic accretionary events in the Mesozoic. These events pre-date the formation of the planation surface and recent uplift of the Andes.

The Sub-Andean Zone and Oriente

East of the Eastern Cordillera is a low plateau, approximately 500–600 m in elevation, known as the Oriente, with the suggestion of a small plateau or more irregular highland along the Sub-Andean Zone (the Napo Uplift in the north and the Cutucu Uplift in the south) at around 2000 m (Figs 3b and 4). The Oriente Plateau is underlain by a somewhat asymmetrical syncline of rocks dating from Cretaceous to Neogene. The Curaray Fm of the Oriente, underlying the Chambira Fm and, in places, located at over 1000 m (Baldock 1982), contains fresh to brackish water fauna of Late Miocene age, indicating that the Cordillera had still not developed at that time. The Sub-Andean Zone consists of folded and thrust faulted rocks of the same series, thrust from west to east. Kennerly (1980) and Baldock (1982) include the Sub-Andean Zone in the Oriente but the difference in tectonic style, the high elevations, the volcanic activity (i.e. Sumaco, Reventador, Sangay) and the extension in the Eastern Cordillera of Colombia and Peru indicate that it belongs to the Cordillera domain. The main fault escarpment is located at the border between these two domains.

The planation surface

In the Andean Cordillera, the presence of extensive flat or gently rolling areas has been recognized since the early 1900s and described as the 'puna surface' (Bowman 1909; McLaughlin 1924) because they are often preserved at the elevation of the 'puna', the upper vegetational belt of the Cordillera. The term 'altiplano', literally meaning a high plain, might well refer to the planation surface, but confusion arises because some authors use it also to describe the top of the Accumulation Plateau of the Inter-Andean Depression.

Most researchers agree that an erosion surface is formed close to a base-level (sea-level) and later uplifted, creating the present high relief. The mean elevation can be measured only where it is not covered by extensive fields of lava and ignimbrite. It is particularly well preserved on the metamorphic rocks of the Eastern Cordillera in the southern part of the country (near Cuenca) where it lies at 3000–3500 m (Fig. 4). Similar elevations are observed in the planation remnants on Palaeozoic rocks cropping out east of Guamote and Ambato. East of Quito the surface is at 3500–4000 m and similar elevations are recorded northward up to the Colombian border.

The Western Cordillera is similarly bevelled at 3000–3500 m in the southern part of the country where it cuts rocks dating from the Cretaceous and Palaeocene, including some granite and granodiorite batholiths. West of Riobamba, it is on the watershed at elevations close to 3900 m and is cut across folded Cretaceous conglomerates and sandstones. The mean elevation decreases progressively towards the Costa and the Oriente where large bevelled areas lie at c. 2000 m (and below this). This is a similar elevation to that of the higher part of the Napo and Cutucu uplift in the Sub-Andean Zone. Sometimes the planation surface is tilted or displaced by younger faults. In places it is also dissected by glacial erosion (Clapperton 1993) and may be reduced to accordant summits. However, all these modifications, which occurred later in the history of the Cordillera, fail to mask the evidence of planation processes. The lateral continuity of the remnants of the planation surface from elevations of between 2000 and over 4000 m, and the presence of clear, steep and slightly dissected fault escarpments which displace it, delimiting horst and graben, compel us to believe that the Ecuadorian Andes were modelled by a single planation surface. We stress again that the genesis of the planation surface pre-dates the creation of the Inter-Andean Tectonic Depression.

Estimates of the age of the high planation surface come from many parts of the Andes. McLaughlin (1924) stated that the Puna surface in Peru was developed upon the Tertiary volcanic rocks of the Cordillera Negra (northern Peru) which are now known to be as young as 14.6 Ma (Tosdal et al. 1984). Muñoz (1956) wrote that 'At their summit, the Andes show remnants of an ... erosion surface which was elevated during the Pliocene and Quaternary'. Several erosion surfaces in southernmost Peru range in age from early Miocene to Pliocene when planation gave way to canyon incision (Tosdal et al. 1984). The initiation of the canyon stage of erosion in northern Peru is about 6 Ma (Cobbing et al. 1981) but in Ecuador, because this stage followed the creation of the Accumulation Plateau, the age must be much younger. In Bolivia an extensive high-altitude palaeosurface formed between 3 and 12 Ma

(Kennan et al. 1997), when it was uplifted to near its present level.

The planation surface in Ecuador and neighbouring countries is here attributed to the Early Pliocene since the planation surface on the Costa cuts Early Pliocene strata. In the Cordillera it cuts upper Miocene rocks younger than 6 Ma and is covered by Middle Pliocene rocks of the volcanic plateau, described in the following section.

The volcanic plateaus

Volcanic rocks, recently grouped together to create the Pisayambo Fm (Litherland et al. 1993), were deposited extensively on top of the planation surface locally creating a volcanic plateau. The Cuenca area was the centre for the main volcanic plateau in the south. The horizontally layered materials cover an area of over $80 \, km^2$, and can be over 1000 m in thickness. They include ignimbrites, agglomerates, tuffs and lavas of rhyolitic to andesitic composition (Baldock 1982). No centre has yet been located and they may have resulted from fissure eruptions. To the north, on both Cordilleras, the composition changes to pyroclastic and massive flows of basaltic-andesite. These were emplaced before the creation of the Inter-Andean Depression, which, although not as well developed as in the north of the country, is clearly recognizable. The volcanic sheet was deposited across the whole area, including the margin of the Eastern Cordillera, where it is less thick. If the Depression had already existed it would have been filled with thick volcanic deposits.

These volcanics have been K-Ar dated by Barberi et al. (1988) who report Mio-Pliocene ages. The ignimbrites of Cuenca province have to be younger than 6 Ma because all along the Cordillera they cover folded rocks of this age. Volcanic layers of the Chota Fm, which was planated, have K-Ar ages of c. 6.3 Ma (Barberi et al. 1988). Early Pliocene is also the age reported by Litherland et al. (1993) for rocks of the Pisayambo Fm.

The Quaternary volcanoes

Over 70 strato-volcanoes erupted in Ecuador during the Quaternary and many of these remain active (Hall & Wood 1985; Clapperton 1993). They rise up to 5000–6000 m, well above the mean elevation of the planation surface, and most are aligned with the borders of the Inter-Andean Depression, with just a few located within the Depression or in the Eastern Cordillera. Barberi et al. (1988) pointed out that the strato-volcanoes became active after about 1.8 Ma and recognized a long gap between the older volcanic plateaus and these recent strato-volcanoes.

The age of uplift

The Cordillera was uplifted in relatively recent times, with more than 3500 m of uplift occurring since the formation of the planation surface, mostly during the Plio-Pleistocene. Radelli (1967) and Van Der Hammen et al. (1973) describe the occurrence of plant species in Colombia which today live below 500 m and which are now preserved in Pliocene deposits located at 2700 m, indicating substantial uplift during the Plio-Quaternary. More recently fission-track and K-Ar dating as well as geological and geomorphological evidence have helped to establish the age and uplift rates for the Colombian Andes (Kroonenberg et al. 1990). They also conclude that the Cordilleras started to rise in the Miocene, but most uplift occurred during the Plio-Pleistocene, with some difference in age between the tectonic blocks.

On the other hand, what was once supposed to be a local sea intrusion in the Oriente of Ecuador during the Upper Miocene (Curaray Fm: Kennerly 1980; Baldock 1982) has also recently been observed in Colombia and Bolivia (Hoorn 1993; Hoorn et al. 1995) confirming that the present-day Cordillera was uplifted during the Plio-Quaternary. This has also been suggested by previous geological evidence (for a review see Audebaud et al. 1973).

The nature of uplift

The interpretation advanced in this paper is shown in simplified form in Figs 3a and 5. Two outer erosion surfaces on gently folded Cretaceous rocks bound two higher plateaus of intensely folded rock, which are separated by a graben of mainly horizontal Cenozoic sediments. The recent simple horst and graben structure is very like that described for Peru by Myers (1975) (Fig. 6a). This pattern is consistent with simple vertical uplift, usually associated with tension rather than compression.

Strike-slip movement and exotic terranes

If exotic terranes are present in Ecuador they imply great horizontal movements, with major strike-slip faults. It is possible that horizontal movements preceded the erosion of the planation

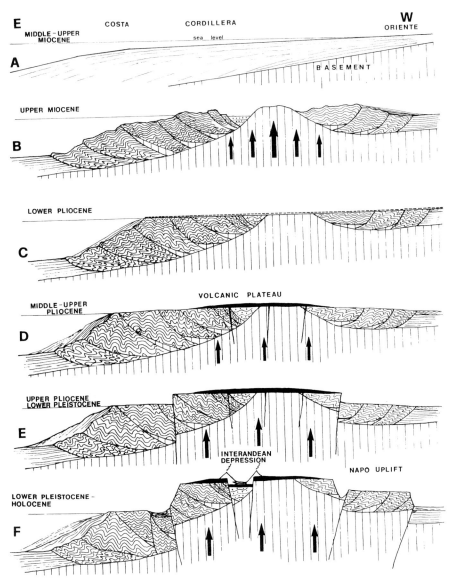

Fig. 5. Sequence of events which led to the build-up of the Cordillera. (A) Cratonic area bordering a sedimentary platform on the Pacific side of south America (Middle Miocene). (B) Beginning of the uplift of the Eastern Cordillera (Miocene). Tectonic denudation of the sedimentary cover and activation of detachment movements. As a consequence, folded and thrusted rocks were created towards the Amazonian basin as well as towards the Pacific. The trench on the western side was the site of continental sedimentation which was further deformed (progressive unconformities: Hungerbuhler et al. 1995; Lavenu et al. 1995) due to the continuation of the detachment movements. (C) Modelling of the planation surface across the Costa, the Cordillera and part of the Oriente (Lower Pliocene). (D) Widespread effusive activity led to the emplacement of a Volcanic Plateau (Middle–Upper Pliocene). (E) Progressive uplifting in the order of 3000–4000 m in c. 4 Ma. (F) Activation of lateral spreading in the Cordillera with creation of the Inter-Andean tectonic depression. Another huge trench formed since the Pliocene along the Pacific side with the creation of a depression hosting continental and estuarine sedimentation (Guayaquil Gulf and Guayaquil Babahoyo line). Activation of the volcanoes in tectonic depression.

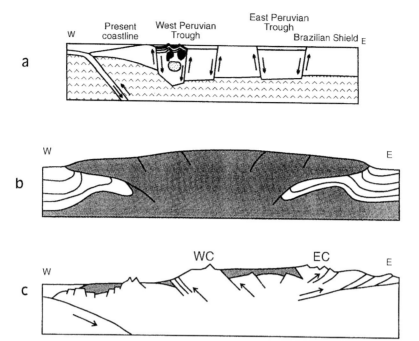

Fig. 6. (a) Diagrammatic cross-section of the Andes of Peru, as a series of fault blocks (after Myers 1975). (b) Sketch cross-section of 'mushroom tectonics', as applied to the Front Range, Colorado, by Jacob (1983), with Precambrian rocks (shaded) spreading over younger rocks on both sides. (c) Sketch cross-section of the fan-like arrangement of faults in the Chilean Andes, with low-angle outer faults, as shown by authors such as Munoz & Charrier (1996) and Reutter et al. (1988). Some of these authors state, or imply, subduction under the Andes from both sides.

surface and vertical uplift. Feininger (1987) presents a strong case for such terranes. The Costa, which equates to his 'Pinon Terrane', has a basaltic basement and a very high Bouguer anomaly; the Santiago Terrane which occupies the south Western Cordillera is the only site where Lower Jurassic limestones occur. The northern Sierra and the Oriente are both on Feininger's South American Craton. Feininger (1987) suggests that the terranes were emplaced from Middle Jurassic to Late Eocene, long before the planation surface was eroded. It is not surprising that, as Litherland & Aspen (1992) suggest, there was reactivation of terrane-bordering faults. These authors provide a good map of the Inter-Andean Graben, showing numerous patches of ophiolite on the flanks of the Cordillera, and the location of a number of major volcanoes on terrane-bounding faults.

Compression and crustal shortening or uplift and spreading

Compression is the most common concept used to explain mountain formation and is generally related to crustal shortening (Muñoz & Charrier 1996). Its action is usually reflected in rock structures such as folds and thrust faults. This concept forms an important part of a series of models and interpretations based on the plate-tectonic assumption that the Andes are the result of the collision between oceanic and continental crust (Suarez et al. 1983; Isack 1988; Wdowinski & Bock 1994a,b).

Most authors forget that a 'planation surface' is an important feature in Earth history that marks the end of a tectonic regime. In elementary diagrams compression is shown causing corrugation at the ground surface thereby resulting in folds and mountains simultaneously. But in many orogenic mountain ranges the ground surface is not folded, and plateaus are commonly preserved in almost horizontal position. Compressive folds and thrusts can also be caused by gravity tectonics. Can this interpretation be applied to the Andes? Unlike the situation shown in Fig. 3a, the main faults in the Ecuadorian Andes (and in many other parts of the Andes) diverge from the Inter-Andean Depression (Kennerly 1980; Baldock 1982; Suarez et al. 1983). This is depicted as a fan-like distribution

of faults reported by Lonsdale (1978) and also on the cross-section accompanying the geological map of Ecuador (Litherland *et al.* 1993). As the faulting has not destroyed the uplifted planation surface, we believe that the faults are divergent because of post-uplift spreading.

Post-uplift spreading has been described around numerous uplifted plateaus. The Colorado Plateau uplift shed sedimentary rocks in opposite directions (Jacob 1983). Clearly an uplifted block could not apply a lateral thrust in opposite directions except by spreading. Many other North American examples are known of uplifted blocks that have spread, with kilometres of thrusting in opposite directions from an uplifted plateau, including the Rocky Mountain area (Lowell 1983), and New Mexico (Woodward 1983). Jacob (1983, p. 235) used the phrase 'mushroom tectonics' to describe this phenomenon, and illustrated it by an example from the Front Range between South Park and the Denver Basin (Fig. 6b). The symmetry of this section is reminiscent of the symmetry of the Andes.

The interpretation of gravity spreading after vertical uplift has also been applied to uplifted blocks in Australia (Ollier & Wyborn 1989). When a fault scarp runs almost straight for tens or even hundreds of kilometres, it must be associated with a high-angle fault, because a low-angle thrust fault in an area of high relief would have a sinuous outcrop. Examples of such straight faults are the Tawonga Fault and Long Plain Fault in southwest Australia. Yet detailed examination in tunnels cut through these faults shows that they are low-angle thrusts, with granite or Palaeozoic sediments thrust over Quaternary alluvium. The simplest explanation for this is gravity spreading of an uplifted block over the alluvium. Similarly in the Andes, we believe that the straightness of the Cordillera margins on the large scale indicates dominantly vertical faulting, but detailed studies indicate thrusting, which is a late-stage modification as uplifting blocks spread. If uplift is high enough spreading is inevitable, because of the low strength of rock, as noted by Jeffreys (1931). Whenever an elevation of over 3 km has been attained, failure by fracture or flow is to be expected.

We believe that the Andean Cordillera was uplifted by so much that gravitational (body) forces exceeded the bearing strength of the rocks, which had to fail by flow or faulting, with consequent spreading. Elliot & Johnson (1978) coined the term 'gravity spreading', rather than gravity sliding, to explain the usual observation that the decollement plane dips towards the hinterland.

The Plio-Pleistocene dynamics recorded in the Cordillera by uplift of a Pliocene planation surface indicates great vertical displacement. After uplift, gravity spreading or 'mushroom tectonics' led to the formation of divergent thrusts. In Ecuador and Peru the spreading of the Cordillera affected the neighbouring areas of the Costa and the Oriente where important basins were created and remain active (i.e. Guajas Basin). As a result of the lateral spreading the central part of the Cordillera was affected by extensional movements with the creation of the Inter-Andean Depression, and the activity of the many young and active volcanoes can reasonably be associated with this extension. Global Positioning System data obtained for present-day crustal shortening at a locality in northern Peru (Leffler *et al.* 1997) could indicate superficial movements which can also be associated with spreading.

Subduction

The ruling theory of plate tectonics and subduction is routinely applied to the Andes, despite the caution expressed by authors such as Gansser (1973), Myers (1975) and Zeil (1979) and contradictory evidence. The crudest models show collision of the Pacific Plate with the South American Plate to form the Andean 'fold mountains'. A more sophisticated approach starts with details of structures within the Andean rocks, relates them to plate tectonics, and assumes that mountain building is somehow associated with this. Megard's (1987) and Isacks' (1988) descriptions of the Andes are good examples of this. However, the presence of a planation surface along the Cordillera is recognized by workers who have examined the large-scale topography (Audebaud *et al.* 1973; Myers 1975; Tosdal *et al.* 1984; Kroonenberg *et al.* 1990; Kennan *et al.* 1997), indicating that the land surface is not folded. It cannot be over-stressed that although some of the rocks are folded, the Andes were formed by vertical uplift following planation.

A variation on this theme is the assumption that the subduction of material, which somehow becomes transformed into granite, ultimately causes uplift. The granites emplaced all along the Andean chain form the most continuous body of granite and granodiorite batholiths in the world (Gansser 1973) but these granites are mostly Mesozoic, and it is difficult to see why uplift should lag so far behind their emplacement. Pitcher & Bussel (1977) suggest that Andean granites rose over a period of 70 Ma but Shackleton (1977) goes further, stating that

'The thickening of crust under the Andes and the recent uplift must be attributed to large accession of material from the mantle, since crustal shortening during Mesozoic and Tertiary time was clearly trivial and could not account for the thickening'. Of course attribution of uplift to mantle processes does not account for the recent uplift.

Even when the existence of a high planation surface is accepted, it is still fashionable to retain a subduction scenario, for example Kennan et al. (1997), who noted, however, that the complete lack of regional palaeosurface tilt is inconsistent with late, ramp-related folding. Similarly, Muñoz & Charrier (1996) claimed underthrusting corresponding to a crustal shortening of 210 km and concluded 'that the Altiplano is an uplifted block controlled by thrust systems at both sides' and, therefore, they suggest subduction from both sides of the Andes. They also reported thrusting of Precambrian metamorphic rocks over late Tertiary sediments in Chile but they note that 'the uplifted Altiplano block reacted to these stresses without a serious compressive deformation on its surface' (Muñoz & Charrier 1996, p. 180). This is more easily explained by gravity spreading with activation of large detachments (slides) in the upper crust during the Upper Miocene (similar to Ecuador) generated by a body force.

Our work in Ecuador emphasizes the symmetry in the arrangement of the Cordillera around the Inter-Andean Depression of Ecuador. Russo & Silver (1996) highlighted an apparent symmetry in the entire Andes as seen on a computer-generated map of South America (illustrated on the front cover of *Geology* in the same issue as their paper). They suggested that this symmetry might be explained by a variation on the Wilson cycle, but if our proposed timing for the Andean tectonics is correct, there would be no time for several Wilson cycles.

Many authors have depicted the symmetry of fan-like divergence of faults (e.g. Megard 1987; Litherland et al. 1993) and some, including Reutter et al. (1988) and Kennan et al. (1997), describe or depict low-angle faults (Fig. 6c) that imply subduction from the east as well as the west. Kennan et al. (1997) state that 'Thin-skinned shortening in the Sub-Andes accommodated c. 140 km of underthrusting of the Brazilian shield beneath the Cordillera Oriental', while Reutter et al. (1988) describe 'underthrusting of the (eastern) foreland under the mobilized crust of the central part of the orogen'.

If the symmetry noted in Ecuador persists along much of the rest of the Andes, the whole question of subduction as a mechanism for creating the Andes will need to be re-evaluated. Gravity spreading can make a symmetrical group of structures, but to make a symmetrical Andean structure by subduction, underthrusting from both sides would be required, which is not part of the general paradigm.

Plate tectonics fail to explain how granite intrusion, planation and uplift have occurred at different times. Nor does it provide a mechanism for vertical uplift followed by gravitational spreading. Above all, subduction is seen as a continuous process, but the geomorphology of the Andes indicates a time of stability, when a wide area including granites and exotic terranes was worn down to form a plain, and a later period of rapid vertical uplift.

Pre-planation and post-planation tectonics

The planation surface divides pre-planation tectonics from post-planation tectonics. This idea seems so obvious that it is hardly worth stating, yet often it is ignored.

The African Planation Surface in Uganda (King 1976) is about as flat as any erosional surface gets, and it is underlain by very complex structures including thrusts and mylonite zones. Nobody would suggest that these structures formed the planation surface. Yet when similar structures are found beneath mountains it is often assumed that the structures formed the mountains. Thus, structures in the Appalachian Mountains were commonly thought of as being related to the Appalachian orogeny that formed the Appalachian Mountains. In reality the Palaeozoic structures were planated, and it was later uplift of the erosion surface and subsequent erosion that created the Appalachians of today. Similarly, the mountains of Scandinavia were planated after the Caledonian orogeny that folded the bedrock, and the Caledonoian orogeny did not make the mountains of today. The same is true in the Andes of Ecuador, and elsewhere. The rock structures in the Mesozoic and older rocks are pre-planation structures, and the 'orogenies' that formed them did not make the present Andes. The fold and thrust belt is the result of Upper Miocene decollements. Decollements, on the other hand, are common features in many 'orogens' (Cook & Varzek 1994; Harry et al. 1995).

The post-planation faults that lifted the horsts of the Eastern and Western Cordillera made the Andes! If our postulated spreading structures are real, they too are post-planation structures. Most accounts of the structure of the Andes take no account of the planation surface, and totally miss

the distinction between pre-planation and post-planation structures. This results in the mistaken belief that all compressive structures 'made the Andes' but the truth is that only the Plio-Pleistocene tectonics are really mountain building. The full significance of the long-recognized erosion surface is still to be appreciated in literature discussing the origin of the Andes.

Conclusions

Remnants of a planation surface formed during the early Pliocene are preserved on the Cordillera of Ecuador. This surface cuts across rocks and structures ranging in age from Palaeozoic to earliest Pliocene.

The driving force for the tectonic movements that created the Andes of Ecuador is vertical uplift, with over 3000 m of uplift in 4–5 million years. Uplift is much younger than the intrusion of granites, and older than the strato-volcanoes that now follow the Andean chain. The fundamental cause of uplift is unknown, but the symmetry of the Andes is undoubtedly a constraint on any possible hypotheses.

In addition to the problem of uplift, there is a problem of erosion: how could this region remain sufficiently tectonically inactive long enough for a regional-scale flat erosion surface to develop, and how could such a surface form in a few millions of years (between late Miocene and Early Pliocene).

One of the greatest of Andean geologists, Bowman (1916, p. 190), stated that 'Proof of the rapid and great uplift of certain now lofty mountain ranges in late geological time is one of the largest contributions of physiography to geologic history'. To this we would add that the distinction between pre-planation structures and post-planation structures adds significantly to both tectonic and geomorphic interpretation.

References

ANDRIESSEN, P. A. M., HELMENS, K. F., HOOGHIEMSTRA, H., RIEZBOS, P. A. & VAN DER HAMMEN, T. 1994. Pliocene–Quaternary chronology of the sediments of the high plain of Bogotá, Eastern Cordillera, Colombia. *Quaternary Science Review*, **12**, 483–503.

ASPEN, J. A. & LITHERLAND, M. 1992. The geology and Mesozoic collisional history of the Cordillera Real, Ecuador. *Tectonophysics*, **205**, 187–204.

AUDEBAUD, E., CAPDEVILLA, R., DALMAYRAC, B. et al. 1973. Les traits geologiques essentiels des Andes centrales (Perou-Bolivia). *Revue de Géographie Physique et Géologie Dynamique*, **15**, 73–114.

BALDOCK, J. W. 1982. *Geology of Ecuador*. Institute of Geological Sciences, London.

BARBERI, F., COLTELLI, M., FERRARA, G., INNOCENTI, F., NAVARRO, J. M. & SANTACROCE, R. 1988. Plio-Quaternary volcanism in Ecuador. *Geological Magazine*, **125**(1), 1–14.

BOWMAN, I. 1909. The Physiography of the Central Andes. *American Journal of Science*, **40**, 197–217.

—— 1916. *The Andes of Southern Peru*. American Geographical Society, New York.

BRISTOW, C. R. & HOFFSTETTER, R. 1977. *Lexique Stratigraphique International: Ecuador*. Centre National de la Recherche Scientifique, Vol. 5.

CLAPPERTON, C. M. 1993. *The Quaternary Geology and Geomorphology of South America*. Elsevier, Amsterdam.

COBBING, E. J., PITCHER, W. S., WILSON, J. J., BALDOCK, W. P., MCCOURT, W. & SNELLING, N. J. 1981. *The geology of the Western Cordillera of northern Peru*. Institute of Geological Sciences, London, Overseas Memoir **5**.

COOK, F. A. & VARZEK, J. L. 1994. Orogen-scale decollements. *Reviews of Geophysics*, **31**(1), 37–60.

ELLIOT, D. & JOHNSON, M. R. W. 1978. Discussion on structures found in thrust belts. *Journal of the Geological Society, London*, **135**, 259–260.

FALSINI, F. 1995. *Rilevamento geologico e geomorfologico dell'area compresa tra Tumbaco e Guayallabamba (Depressione Interandina, Ecuador Centrale)*. Tesi di Laurea, Università di Siena.

FAUCHER, B. & SAVOYAT, E. 1973. Esquisse géologique des Andes de l'Equateur. *Revue de Géographie Physique et Géologie Dynamique*, **15**, 115–142.

FEININGER, T. 1987. Allochtonous terranes in the Andes of Ecuador and northwestern Peru. *Canadian Journal of Earth Sciences*, **24**, 266–278.

FICCARELLI, G., AZZAROLI, A., BORSELLI, V., COLTORTI, M., DRAMIS, F., FEJFAR, O., HIRTZ, A. & TORRE, D. 1992. Stratigraphy and palaeontology of the Upper Pleistocene deposits in the Inter-Andean Depression, Northern Ecuador. *Journal of South America Earth Science*, **6**(3), 145–150.

GANSSER, A. 1973. Facts and theories on the Andes. *Journal of the Geological Society, London*, **129**, 93–131.

HALL, M. & WOOD, C. A. 1985. Volcano-tectonic segmentation of the Northern Andes. *Geology*, **13**, 203–207.

HARRY, D. L., OLDOW, J. S. & SAWYER, D. S. 1995. The growth of orogenic belts and the role of crustal heterogeneities in decollement tectonics. *Geological Society of America Bulletin*, **107**(12), 1411–1426.

HOORN, C. 1993. Marine incursions and the influence of Andean tectonics on the Miocene depositional history of northwestern Amazonia: results of a palynostratigraphic study. *Palaeogeography, Palaeoclimatology, Palaeoecology*, **105**, 267–309.

——, GUERRERO, J., SARMIENTO, G. A., & LORENTE, M. A. 1995. Andean tectonics as a cause for changing drainage patterns in Miocene northern South America. *Geology*, **23**(3), 237–240.

HUNGERBUHLER, D., STEINMANN, M., WILKLER, W., SEWARD, D., EGUEZ, A., HELLER, F. & FORD, M. 1995. An integrated study of fill and deformation in the Andean Intermontane basin of Nabon (Late Miocene), southern Ecuador. *Sedimentary Geology*, **96**, 257–279.

ISACKS, B. L. 1988. Uplift of the Central Andean Plateau and bending of the Bolivian Orocline. *Journal of Geophysical Research*, **93**(B4), 3211–3231.

JACOB, A. F. 1983. Mountain front thrust, southeastern Front Range and northeastern Wet Mountains, Colorado. *In*: LOWELL, J. D. (ed.), *Rocky Mountain Foreland Basins and Uplifts*. Rocky Mountain Association of Geologists, Denver, 229–244.

JEFFREYS, H. 1931. On the mechanics of mountains. *Geological Magazine*, **68**, 433–442.

KENNAN, L., LAMB, S. H. & HOOKE, L. 1997. High-altitude palaeosurfaces in the Bolivian Andes: evidence for late Cenozoic surface uplift. *In*: WIDDOWSON, M. (ed.), *Palaeosurfaces: Recognition, Reconstruction and Palaeoenvironmental Interpretation*. Geological Society, London, Special Publications, **120**, 307–323.

KENNERLY, J. B. 1980. *Outline of the geology of Ecuador*. Institute of Geological Sciences, London, Overseas Geology and Mineral Resources, **55**.

KING, L. 1976. Planation remnants upon highlands. *Zeitschrift für Geomorphologie*, *N.S.*, **20**(2), 133–148.

KROONENBERG, S. B., BAKKER, J. G. M. & VAN DER WIEL, M. 1990. Late Cenozoic uplift and paleogeography of the Colombian Andes: constraints on the development of the high-Andean biota. *Geologie en Mijnbouw*, **69**, 279–290.

LAVENU, A., NOBLET, C. & WINTER, T. 1995. Neogene ongoing tectonics in the Southern Ecuadorian Andes: analysis of the evolution of the stress field. *Journal of Structural Geology*, **17**, 47–58.

LEFFLER, L., STEIN, S., MAO, A., DIXON, T., ELLIS, M. A., OCOLA, L. & SACKS, I. S. 1997. Constraints on present-day shortening rate across the central Eastern Andes from GPS data. *Geophysical Research Letters*, **24**, 1031–1034.

LITHERLAND, M. & ASPEN, J. A. 1992. Terrane bordering reactivation. *Journal of South American Earth Science*, **5**, 71–6.

—, ZAMORA, A. & EGUEZ, A. 1993. *Mapa geologico del Euador, escala 1:1,000,000*. CODIGEM-BGS, Quito.

LONSDALE, P. 1978. Ecuadorian subduction system. *AAPG Bulletin*, **62**, 2454–2477.

LOWELL, J. D. (ed.) 1983. *Rocky Mountain Foreland Basins and Uplifts*. Rocky Mountain Association of Geologists, Denver.

MCLAUGHLIN, D. H. 1924. Geology and physiography of the Peruvian Cordillera, Departments of Junin and Lima. *Bulletin of the Geoogical Society of America*, **35**, 591–632.

MEGARD, F. 1987. Structure and evolution of the Peruvian Andes. *In*: SCHAER, J. P. & ROGERS, J. (eds) *The Anatomy of Mountain Ranges*. Princeton University Press, 179–210.

MUÑOZ, J. 1956. *In*: JENKS, W. F. (ed.) *Handbook of South American Geology*. Geological Society of America, Memoir, **65**, 187–214.

MUÑOZ, N. & CHARRIER, R. 1996. Uplift of the western border of the Altiplano on a west-vergent thrust system, Northern Chile. *Journal of South American Earth Science*, **9**, 171–181.

MYERS, J. S. 1975. Vertical crustal movement of the Andes in Peru. *Nature*, **254**, 672–644.

OLLIER, C. D. & WYBORN, D. 1989. Geology of alpine Australia. *In*: GOOD, R. (ed.) *The Scientific Significance of the Australian Alps*. Australian Alps, National Parks Liaison Committee and Australian Academy of Science, Canberra, 35–53.

PILGER, R. 1981. Plate reconstructions, aseismic ridges, and low-angle subduction beneath the Andes. *Geological Society of America, Bulletin*, **92**, 448–456.

PITCHER, W. A. & BUSSEL, M. A. 1977. Structural control of batholithic emplacement in Peru: a review. *Journal of the Geological Society, London*, **133**, 249–256.

RADELLI, L. 1967. *Géologie des Andes Colombiennes*. Travaux du Laboratoire de la Faculté des Sciences de Grenoble, Memoir No. 6.

REUTTER, K. J., GIESE, P., GOTZE, H. J., SCHEUBER, E., SCWAB, K., SCHWARZ, G. & WIGGER, P. 1988. Structures and crustal development of the central Andes between 21° and 25°S. *In*: BAHLBURG, H., BREITKREUZ, C. & GIESE, P. (eds), *The Southern Andes*. Springer, Berlin, 231–261.

ROSANIA, S. G. 1989. Petroleum prospects of the sedimentary basins of Ecuador. *In*: ERICKSEN, G. E., PINOCHET M. T. & REINEMUND, J. A. (eds) *Geology of the Andes and its relation to hydrocarbon and mineral resources*. Circum-Pacific Council for Energy and Mineral Resources. Houston, Texas, Earth Science Series, **11**, 415–430.

RUSSO, R. M. & SILVER, P. G. 1996. Cordillera formation, mantle dynamics, and the Wilson Cycle. *Geology*, **24**, 511–514.

SHACKLETON, R. S. 1977. Discussion after Pitcher and Bussel. *Journal of the Geological Society, London*, **133**, 255.

SUAREZ, G., MOLNAR, P. & BURCHFIEL, B. C. 1983. Seismicity, fault plane solution, depth of faulting and active tectonics of the Andes of Peru, Ecuador and Southern Colombia. *Journal of Geophysical Research*, **88**(B12), 10403–10428.

TOSDAL, R. M., CLARK, A. H. & FARRAR, E. 1984. Cenozoic polyphase landscape and tectonic evolution of the Cordillera Occidental, southernmost Peru. *Geological Society of America Bulletin*, **95**, 1318–1332.

VAN DER HAMMEN, T., WERNER, J. H. & VAN DOMMELEN, H. 1973. Palynological record of the upheaval of the Northern Andes: a study of the Pliocene and Lower Quaternary of the Colombian Eastern Cordillera and the early evolution of its high-Andean biota. *Palaeogeography, Palaeoclimatology, Palaeoecology*, **16**, 1–22.

WDOWINSKI, S. & BOCK, Y. 1994a. The evolution of deformation and topography of the high elevated plateaus. 1. Model, numerical analysis and general results. *Journal of Geophysical Research*, **99**, 7103–7119.

—— & —— 1994b. The evolution of deformation and topography of the high elevated plateaus. 2. Application to the Central Andes. *Journal of Geophysical Research*, **99**, 7121–7130.

WOODWARD, L. A. 1983. Potential oil and gas traps along the overhang of the Nacimiento Uplift, northwestern New Mexico. *In*: LOWELL, J. D. (ed.) *Rocky Mountain Foreland Basins and Uplifts*. Rocky Mountain Association of Geologists, Denver, 213–218.

ZEIL, W. 1979. *The Andes. A geological review*. Gedruder Borntraeger, Berlin.

Cosmogenic isotope data support previous evidence of extremely low rates of denudation in the Dry Valleys region, southern Victoria Land, Antarctica

M. A. SUMMERFIELD[1], D. E. SUGDEN[1], G. H. DENTON[2], D. R. MARCHANT[3], H. A. P. COCKBURN[1] & F. M. STUART[4]

[1] *Department of Geography, University of Edinburgh, Edinburgh EH8 9XP, UK*
[2] *Department of Geological Sciences and Institute for Quaternary Studies, University of Maine, Orono, Maine 04469, USA*
[3] *Department of Earth Sciences, Boston University, 675 Commonwealth Avenue, Boston, Massachusetts 02215, USA*
[4] *Isotope Geosciences Unit, Scottish Universities Research and Reactor Centre, East Kilbride, G75 0QF, UK*

Abstract: Quantitative estimates of denudation rates in different tectonic and climatic environments are of fundamental importance to an understanding of long-term landscape development. Little quantitative information is currently available on minimum denudation rates, although this is important in placing constraints on the maximum survival potential of individual landforms or erosion surfaces in terrestrial environments. The persistence for up to >15 Ma of a hyper-arid, polar climate in the ice-free Dry Valleys area of southern Victoria Land, Antarctica, makes this a highly probable environment for minimum terrestrial denudation rates. ^{40}Ar/^{39}Ar dating constraints on the mimimum formation ages of individual landforms indicates generally little modification of even minor features over the past few million years. Relatively, the highest rates of denudation occur at low elevations near the Ross Sea coast. By contrast, rates of landscape change are exceedingly slow at elevations above c. 1500 m in the western Dry Valleys where landforms with a relief of a only a few metres have survived with minimal modification since the mid-Miocene. Measurements of *in situ*-produced cosmogenic isotopes indicate that even on rectilinear slopes maximum denudation rates may be only $c.$ 1 m Ma^{-1}, with rates falling to <0.2 m Ma^{-1} at some locations on low relief surfaces at high elevation.

Quantifying rates of denudation is a prerequisite to the comprehensive understanding of landscape evolution. Over the past decade a range of new geochronological techniques has been applied to the problem of rates of long-term landscape change and there are now increasing opportunities to compare this growing body of data with estimates of modern denudation rates based on sediment and solute load data (Summerfield & Hulton 1994). Although the data now available indicate that long-term, natural denudation rates can range up to $c.$ 10 000 m Ma^{-1} in active orogenic belts, such as over parts of the Southern Alps of New Zealand (Tippett & Kamp 1995), the lower limit for denudation rates in terrestrial environments is rather poorly constrained. Obviously in some environments long-term deposition has occurred, and the net excess of sedimentation over denudation may persist over extensive areas for tens of millions of years in cratonic basins in continental interiors, such as the Kalahari Basin of southern Africa. However, quantitative data on mimimum rates of denudation in environments not subject to significant or persistent sediment accumulation are important because the idea of the 'age' of a landform or landsurface is predicated upon the notion of minimal modification by denudational processes since its creation.

The combination of persistent low temperatures and hyper-aridity suggests that the ice-free 'oasis' of the Dry Valleys area of southern Victoria Land, Antarctica, is a prime candidate for an environment that has experienced very low rates of denudation and extreme landscape stability. Previous work on tors, rock slopes and associated landforms in this region (Selby 1971, 1974) has indicated low rates of landscape change, an interpretation which echoed ideas of some of the early explorers of the Dry Valleys and adjacent ice-free areas within the Transantarctic Mountains (Priestley 1909; Taylor 1922). However, none of these studies was able to call upon an adequate body of geochronological data to support their interpretations of

rates of landscape change which were derived largely from morphological and stratigraphic relationships in the landscape. Here we review a range of geochronological data which, in combination with information from morphological and depositional evidence, have provided quantitative estimates of rates of denudation and landscape change in the Dry Valleys region. We also show how data from cosmogenic isotope analysis can provide a means, hitherto unavailable, of constraining site-specific denudation rates in this extreme morphoclimatic environment.

Field setting

The Dry Valleys region consists of a number of ice-free, or partially ice-free, transverse valleys and intervening mountain ridges in the Transantarctic Mountains flanking McMurdo Sound in the Ross Sea embayment between 77°15′S and 77°45′S and 160°E to 164°E (Fig. 1). Inland lies the Polar Plateau formed by the East Antarctic Ice Sheet from which outlet glaciers have drained at various times in the past through the Dry Valleys. The majority of the 4000 km^2 ice-free area is accounted for by the Taylor, Wright and Victoria Valleys and the intervening 1500–2400 m high Asgard and Olympus Ranges. These transverse valleys cut across the compound escarpment that marks the Ross Sea margin of the topographic upwarp on the rift-flank uplift forming the Transantarctic Mountains.

Geological and tectonic setting

Exposed bedrock in the Dry Valleys region comprises a basement complex of Precambrian igneous and meta-igneous rocks overlain by gently dipping sandstones, siltstones and conglomerates of the Devonian–Triassic Beacon Supergroup. Both the basement and overlying sedimentary strata are intruded by dolerites of Jurassic age (Ferrar dolerites) which form extensive sills up to several hundred metres thick. Highly localized volcanic cones and ash deposits of late Cenozoic age are also extensively distributed across the region. The major structural components of the Transantarctic Mountains in the Dry Valleys area were established by asymmetric rifting in the Eocene, with the lower plate forming the Ross Sea Embayment and a combination of block-tilting and flexure in the upper-plate margin producing an upwarped rim to the extensive interior plateau and the modest inland dip observed in the Beacon Supergroup strata (Fitzgerald *et al.* 1986; Fitzgerald 1992).

Morphology of the Dry Valleys region

The Dry Valleys region is characterized by two major landscape components: high-elevation, low-relief surfaces, and rectilinear slopes (Sugden *et al.* 1995a). The high-elevation surfaces in general lie above an elevation of 1800 m and occur both as flat-topped ridges in the western sectors of the Asgard and Olympus Ranges and, more extensively, on the more elevated terrain between 2000 and 2400 m lying at the head of the Dry Valleys and separating them from the ice-cover of the Polar Plateau. Rectilinear slopes, typically at angles of 26–37°, are prominent both on the flanks of the high surfaces adjacent to the Polar Plateau and on the sides of the Dry Valleys themselves. At higher elevations in the western sector of the Dry Valleys the rectilinear slopes lead up to strength-equilibrium free face slopes at angles of >60° and merge down slope into colluvium-mantled footslopes at an angle of around 18° (Selby 1971). At lower elevations in the eastern part of the Dry Valleys, free faces are less common and the rectilinear slopes of adjacent valleys meet as sharp-edged interfluves. These rectilinear forms have been interpreted as Richter denudation slopes by Selby (1971, 1974, 1993). They are cut in bedrock but partly covered by a thin veneer of generally coarse rock debris up to 1 m thick.

Present and past climates

The present climate of the Dry Valleys region is cold and hyper-arid. Mean annual temperatures at lower elevations in the central and eastern sectors of the valleys are around −20°C, although midday temperatures in summer may exceed +5°C for several days (Bull 1966; Schwerdtfeger 1984). The presence of active debris and mud flows, together with episodic flow in stream channels, attests to temperatures above 0°C below elevations of *c.* 800 m under the present climatic regime. Meteorological data for higher inland locations are limited but, on the assumption of a lapse rate of 10°C km^{-1}, mean annual temperatures at high elevations fringing the Polar Plateau and in the western Asgard and Olympus Ranges must drop to −30°C to −40°C (Schwerdtfeger 1984; Fortuin & Oerlemans 1990). Mean annual precipitation ranges from around 100 mm water equivalent

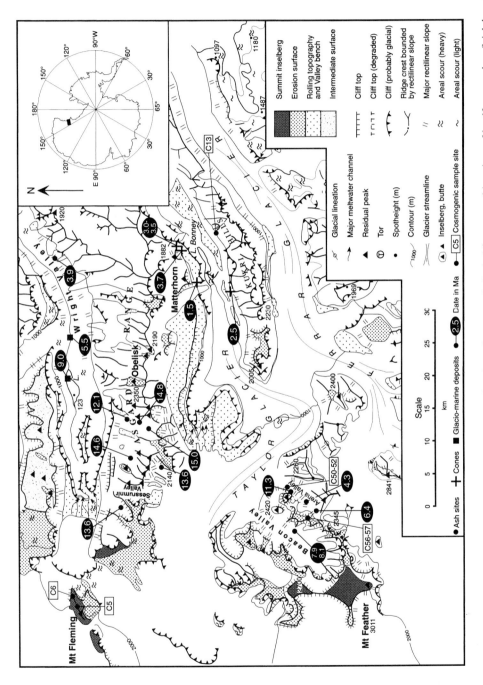

Fig. 1. Morphological map of Beacon, Taylor and Wright valleys showing the main geomorphological features and locations of landforms and surficial deposits dated by association with volcanic deposits together with cosmogenic ^{21}Ne sample sites.

near the coast to <10 mm at high elevations in the west, and this trend is mirrored by relative humidity which drops from around 75% immediately inland from the Ross Sea to as low as 5% on the Polar Plateau. Throughout the year strong, low-humidity (5–60%) katabatic winds blow from the Polar Plateau, but during the summer they are largely confined to the western ends of the Dry Valleys and to the valley bottoms. Snow accumulation is sufficient to sustain small, cold-based glaciers which flow down the sides of Taylor, Victoria and Wright Valleys.

The systematic trend in climatic conditions in response to altitude and distance from the coast suggests three morphoclimatic zones based on temperature, precipitation, relative humidity and soil-moisture content (Marchant & Denton 1996). In the coastal zone, with a mean annual temperature of $c. -17°C$ and mean relative humidity of $c. 75\%$, there are active gelifluction terraces and lobes, streams, levées, debris and mud flows and sub-xerous soils. In the intermediate zone between $c. 40$ and 55 km inland, where the mean annual temperature is $c. -27°C$ and mean relative humidity is $c. 45\%$, there is little evidence of active slope processes, with streams, gelifluction lobes and debris flows being mostly confined to warmer north-facing slopes with more prolonged availability of liquid water on the surface and in the immediate substrate. Slope forms indicative of liquid water are essentially absent from the third morphoclimatic zone distributed across the elevated terrain of the western Dry Valleys in the Quartermain Mountains and the higher, western sectors of the Asgard and Olympus Ranges.

Late Cenozoic climates in the Dry Valleys region are a subject of dispute. According to one view, based primarily on the interpretation of Pliocene-age marine diatoms present in glaciogenic Sirius Group deposits located in the Transantarctic Mountains, the East Antarctic Ice Sheet experienced deglaciation in the early Pliocene (c. 4.5 Ma BP), with deglacial conditions probably continuing into the mid-Pliocene (Webb et al. 1984; Webb & Harwood 1991; Barrett et al. 1992; Harwood 1994). This model postulates the existence of seaways across the Wilkes-Pensacola Basin in the interior of Antarctica during this period, with these marine incursions providing the source of the diatoms found in the Sirius Group tills. Although it has been argued that such dramatic changes in the extent of the East Antarctic Ice Sheet could be achieved with a modest increase in temperature compared with the present (Webb & Harwood 1991; Harwood 1994), ice-sheet modelling studies indicate that temperatures 17–20°C above those of the present would be required to remove ice from the Wilkes-Pensacola Basin (Huybrechts 1993). Furthermore, there is a growing consensus that the marine diatoms that have been used to date the Sirius Group deposits are allogenic contaminants emplaced by aeolian deposition (Burckle & Potter 1996; Kellogg & Kellogg 1996), or possibly derived from fall-out from a bolide impact (Gersonde et al. 1997), and that they are therefore not capable of constraining the age of deposition to the Pliocene.

A contrasting interpretation of the history of the East Antarctic Ice Sheet is that it has experienced long-term stability under a hyper-arid, polar climatic regime similar to that of the present for at least the past 15 Ma (Clapperton & Sugden 1990; Denton et al. 1993; Marchant et al. 1993a; Sugden et al. 1995a). This view is based on a range of evidence including marine stratigraphic data from the Southern Ocean (Kennett & Hodell 1993), modelling studies of ice-sheet dynamics (Huybrechts 1993), and morphological, stratigraphic and $^{40}Ar/^{39}Ar$ data from the Dry Valleys region and neighbouring areas (Denton et al. 1993; Marchant et al. 1993a, 1996). Irrespective of the veracity of these contrasting interpretations of the longer term climatic history of Antarctica, there is agreement that there has been remarkable climatic stability in the Dry Valleys area for the last 2–3 Ma, with only modest fluctuations in the extent of ice cover in the outflow and valley-side glaciers during this period. This morphoclimatic regime contrasts starkly with the large-magnitude climatic fluctuations driven by glacial–interglacial oscillations that have characterized virtually all other morphoclimatic zones on Earth throughout the Quaternary.

Morphological and depositional evidence of rates of landscape change

The last few years have seen significant progress in the geochronometric dating of landforms in the Dry Valleys area and this has greatly enhanced our knowledge of the glacial history and long-term landscape evolution of the area (Denton et al. 1993). The main conclusion arising from this programme of landform dating is that a number of minor landforms, including small cinder cones, surficial forms associated with volcanic ash falls, raised marine features and glacial forms, have survived without significant modification since the mid-Miocene. The implication of this is that surface geomorphic processes have operated at a very low level of

activity over the past c. 15 Ma. Here we collate observations on these various landforms and deposits in order to assess rates of landscape change; a summary of observations is presented in Table 1 and site locations are shown in Fig. 1.

Cinder cones

In central Taylor Valley the glacially moulded basement surface is dotted with a series of small cinder cones, typically 200-1000 m across, and associated lava flows (Wilch et al. 1993). Fourteen flows have been dated by whole-rock $^{40}Ar/^{39}Ar$ analysis, and have been shown to range in age from 1.5 to 3.9 Ma. Some of the cones retain their original form, whereas others have been partially eroded. In the latter cases, trails of dark volcanic rocks may be seen in the regolith extending several tens of metres directly downslope of the volcanic outcrop. There is insufficient evidence to quantify precisely the amount of denudation involved, but there is a relationship between the date of eruption and degree of erosion (Wilch et al. 1993). For example, the youngest 1.5 Ma old cone east of Rhone Glacier has essentially retained its initial form, whereas the older West Matterhorn cone of 3.0-3.5 Ma is heavily eroded (Table 1). The East Borns cone, with an age of 2.53 Ma, shows an intermediate degree of erosional modification. Within the altitudinal range of the cinder cones in Taylor Valley (209 to 828 m), denudational processes have been capable of eroding small cinder cones, but only slowly and over time scales of millions of years. A further point is that all the cones have been erupted subaerially, and their age and altitudinal relationships demonstrate recent tectonic stability and constrain surface uplift in central Taylor Valleys to less than c. 300 m in the last 2.57 Ma (Wilch et al. 1993).

Volcanic ash deposits

Volcanic ash falls, often of considerable antiquity, are intimately related to several types of deposit in the Dry Valleys area (Table 1). In Arena Valley at the western extremity of Taylor Valley there are two avalanche tongues where 30% of the matrix consists of volcanic ash (Marchant et al. 1993a). The unconsolidated ash is covered by a protective lag of pebbles, cobbles and boulders. One avalanche tongue, containing ash 11.3 Ma old, forms an elongated mound rising 3 m above the surrounding slope, the overall angle of which is 28°. In the western Dry Valleys area volcanic ashes are found at high altitudes in association with relict frost wedges (Marchant et al. 1993a,b), the ash having collected in the polygonal cracks much as snow does today (Fig. 2). The high concentration of ash and its uniform geochemistry, together with the bimodal grain-size distribution of glass shards, stratification by grain size, intact bubble vesicles and angular glass shards point to deposition by primary air falls with minimum subsequent disturbance. As the crack patterns have evolved, ash has become preserved in pockets in relict wedges (Fig. 3). Some 50 ashes associated with tills, slope regolith, lag deposits, lake sediments and, in one case, a moraine overlying relict glacier ice (Table 1), have been dated by $^{40}Ar/^{39}Ar$ single/multiple crystal analysis, the oldest yielding an age of c. 15.0 Ma. In a number of other locations, such as in Wright, Arena and Nibelungen valleys, ashes ranging in age from 3.9 to 14.8 Ma have been located overlying old desert pavement surfaces (Table 1).

Various inferences concerning surface geomorphic processes can be drawn from this preservation of volcanic ashes dating back to 15 Ma BP. For instance, it is evident that regolith, particularly at high altitudes, has undergone little weathering or transport since the mid-Miocene. This is particularly true of the ash-filled wedges which have retained their essential integrity for millions of years, thereby demonstrating that during this period there has been little disturbance or differential movement of the parent material in which they lie, be it till, slope regolith or moraine-covered glacial ice. This point is further supported by an avalanche tongue in the Arena Valley, which has retained its form on a slope of 28° for 11.3 Ma (Fig. 4). It follows from this that the rectilinear slopes on which the ashes are preserved must also be old. Volcanic ashes occur on valley floors and, in one case, in a lake deposit associated with a glacially scoured rock basin (Table 1). The age and spatial extent of the ashes imply that the landsurface, at least at higher altitudes, is essentially relict from pre-mid-Miocene times. In areas with a protective layer of regolith there has been minimal bedrock lowering in the last 15 Ma.

The preservation of landforms with which the ashes are associated is itself evidence of a remarkable degree of climatic stability at higher elevations in the Dry Valleys area since at least the mid-Miocene. Buried lag deposits reflect arid conditions as long ago as 14.8 Ma BP, while the presence of frost cracks to trap the ashes suggests

Table 1. *Evidence demonstrating slow rates of landscape change*

Location	Site/stratigraphy	Altitude (m)	Basis of dating	$^{40}Ar/^{39}Ar$ age (Ma) (unless stated otherwise)	Comments
Cinder cones in Taylor Valley (Wilch *et al.* 1993)					
East Rhone	Volcanic cone	671	Whole-rock basalt	1.5	Intact pristine cone
East Borns	Volcanic cone	587	Whole-rock basalt	2.53	Partially eroded cone
West Matterhorn	Volcanic outcrop	602	Whole-rock basalt	3.0–3.5	Heavily eroded volcanic outcrop
East Matterhorn	Volcanic outcrop	828	Whole-rock basalt	3.74	Eroded volcanic outcrop
Volcanic ash avalanches (Marchant *et al.* 1993*a*)					
Upper Arena Valley	Ash avalanche tongue, 20 × 200 m in size	1380	Sanidine; 3 crystals	6.4	Deposit survived intact for 6 Ma, 1–1.5 m lowering of surrounding, exposed Arena sandstone
Lower Arena Valley	Ash avalanche tongue, 50 × 350 m in size, standing 3 m proud of slope of 28°	1625–1650	Sanidine; 3 crystals	11.3	Preservation of avalanche tongue morphology on steep rectilinear slope for >11.3 Ma
Volcanic ash in unconsolidated tills and regolith (Marchant *et al.* 1993*b*)					
Nibelungen Valley	Ash in wedge in weathered regolith at foot of rectilinear slope	1450	Sanidine; 5 crystals	12.1	Ancient sand wedge; regolith older than 12 Ma; cold desert climate shown by existence of wedge
Inland Forts, Asgard Range	Ash in wedge in weathered till	1650	Glass	13.6	Till and rectilinear slope older than 13.6 Ma; cold desert conditions
Koenig Valley, Asgard Range	Ash in wedge in till on valley floor	1800	Sanidine; 5 crystals	13.6	Till and rectilinear slope older than 13.6 Ma; cold desert conditions
Nibelungen Valley, Asgard Range	Ash in wedge	1600	Sanidine; 10 crystals	15.0	Till at base of rectilinear slope has survived for 15.0 Ma; cold desert climate
Njord Valley, Asgard Range	Grey, stratified with dropstones	1600	Sanidine; 3 crystals	14.6	Small pond deposit and bedrock hollow survived 14.6 Ma
Volcanic ash above glacier ice (Sugden *et al.* 1995*b*)					
Beacon Valley, Quartermain Mountains	Ash in wedge overlying glacier ice	1380	Sanidine; 14 crystals	7.9, 8.1	Glacier ice >8 Ma old; persistent cold conditions for >8 Ma

Location	Description		Method	Age (Ma)	Interpretation
Volcanic ash burying desert pavements					
West Central Arena Valley, Quartermain Mountains	Ash overlies pavement and is covered by regolith	1800	Anorthoclase	4.3	Minimal slope development in 4.3 Ma; desert conditions at the time (Marchant et al. 1993a)
Nibelungen Valley, Asgard Range	Ash overlies desert pavement at base of steep rectilinear slope; covered by till	1500	Sanidine	14.8	Rectilinear slope older than 14.8 Ma; desert conditions at the time (Marchant et al. 1993b)
Wright Valley between Hart and Goodspeed glaciers	Ash overlies oxidized regolith; overlain by till	378–526	Glass	3.9	Minimum age of regolith on side of Wright Valley; desert conditions at the time (Hall et al. 1993)
Preservation of raised marine deposits (Prentice et al. 1993)					
Central Wright Valley	Glacio-marine Jason diamicton near Lake Vanda	3–250	^{87}Sr/^{86}Sr on shell fragments	9 ± 1.5	Fjord sediments have survived in valley bottom since Miocene
Central Wright Valley	Shells in growing position on valley floor and in Prospect Mesa section	165	^{87}Sr/^{86}Sr on *Chlamys tuftsensis*	5.5 ± 0.4	Details of bed of fjord have survived for 5.5 Ma
Modification of glacial landforms					
Sessrumnir Valley, Asgard Range	Meltwater channels and potholes	1550	Association with ash-dated till deposits	>13.6	Cavernous weathering and undercutting of 4 m in >13.6 Ma (Sugden et al. 1991)
Central Wright Valley	Alpine IV moraine ridges	400–800	Whole-rock basalt	>3.7	Weathered stones but ridge forms intact for over 3.7 Ma (Hall et al. 1993)
Central Wright Valley	Feather edge of Peleus till occurs on 25–35° slope	<1150	Minimum age of till	>3.8	Upper till limit survived on steep slope for over 3.8 Ma (Hall et al. 1993)

Fig. 2. Polygonal cracks in regolith in McKelvey Valley, central Dry Valleys, picked out by snow. The polygons range from 5 to 15 m across.

Fig. 3. Volcanic ash preserved in a relict wedge in till, Beacon Valley.

Fig. 4. Volcanic ash avalanche tongue (arrowed) in Arena Valley, $^{40}Ar/^{39}Ar$ dated at 11.3 Ma BP. Moraines associated with an enlarged Taylor Glacier have been deposited on the tongue without deforming or displacing it.

that the temperatures have been typical of a cold polar climate since the mid-Miocene. In Beacon Valley at the head of Taylor Valley, the preservation of glacier ice beneath a thin moraine cover incorporating 8 Ma old volcanic ash deposits is difficult to explain unless a polar climate similar to that of today has persisted since this time (Sugden et al. 1995b). The very freshness of the ash, which permits its dating, also demonstrates the lack of chemical weathering (Marchant et al. 1993b), and this, in turn, implies the absence over the past several million years of conditions significantly milder and moister than those of the present. In Arena Valley, the clay content of ashes up to 11.3 Ma old is less than 5%, and this can be compared to a figure of 60% attained in just 50 ka under the temperate climate of New Zealand through highly active chemical weathering (Birrell & Pullar 1973). More generally, the lack of gelifluction lobes and terraces, rills and stream channels in dated Miocene and Pliocene deposits above c. 1200 m in the far western Dry Valleys area demonstrates that the temperature and moisture conditions characteristic of lower elevations closer to the Ross Sea to the east have not extended into these high-elevation areas over the past 15 Ma or so.

Raised marine features

Miocene and Pliocene raised marine deposits in the bottom of Wright Valley demonstrate slow rates of denudation even at low altitudes in the Dry Valleys region. The Jason glacio-marine diamicton occurs at an altitude of 3 to 250 m on the northern shore of Lake Vanda in Wright Valley, and $^{87}Sr/^{86}Sr$ dating of shell fragments demonstrates a late Miocene (9.5 ± 1.5 Ma) age for this deposit. In the vicinity of Prospect Mesa, a few kilometres along the valley to the east, there is a shell fauna with $^{87}Sr/^{86}Sr$ ages of 5.5 ± 0.4 Ma which, remarkably, is exposed in growth positions on parts of the present floor of the valley. The implication here is that the bed of a former fjord has survived since emerging from the retreating sea around 3.4 Ma ago (Hall et al. 1993), although there are some uncertainties surrounding the significance of these raised shell deposits for estimating rates of landscape

change. The date of emergence is based on extrapolation from a marine core in an adjacent valley (Ishman & Rieck 1992) and Wright Valley fjord would have been isolated as a lake of unknown size and for an unknown duration before finally drying out. Indeed, the present Lake Vanda is a direct descendant of the isolated fjord and is known to have varied in extent sufficiently to cover both sites on different occasions in the past. Furthermore, one cannot exclude the possibility that the fjord bed, including the *in situ* shells, was buried by sediment that has subsequently been eroded. Nonetheless, it remains clear that unconsolidated marine sediments have remained intact in the bottom of one of the larger of the Dry Valleys for c. 9 Ma. Furthermore, the preservation of fragile shells in growth positions on the valley floor is consistent with minimal amounts of landscape modification over this period.

Glacial landforms

The modification of glacial landforms of known age can also be used to estimate rates of landscape change. For example, weathering has affected the walls of meltwater channels and associated potholes at an altitude of 1550 m in the Asgard Range (Table 1). Cavernous weathering forms are common on former glacial surfaces, and in places fine-grained sandstone beds have been exploited to form overhangs up to 4 m high (Sugden *et al.* 1991). This represents the sum of subaerial denudation in the >13.6 Ma since the last warm-based glaciers overrode the mountains (Marchant *et al.* 1993*b*). Another example concerns Alpine IV moraines flanking the glaciers flowing into the side of Wright Valley at altitudes of 400–800 m. Ages for basalt clasts show the moraines to be over 3.7 Ma old and yet they retain their ridge form. There is evidence of some weathering; former upstanding boulders have been planed off and shattered and cavernously weathered, and ventifacts mantle the surface. Soils are oxidized to a depth of 0.42 m and contain visible salt horizons (Hall *et al.* 1993). Finally, one can point to the presence of Peleus Till in Wright Valley, whose upper feather edge is preserved on slopes of 25–35°. This deposit is at least 3.8 Ma old, and could well be the equivalent of the Asgard Till which is at least 13.6 Ma old (Hall *et al.* 1993). The implication here is that surface geomorphic processes acting over several million years have been unable to remove a till deposit from slopes of 25–35° at an altitude of 1150 m in Wright Valley and that therefore rates of denudation have been extremely low.

Denudation rate estimates from cosmogenic isotope data

The semi-quantitative estimates of rates of landscape change for the Dry Valleys area outlined above can be supplemented by denudation rates estimated from the accumulation of *in situ* produced cosmogenic isotopes in surface rock samples (Bierman 1994; Kurz & Brook 1994). A number of studies have previously reported denudation rate estimates for sites in the Dry Valleys and neighbouring ice-free areas in the Antarctic 'oasis' based on cosmogenic nuclide measurements. Nishiizumi *et al.* (1991), for example, have modelled maximum denudation rates over the past few million years on the basis of measurements of the cosmogenic radionuclides ^{10}Be and ^{26}Al. Nine samples from Allan Nunatak (Allan Hills), located about 75 km northwest of the northern limit of the Dry Valleys area, yielded denudation rates of 0.24–1.31 m Ma^{-1} (mean: 0.65 m Ma^{-1}), while four further samples from Wright Valley gave a range of 0.54–1.31 m Ma^{-1} (mean: 0.93 m Ma^{-1}). Somewhat lower denudation rates of 0.06–0.27 m Ma^{-1} over the past 2–3 Ma, based on cosmogenic ^{10}Be measurements, have been reported by Brook *et al.* (1995) for quartz arenite boulders and cobbles in glacial deposits in the Quartermain Mountains and Asgard Range. Extremely low maximum denudation rates based on cosmogenic ^{26}Al and ^{10}Be data have also been reported by Ivy-Ochs *et al.* (1995). Their samples from boulders in Sirius Group deposits on Table Mountain at the head of Ferrar Glacier and at Mount Fleming at the western end of Wright Valley, together with high-elevation outcrop samples of Beacon Supergroup sandstone, yielded denudation rates ranging only up to 0.7 m Ma^{-1}. Denudation rates of less than 0.1 m Ma^{-1} over the past several million years have also been revealed by measurements of the cosmogenic ^{21}Ne in pyroxene, quartz and whole-rock dolerite samples from Sirius Group tillites and associated bedrock from high-elevation sites in the Mount Fleming and Table Mountain areas (Bruno *et al.* 1997).

In order to supplement these data and provide estimates of denudation rates in contrasting morphological settings, we have measured cosmogenic ^{21}Ne in quartz in samples from both low-relief, plateau surfaces and rectilinear slopes in the Dry Valleys. Four samples from *in situ* bedrock outcrops of Beacon Sandstone

(Beacon Supergroup) were collected from sites on high-elevation (>2000 m) surfaces in the western Dry Valleys: two from the eastern flank of Mount Fleming (C5, C6) and two from the flat-topped interfluve between Arena and Beacon valleys at head of Taylor Glacier (C56, C57). Three further samples (C50–C52) of Arena Sandstone (Beacon Supergroup) were collected from a 36–38° rectilinear slope in the Quatermain Mountains on the western side of Arena Valley, this site being selected because of its proximity to the ^{40}Ar/^{39}Ar-dated ash avalanche deposits described above. The final sample analysed (C13) comprised exfoliating fragments of granite from a rectilinear slope in lower Taylor Valley adjacent to a ^{40}Ar/^{39}Ar-dated, partially eroded pyroclastic and lava-flow deposit. Further details of sample locations together with documentation of analytical methods and procedures for calculating denudation rates are given elsewhere (Summerfield et al. 1999).

Of the eight samples analysed, the highest denudation rate recorded of $1.02\,\text{m}\,\text{Ma}^{-1}$ over at least the past 1 Ma was from a rectilinear slope in lower Taylor Valley at an elevation of $c.\,800\,\text{m}$ (C13) (Fig. 1). All four samples from the high-elevation surfaces (>2000 m) in the western Dry Valleys display much lower rates of 0.135–$0.165\,\text{m}\,\text{Ma}^{-1}$ (over at least the past 4 Ma). Samples from the rectilinear slope on the western side of Arena Valley (C50–C52) yielded intermediate denudation rates of 0.23–$0.60\,\text{m}\,\text{Ma}^{-1}$ (over the past $c.\,2$–$4\,\text{Ma}$). These rates accord with the range of morphological and stratigraphic evidence outlined above; in particular, the rates for the rectilinear slope in Arena Valley are compatible with the preservation of the nearby >11 Ma old ash avalanche deposit, and the somewhat higher rate for the lower Taylor Valley site is consistent with the partially degraded state of the adjacent 3.57 Ma old pyroclastic and lava-flow deposit (Wilch et al. 1993).

Discussion and conclusions

The diverse range of data collated here creates a coherent picture of extremely low rates of denudation of the order of $1\,\text{m}\,\text{Ma}^{-1}$ or less over the past several million years in the hyper-arid, polar environment of the Dry Valleys region. Rates in the relatively warmer environments at lower elevations nearer the Ross Sea appear to be somewhat higher than those on the elevated surfaces above 1500–2000 m in the western sector of the Dry Valleys, presumably due to the seasonal presence of liquid water and more active physical and biological weathering. These rates can be compared with a much higher mean denudation rate for the Dry Valleys region of 40–$100\,\text{m}\,\text{Ma}^{-1}$ over the past $c.\,50\,\text{Ma}$ derived from fission-track thermochronology (Fitzgerald 1992).

Direct comparison with rates over the past few million years for other morphoclimatic and tectonic environments is problematic since the extreme stability of the Dry Valleys region and the resulting long-term preservation of landforms and deposits means that average denudation rates here apply to extensive periods of several millions of years. In more active environments such features survive for much less time and, consequently, rates based on the degree of modification of minor landforms and deposits apply to shorter periods. This situation also applies to data from cosmogenic isotopes, since they effectively measure the residence time of the top $c.\,2\,\text{m}$ of the Earth's surface. Two of the few sets of comparable cosmogenic isotope data are the estimate of maximum denudation rates as low as $0.6\,\text{m}\,\text{Ma}^{-1}$ for granitic domes in the arid Eyre Peninsula in South Australia (Bierman & Turner 1995), and the much higher rate of 5–$11\,\text{m}\,\text{Ma}^{-1}$ reported for volcanic ash-flow tuffs in New Mexico, USA (Albrecht et al. 1993).

Mean denudation rates for large drainage basins estimated from modern sediment and solute load data range from $5\,\text{m}\,\text{Ma}^{-1}$ to $688\,\text{m}\,\text{Ma}^{-1}$ and show that rates for the Dry Valleys area lie well below the average for low-relief basins (Summerfield & Hulton 1994). Mean denudation rates for individual catchments may be as low as $1\,\text{m}\,\text{Ma}^{-1}$ (Summerfield 1991), but these figures include large areas of net deposition, and thus under-estimate local rates of denudation comparable to the site-specific rates provided by cosmogenic isotope analysis and the modification of minor landforms and deposits.

We have demonstrated how limited modification of a range of landforms and deposits in the Dry Valleys area, whose age is constrained by associated radiometrically dated volcanic rocks, indicates very slow rates of denudation and landscape change in this hyper-arid, polar environment, probably for the past 15 Ma or more. Cosmogenic isotope analyses of surface samples provide additional quantitative data on denudation rates and indicate that over the past several million years these have ranged from around $1\,\text{m}\,\text{Ma}^{-1}$ down to $0.1\,\text{m}\,\text{Ma}^{-1}$ with the lowest rates occurring at high elevations inland where temperatures appear to be always too low for liquid water or biological activity to exist.

We conclude that rates of denudation of 0.1–1.0 m Ma^{-1} represent a minimum in terrestrial subaerial environments. At such low rates not only major landforms, such as erosion surfaces, but also minor forms are able to survive for several million years. In environments where liquid water is present for at least some period of time each year it is likely that long-term denudation rates will be above these values. Further applications of cosmogenic isotope analysis in a range of other environments should provide valuable constraints on long-term rates of landscape change and the survival potential of erosion surfaces and other landforms used to reconstruct landscape history.

Field support for much of the work reported here was provided by the U.S. Division of Polar Programs of the National Science Foundation through grants to the University of Maine and the Berkeley Geochemistry Center. The cosmogenic isotope analysis was supported by the Natural Environment Research Council through grant no. GR3/9128.

References

ALBRECHT, A., HERZOG, G. F., KLEIN, J., DEZFOULY-ARJOMANDY, B. & GOFF, F. 1993. Quaternary erosion and cosmic-ray-exposure history derived from ^{10}Be and ^{26}Al produced in situ – An example from Pajarito plateau, Valles caldera region. *Geology*, **21**, 551–554.

BARRETT, P. J., ADAMS, C. J., MCINTOSH, C. J., SWISHER III, C. C. & WILSON, G. S. 1992. Geochronological evidence supporting Antarctic deglaciation three million years ago. *Nature*, **359**, 816–818.

BIERMAN, P. R. 1994. Using in situ cosmogenic isotopes to estimate rates of landscape evolution: A review from the geomorphic perspective. *Journal of Geophyscial Research*, **99**, 13885–13896.

—— & TURNER, J. 1995. ^{10}Be and ^{26}Al evidence for exceptionally low rates of Australian bedrock erosion and the likely existence of pre-Pleistocene landscapes. *Quaternary Research*, **44**, 378–382.

BIRRELL, K. S. & PULLAR, W. A. 1973. Weathering of paleosols in Holocene and late Pleistocene tephras in central North Island, New Zealand. *New Zealand Journal of Geology and Geophysics*, **16**, 687–702.

BROOK, E. J., BROWN, E. T., KURZ, M. D., ACKERT, R. P. JR, RAISBECK, G. M. & YIOU, F. 1995. Constraints on age, erosion, and uplift of Neogene glacial deposits in the Transantarctic Mountains determined from in situ cosmogenic ^{10}Be and ^{26}Al. *Geology*, **23**, 1063–1066.

BRUNO, L. A., BAUR, H., GRAF, T., SCHLÜCHTER, C., SIGNER, P. & WIELER, R. 1997. Dating of Sirius Group tillites in the Antarctic Dry Valleys with cosmogenic ^{3}He and ^{21}Ne. *Earth and Planetary Science Letters*, **147**, 37–54.

BULL, C. 1966. Climatological observations in ice-free areas of southern Victoria Land, Antarctica. *American Geophysical Union Antarctic Research Series*, **9**, 177–194.

BURCKLE, L. H. & POTTER, N. JR. 1996. Pliocene–Pleistocene diatoms in Paleozoic and Mesozoic sedimentary and igneous rocks from Antarctica: A Sirius problem solved. *Geology*, **24**, 235–238.

CLAPPERTON, C. M. & SUGDEN, D. E. 1990. Late Cenozoic glacial history of the Ross Sea embayment, Antarctica. *Quaternary Science Reviews*, **9**, 253–272.

DENTON, G. H., SUGDEN, D. E., MARCHANT, D. R., HALL, B. L. & WILCH, T. I. 1993. East Antarctic Ice Sheet sensitivity to Pliocene climatic change from a Dry Valleys' perspective. *Geografiska Annaler*, **75A**, 155–204.

FITZGERALD, P. G. 1992. The Transantarctic Mountains of southern Victoria Land: The application of apatite fission track analysis to a rift shoulder uplift. *Tectonics*, **11**, 634–662.

——, SANDIFORD, M., BARRETT, P. J. & GLEADOW, A. J. W. 1986. Asymmetric extension associated with uplift and subsidence of the Transantarctic Mountains and Ross Embayment. *Earth and Planetary Science Letters*, **81**, 67–78.

FORTUIN, J. P. F. & OERLEMANS, J. 1990. Parameterization of the annual surface temperature and mass balance of Antarctica. *Annals of Glaciology*, **14**, 78–84.

GERSONDE, R., KYTE, F. T., BLEIL, U. et al. 1997. Geological record and reconstruction of the late Pliocene impact of the Eltanin asteroid in the Southern Ocean. *Nature*, **390**, 357–363.

HALL, B. L., DENTON, G. H., LUX, D. R. & BOCKHEIM, J. R. 1993. Late Tertiary Antarctic paleoclimate and ice-sheet dynamics inferred from surficial deposits in Wright Valley. *Geografiska. Annaler*, **75A**, 239–267.

HARWOOD, D. M. 1994. The continuing debate on Pliocene Antarctic deglaciation. *In*: VAN DER WATEREN, F. M., VERBERS, A. L. L. M. & TESSENSOHN, F. (eds) *LIRA Workshop on Landscape Evolution: A Multidisciplinary Approach to the Relationship Between Cenozoic Climate Change and Tectonics in the Ross Sea Area, Antarctica*. Rijks Geologische Dienst., Haarlem, 101–105.

HUYBRECHTS, P. 1993. Glaciological modelling of the late Cenozoic East Antarctic Ice Sheet: Stability or dynamism? *Geografiska Annaler*, **75A**, 221–238.

ISHMAN, S. E. & RIECK, H. J. 1992. A late Neogene Antarctic glacio eustatic record, Victoria Land basin margin, Antarctica. *In*: KENNETT, J. P. & WARNKE, D. A. (eds) *The Antarctic Paleoenvironment: A Perspective on Global Change*. Antarctic Research Series, **56**, 327–347.

IVY-OCHS, S., SCHLÜCHTER, C., KUBIK, P. W., DITTRICH-HANNEN, B. & BEER, J. 1995. Minimum ^{10}Be exposure ages of early Pliocene for the Table Mountain plateau and the Sirius Group at Mount Fleming, Dry Valleys, Antarctica. *Geology*, **23**, 1007–1010.

KELLOGG, D. E. & KELLOGG, T. B. 1996. Diatoms in South Pole ice: Implications for eolian contamination of Sirius Group deposits. *Geology*, **24**, 115–118.

KENNETT, J. P. & HODELL, D. A. 1993. Evidence for relative climatic stability of Antarctica during the early Pliocene: A marine perspective. *Geografiska Annaler*, **75A**, 205–220.

KURZ, M. D. & BROOK, E. J. 1994. Surface exposure dating with cosmogenic nuclides. *In*: BECK, C. (ed.) *Dating in Exposed and Surface Contexts*. University of New Mexico, Alberquerque, 139–159.

MARCHANT, D. R. & DENTON, G. H. 1996. Miocene and Pliocene paleoclimate of the Dry Valleys region, Southern Victoria land: a geomorphological approach. *Marine Micropaleontology*, **27**, 253–271.

——, —— & SWISHER, C. C. III 1993a. Miocene Pliocene Pleistocene glacial history of Arena Valley, Quartermain Mountains, Antarctica. *Geografiska Annaler*, **75A**, 269–302.

——, SUGDEN, D. E. & SWISHER, C. C. III 1993b. Miocene glacial stratigraphy and landscape evolution of the western Asgard Range, Antarctica, *Geografiska Annaler*, **75A**, 303–330.

——, ——, SWISHER, C. C. III & POTTER, N. JR. 1996. Late Cenozoic Antarctic paleoclimate reconstructed from volcanic ashes in the Dry Valleys region of southern Victoria Land. *Geological Society of America Bulletin*, **108**, 181–194.

NISHIIZUMI, K., KOHL, C. P., ARNOLD, J. R., KLEIN, J., FINK, D. & MIDDLETON, R. 1991. Cosmic ray produced ^{10}Be and ^{26}Al in Antarctic rocks: exposure and erosion history. *Earth and Planetary Science Letters*, **104**, 440–454.

PRENTICE, M. L., BOCKHEIM, J. G., WILSON, S. C., BURCKLE, L. H., HODELL, D. A., SCHLÜCHTER, C. & KELLOGG, D. E. 1993. Late Neogene Antarctic glacial history: evidence from central Wright Valley. *In*: KENNETT, J. P. & WARNKE, D. A. (eds) *The Antarctic Palaeoenvironment: A Perspective on Global Change*. Antarctic Research Series, **56**, 207–250.

PRIESTLEY, R. E. 1909. Scientific results of the western journey. *In*: SHACKLETON, E. H. (ed.) *The Heart of the Antarctic*, Vol. 2. Heinemann, London, 315–333.

SCHWERDTFEGER, W. 1984. *Weather and Climate of the Antarctic*. Elsevier, Amsterdam.

SELBY, M. J. 1971. Slopes and their development in an ice-free, arid area of Antarctica. *Geografiska Annaler*, **53A**, 235–245.

—— 1974. Slope evolution in an Antarctic oasis. *New Zealand Geographer*, **30**, 18–34.

—— 1993. *Hillslope Materials and Processes*, 2nd edn. Oxford University, Oxford.

SUGDEN, D. E., DENTON, G. H. & MARCHANT, D. R. 1991. Subglacial meltwater channel systems and ice sheet overriding, Asgard Range, Antarctica. *Geografiska Annaler*, **73A**, 109–121.

——, —— & —— 1995a. Landscape evolution of the Dry Valleys, Transantarctic Mountains: Tectonic implications. *Journal of Geophysical Research*, **100**, 9949–9967.

——, MARCHANT, D. R., POTTER, N. L. JR., SOUCHEZ, R. A., DENTON, G. H., SWISHER, C. C. III & TISON, J-L. 1995b. Preservation of Miocene ice in East Antarctica. *Nature*, **376**, 412–414.

SUMMERFIELD, M. A. 1991. *Global Geomorphology*. Longman, London.

—— & HULTON, N. J. 1994. Natural controls of fluvial denudation rates in major world drainage basins. *Journal of Geophysical Research*, **99**, 13871–13883.

——, STUART, F. M., COCKBURN, H. A. P., SUGDEN, D. E., DUNAI, T., MARCHANT, D. R. & DENTON, G. H. 1999. Long-term rates of denudation in the Dry Valleys region of the Transantarctic Mountains, southern Victoria Land, Antarctica: Preliminary results based on *in-situ*-produced cosmogenic ^{21}Ne. *Geomorphology*, **27**, 113–129.

TAYLOR, G. 1922. The physiography of McMurdo Sound and Granite Harbour Region. *In: British Antarctic (Terra Nova) Expedition, 1910–1913*. Harrison, London.

TIPPETT, J. M. & KAMP, P. J. J. 1995. Geomorphic evolution of the Southern Alps, New Zealand. *Earth Surface Processes and Landforms*, **20**, 177–192.

WEBB, P. N. & HARWOOD, D. M. 1991. Late Cenozoic glacial history of the Ross Embayment, Antarctica. *Quaternary Science Reviews*, **10**, 215–223.

——, ——, MCKELVEY, B. C., MERCER, J. H. & STOTT, L. D. 1984. Cenozoic marine sedimentation and ice-volume variation on the East Antarctic craton. *Geology*, **12**, 287–291.

WILCH, T. I., DENTON, G. H., LUX, D. R. & MCINTOSH, W. C. 1993. Limited Pliocene glacier extent and surface uplift in middle Taylor Valley, Antarctica. *Geografiska Annaler*, **75A**, 331–351.

Index

Note: page numbers in *italic* refer to illustrations; those in bold refer to **tables**

Accumulation Plateau 240, *241*, 242, *243-244*, 244
Adriatic Sea 119
  western sector, seismo-stratigraphy surveys 121
African Planation Surface, Uganda 250
Ahtme Elevation 75, 77, *77*, *78*, 81
Ain Grab Formation *130*, *131*, 132, 140
Allan Nunatak 264
alluvial fans
  Guide Dong Shan 190, *190*, 196
  Nemegt Uul *203*, 205, *206*, 208, *214*, 217
  and terraces, Guide Basin, Tibet 190, *190*, 192, 196
Alpine Storm, outer ripples 3
altiplano, confusion over term 245
Altyn Tagh Fault 184-185, *184*
Andes 248-250
Andes, Chilean, fan-like faults *248*
Andes, Ecuadorian 249-251
  Cordillera and gravity spreading 249
  fold and thrust belt results from Upper Miocene decollements 250
  formed by vertical uplift following planation 249, 251
  granite/granodiorite batholiths, mainly Mesozoic 250
  modelled by single planation surface 245
  morphotectonic units 239-246
  symmetry round Inter-Andean Depression 250
  uplift 246-251
Andes, Peruvian, block faults *248*
angiosperms 55
Anglesey 70
Anglesey–Arfon block *66*, 70
Anglesey–Snowdonian landscape contrast 71
Anglesey/adjacent Welsh mainland, Menaian Surface/Platform 48-49
Antarctica, extreme landscape stability 255-267
  denudation rate estimates from cosmogenic isotope data 264-265
  Dry Valleys region 256-258
  rates of landscape change 258-264
anticlines, crustal ramp 184
Apaqui River Gorge 240, 242
apatite fission-tracking studies (AFT) 14-15, 71
  indicates major erosional event, eastern flank of Southern Scandes 88
  Northern Apennines 121
Apennine thrust belt, extension and uplift, post-Miocene 109

Apennines
  Northern, Pliocene to present-day uplift and denudation rates 119-125
  remnant landsurfaces 110-116
Appalachian Mountains 250
Arena Sandstone 265
Arena Valley
  avalanche tongues in 259, *263*
  clay content of ash 263
Argentina, palaeo-landsurfaces and neotectonics 229-238
  geological setting 229-232
  uplift characteristics 235-236
Asgard Range, glacial landforms 264
Ashdown Forest/Crowborough–Burwash–Mountfield–Fairlight axis *26*, 29
Asian lithosphere, underlying northeastern margin of Tibetan Plateau 184
Aterno River, uppermost reach, case study 112-113
Atlas Mountains 127
Atlas Thrust Front *128*
Australia, gravity spreading after vertical uplift 249
avalanche tongues, Arena Valley, retain form on slopes 259, *263*

badlands
  Gobi desert 205, *206*
  Irangi Hills 158
bajadas *see* alluvial fans
Bala Fault *66*, 67
Ballyadams Formation 53
Ballygaddy Outlier, Co. Offaly, saprolite fill 53
Baltic Oil-Shale Basin, pre-Devonian landscape 75-83
Baltic Syneclise 82
Barrow Valley, Co. Carlow 50
Barton Basement Bed, chert in 34-35
basement
  British Isles, a complex mosaic of differing terranes 67
  Caledonian, Scandinavia 85
  crystalline, Sudetic Foreland 95
  Lower Cretaceous, cut by planation surface 239
  Precambrian
    Baltic Oil-Shale Basin 76
    Scandinavia 85
basement control 13, 28
basin inversion, south of the Variscan Front 13
beach gravels, St Agnes sediments 47
Beacon Cottage Farm Outlier 48
  Oligocene sediments 48, 56
  and St Agnes Outlier 56
  sediments rest on more irregular surface 48, 56
Beacon Member 56
Beacon Supergroup 256, 264
Beacon Valley 263

bedrock
  Baltic Oil-Shale Basin 76-77, *78*
  deeply weathered, found under Neogene sediments 96-97
  Dry Valleys region, intruded by dolerites 256
  serpentinite, Niemcza Hills, deeply weathered 100
bedrock gorges, rectilinear 185
Bees Nest Pit, Derbyshire, Kenslow Clays in 52
Beglia Formation *131*, 132
Bekaa Valley
  historical evidence for earthquakes 145
  Yammouneh Fault forming western boundary 145
BIRPS, deep seismic-reflection profiling 66
Black Mountains–Fforest Fawr–Brecon Beacons scarp *66*, 68-69
  relief inverted with respect to Palaeozoic landscape 69
Blackdown 25
Bohemian Massif, Sudetes and Sudetic Foreland 93-105
Boiano basin case study 113, 115-116
  altitudinal distribution of terrace sequences 115, *115*
  an inter-montane depression 113
  bounded by normal fault-sets 113
  SW fault controls tectonic evolution of 115, 116
  terrace sequences, anomaly in accordance of 115
Bosherston–Castlemartin Surface 56
  erosional bevel 48
braid plains, Guide Basin 190, *196*
Braintree Line 35
'Brecce di Bisegna' Formation 111
Brevard Fault Zone, appearance on COCORP profile 69-70
British Institutions Reflection Profiling Syndicate *see* BIRPS
British Isles
  palaeoforms and saprolites preserved on low coastal platforms 71
  sharp lateral discontinuities in continental crust beneath 66
British margin
  receives continuous energy input 71
  stress state due to dynamic processes 68
  strong control of pre-rifting structure 65-66, *66*
brittle crust 67
  in front of Snowdonian front 70
  preserves signature of previous tectonic events 68
bulk-shortening 13
Bullers Hill Gravel 11
Bullslaughter Bay syncline, saprolite bodies, chemically weathered mudstone 56

Caesar's Camp Gravels, reassessed 8
Calabrian level, ubiquitous nature of pervasive 49-50
calcrete 208
Cantalupo sequence 115
Carboniferous Limestone, Central Ireland, undated saprolitic/palaeosol masses 58
Cardigan Bay Basin 68
'Celtic Plain', early Neogene 58
Central Europe, epeirogenic and block uplift, Late Tertiary 93
Central Range, Taiwan 170, *172*
Cha Lang Valley 196
  braid plain *187*, 190, 192, *193*, 196
  faults, Guide Formation badlands *193*, *194*
Chalk
  complexity of sedimentation and local variation in 30-31
  preservation of outliers indicates long periods of morphostasis 12
  rapid removal of 18
  timing of denudation of 7
  varying dates for exhumation of 40
  see also Upper Chalk
Chalklands
  divisible into morphotectonic regions 16
  evolutionary geomorphology, a contemporary synthesis 16-20
  morphostasis, late Palaeogene and Neogene 41
  southern 2-3
Chambira Formation 245
Channel High 2
  shows inversion 13
Channel Uplands, isolation of 16
Chattian and Miocene deposits, links with 5-150 m surfaces 55-58
Cherhil Formation *130*
Cherichira Anticline *130*, 135, 136, *136*, 138
Cherichira Blind Thrust *133*, 135, *136*
Cherichira and Grigema drainage systems, evolution of 138-139
Cherichira Sole Thrust 135-137
  tectonic uplift during reactivation of 138
Cherichira Thrust *130*, 133-135, *136*
cinder cones, Dry Valleys region, slow erosion of 259
Clay-with-Flints 7, 11
climate
  climatic environment, Taiwan 173-175
  Eocene 19, 34
  Gobi Desert 204
  Irangi Hills 158
  present and past, Antarctica 256, 258
  regional and global, affected by the raising of Tibet 183
climatic cooling, throughout the Oligocene and Miocene 53
Clogrenan Formation 53
Coastal Range, Taiwan 169-170, *172*

compression
  and crustal shortening 248
  late Silurian 82
conifers 55
continental crust
  anisotropy of 66-8
  imposed stress, brittle and ductile crust, different behaviour 68
  permanent record of geological events affecting it 67
continental plate boundary system 143
Cornubian massif 68
Cornubian terrane 67
Cornwall, west 51
  wide 50-120 m surface, marine/Pliocene view 45, 47
cosmogenic isotope data, Dry Valleys region 264-265
crustal basins, Scandinavia 86
crustal blocks, morphotectonic equilibrium 67
crustal discontinuity(ies)
  major, between Snowdon and Anglesey-Arfon blocks 70
  and the origins of major escarpments 68-69
  strongly influence present landscapes 71
crustal movements, Irangi Hills 162
crustal thickening 184
  associated with active collision 171, 173
cuesta topography, extra-glacial Chalklands 2-3
cuestas, developed on Upper and Lower Greensand 25
Culver Chalk 31
Curaray Formation, Late Miocene fauna 245
Cutucu Uplift 245

Danghe Nan Shan 184, *184*
Danian 11
  creation of Summit Surface implies rapid removal of Chalk 12
Dartmoor, as a monadnock 11
Datong Shan *184*
Dead Sea Transform 143, 145
  and Yammouneh Fault, plate tectonic and regional setting *144*
  Yammouneh Fault not active segment in Lebanon 150
decollements 250, 251
deforestation, Irangi Hills 158
deformation
  compressive 170-171
  Miocene *see* tectonism, Miocene
  Nemegt Uul 215, 217
  neotectonic, poorly understood, southern Argentina 229
denudation
  Central Weald 38-39, *39*
  controlled by bedrock properties 105
  gross, Tertiary, variations, southern England 32-33, *33*

  high rates of induced by extreme precipitation **172**, 174
  lithologically and structurally controlled 103-104
  net, southern England, since end-Cretaceous 33
  North Downs backslope benches 38, 39-40, *40*
  Palaeocene 18
  Palaeogene 32
  Pleistocene 12, 28
  Quaternary, deduced rates of 40-41
  rates in the Northern Apennines 121-122
  removal of Chalk cover, southern England 25
  South Downs crest 38, 39-40, *40*
denudation chronology studies 49-51
denudation rates, catchments in semi-arid areas **165**
desert pavements, Nemegt Uul *207*, 208, 213, 215
Djebel Cherichira
  aspects of stratigraphy 129, 131-132
  location of 127, *128*
  structural uplift 128-129
  structure of thrust front 132-137
  tectonic evolution and structural uplift of thrust front at 132-137
  topographic uplift and erosion of 137-138
Djebel Cherichira and Oued Grigema, uplift and erosional history of 127-142
Djebels, Triassic-cored, evolution of in Tunisia 129, 131-132
Doble Member 56
Dong Gou Grassland, Guide Basin *181*, *187*, 188-189, *189*
Dorking-Penshurst-Tonbridge-Biddenden axis 29
Douhou Louyang river *187*, 190
Dover Straits 30
  discordant, explanations 27
drainage
  discordant 5, 27
drainage patterns
  Irangi Hills, modified 161
  southern England, accordant and discordant *7*, 9
  Weald, concordant and discordant 27, *27*
Drum Hills, planation staircase system 49
Dry Valleys region, Antarctica
  active slope processes in coastal zone 258
  climates past and present 256, 258
  denudation rate estimates from cosmogenic isotope data 264-265
  described 256
  evidence demonstrating slow rate of landscape change **260-261**
  geological and tectonic setting 256
  glacial landforms 264

mean denudation rates very low 265
modification of landforms limited 265
morphological/depositional evidence of rates of landscape change 258-264
morphology of 256
persistent low temperature, hyper-aridity and low rates of landscape change 255-256
progress in geochronometric dating of landforms 255, 258-259
raised marine features 263-264
ductile creep 68
ductile crust, ductility increases with depth 67
dunes, and desert pavement, Nemegt Uul 206, 209, 213, 215, 217
dynamic equilibrium 65

earthquakes
  Aterno River, uppermost reach 112
  Boiano basin 115
  Irangi Hills 161-162, 163, *163*
  Lleyn Peninsula 70
  Qinghai Lake and Gonghe Basin 185
East Africa, suspended-sediment yields from large catchments 166
East Africa Rift System 157, *158*
  data on uplift rates scarce 163
East Anglia, downwarping of 16, 19
East Antarctic Ice Sheet 258
East Devon Plateau
  Combpyne Soil remnants 11
  summit mantled by residual soils 11
Eastern Cordillera (Cordillera Real), Ecuador *241, 242, 243-244*, 244-245
  planation surface 245
Eastern Longitudinal Valley, Taiwan 169, 170, *172*
Ecuador, planation surfaces 239-253
El Bouqaia Depression *148*, 151
El Houfia Extensional Faults *130*
  history of 137
end Neogene datum, position of 35-36, *36*
endogenic energy 68
Eocene
  sub-tropical climate 34
  subaerial erosion in a hot climate 19
Equilibrium Line Altitude (*ELA*) depressions, Qinghai Nan Shan and Laji Shan 196
erosion
  assessment of using suspended sediment data 219
  marine
    Cenozoic 58
    Eocene, link with London Basin Chalkland platforms rejected 8-9
    Weald, pulses of in later Palaeogene 34

erosion cycle 68
erosional relief, East European Craton, development of 75
erosional surface, in broad region from Salisbury to Dartmoor 11-12
escarpments *66*, 68-69
  erosional 152
  Jelania Góra basin 102
  origins of and crustal discontinuities 68-69
  separating Irangi Hills from Maasai Plains 157
  straight, associated with high-angle faults 240
  Walbrzych, Middle Sudetes, lithologically controlled 103
  Wesenberg Escarpment 78-79, 82
  Western Cordillera, Ecuador 239
  see also Snowdonia front scarp
Estonia, flatter hilly relief 75
etchplains, on stable blocks 71
etchplanation 17-18, 19, 105
  dynamic 65
  widely recognized in the tropics 18
etchsurfaces
  stripped, Irangi Hills 157
  sub-Cretaceous 87-88
  sub-Jurassic 87-8
  sub-Mesozoic 87, 87-88, *87*
  Sudetic Foreland 97-101
    a general view 101
Eurasian continental plate 169, *171*
Evolutionary model *37*, 39, *39*

Faille du Pas de Calais 27
fault complexes/systems, strike-slip, left-lateral 184, 202
fault propagation fold *see* Cherichira Anticline
faulting
  co-seismic 112
  dip-slip 155
  extensional, Guide Basin 193-194, 196, 198
  Guide Basin 185, 190
  normal 111
    Guide Basin 194
    Xining Basin 194
  Quaternary, Nemegt Uul 215
  strike-slip 170-171
  thrust, and oblique-slip 215
  transcurrent, Plio-Quaternary, evidence against 147-152
faults 183
  active, central and southern Italy 109
  base Jurassic *26*
  Boiano basin 113, 115
  dip at low angles into reflective lower crust 68
  dip-slip normal 109
  divide Irangi Hills into tectonic blocks 160
  extensional
    Ecuador 239, 244
    El Houfia Extensional Faults *130, 148*, 151

fan-like divergence 250
  geometry of at depth, reconstruction 233, *233*
  normal, Guide Formation 193, *194*
  post-planation 251
  pre-existing, reactivation of in brittle crust 67
  reverse, Sierras Pampeanas 229, 230-231, 232, 235
  sigmoidal curvature 203
  strike-slip 248
  see also thrust faults
Fennoscandian Shield 75, *76*
  basic relief types *87*
  differing amounts of Neogene and Palaeogene uplift 89
Ferrar dolerites 256
Ferrar Glacier *257*
fill terraces 110
flexural subsidence 138
  regional 128-129
flexuring
  early Palaeogene 3
  mid-Tertiary 28
Flimston Outlier/Flimston Pipeclays 48, 56-57
floods, catastrophic, from cylones 174
floras 53, 55
folds/folding
  and faulting, contemporaneous, indications of 13
  large, Nemegt Uul 204
  Oligo-Miocene 36
  plains-type, Baltic Oil-Shale Basin 77
  small en echelon 28
foredeep
  Northern Apennines 119-121, **121**
foreland basin development 140
Fortuna Sandstone Formation *130, 131*
Fucino basin case study 111-112, *112, 114,* 116, 119

Gaohongai River *187*, 188
  incision into basin deposits 194, *197-198*
Gash Breccias 48
geographical information systems 201
Ghab Fault 145
gibbsite 57
glacial landforms, Dry Valleys region 264
glaciation, Quaternary in northeast Tibet, no firm evidence for 185
Gobi Altai Mountains 201, *202*
  North Gobi Altai 202
Gobi desert 204-205
Gobi Platform *184*, 185
Gobi–Tien Shan fault system 201, 202, *202*
goethite 57
Gonghe Lake Basin 184
gravels, siliclastic, covering Bosherston–Castlemartin Surface 48
  age of 57
  possibly a pediment 57

gravity sliding, Ecuadorian Andes 240
gravity spreading 249, 250
gravity tectonics, causing compressive folds and thrusts 249
Great Glen Fault *66*, 67, 68, 70
Grigema Unconformity *130*, 131, 135, *136*
Grochowa Massif 95
Groombridge–Benenden axis 29
Guajaquil–Babahojo–S. Domingo Fault *241*
Guayalabamba Gorge, Ecuadorian Andes 242
Guide Basin, Tibet 185
  alluvial fans and terraces 190, *190*, 192, 196
  base-level lowering, due to Yellow River drainage 193
  deposition of molasse in Miocene 185
  faulting 190
  flysch deposits thrust and folded 185
  loess accumulation, Surface 2 194
  model for development of *197-198*
  molasse deposits suggest closed basin 193
  reactivation of faults in Cenozoic 185
  recent small-scale glaciation and deglaciation 195-196
  Surface 1 192, 193, *197-198*
  surfaces developed in 188-189, 196, 198
Guide Dong Shan 185, *187*, 188, *190*
  high piedmont terrace 190, *190*, *193*
  incised alluvial fan *190*, 196
  scarps attributable to faulting 190, *193*
  Surface 1 at foot of 193, *197-198*
Guide Formation, described 185
Gwna Group 49

Haiyuan Fault *184*
  activity along 184-185
  results of changes in trends 185
Haldon Hills 3, 11
Hampshire–Dieppe Basin 13, 16, 18
Hangay Dome 202
Hardy's multiquadric method 220
Hastings Beds 41
  forming 'High' Weald 25
Haubi basin
  catchment denudation rate calculated 163-164
  highly accelerated soil erosion 166
  increase in apparent sedimentation rate 163
Haubi, Lake
  deep sediment profile 159, *161*
  formation of 161
  severely degraded catchment 158
Headley Sand 3
  fossiliferous ironstone problem 8
high-elevation surfaces, Dry Valleys Region 256
Hindhead 25

Hipparion cf. tchicoicum 185
Hipparion fassatum 185
Hipparion platyodus 185
hog's back ridges 2
Hollymount Outlier, Co. Laois, age of solution-subsidence infill 52-53, 58
homeomorphism 60
Homs Basalt 154
  age of 150-151, **151**
  lies unconformably against Yammouneh Fault 150, 154
  relationship to Yammouneh Fault 145, 147-150
  relationships with shattered carbonates 150
Hythe Beds 25, 36

Iapetus Suture *66*, 67
ignimbrites, Cuenca, Ecuador 246
Indian crust, underthrusting Tibetan Plateau 184
inherited landscapes and uplift, Sudetic Foreland 93-105
Inter-Andean Depression 240, *241*, 242, *243-234*, 244
  Inter-Andean Graben, ophiolites and volcanoes 248
  Quaternary strato-volcanoes 246
Intermediate Atlas *128*
intermontane basins
  central Italy 111
  compressional 184
intrusions
  Andes, Ecuadorian 250
  Sierras Pampeanas 232
  Sudetic Foreland 95
  see also Klodzko-Zloty Stok Granitoid Massif; Strzegom-Sobótka granitoid Massif
inversion structures *see* linear inversion structures; upwarping
inversion tectonics 28-29
  framework for southern England 34
  in the Weald 28, 29
Irangi Hills, Kondoa, Tanzania
  earthquakes, uplift and natural erosions 161-163
  geology and climate 157-158
  N–S fault scarps 157, *159*
  recent sedimentation and denudation rates 163-166
  tectonics and drainage systems 159-161
  topographic transects 159, *160*
ironstone, fossiliferous 8
Isle of Wight Monocline 10, 18
isostatic readjustment/rebound 65, 123, 174
Italy, central, palaeolandsurfaces and recent tectonics 109-117
Izhora Heights 75, 77, *77*

Jabel Akroum ridge *148*, 151-152
Jason glacio-marine diamicton, evidence of 263
Jelania Góra intramontane basin **98**

of complex origin 102
morphological boundary 102
relationships between landforms and bedrock properties 102
Jumbeli–Naranjal Fault 139

K-Ar dating, Homs Basalt 150-151, **151**, 155
Kang Zhou Shan, Guide Basin *187*, 189-190, *189*, *191*
kaolinite, desilication of 57
Kenslow Clays 52
Kenslow flora, and evolution of the Pennines 51
Klodzko region **98**
Klodzko-Zloty Stok Granitoid Massif 103, *104*
  bedrock-controlled topography 103, *104*
Kredija Backthrust *130*, *131*, *136*, 137
Kredija Strike-Slip Fault *130*
  age of 140
  truncates El Houfia Extensional Faults and Cherichira Thrust 137

Laga Formation 119, *120*
Laji Shan 185
land degradation, Irangi Hills 157
landform evolution, long-term, neglected 65
landforms
  may help to unravel uplift and subsidence histories 93
  pre-Devonian, Baltic Oil-Shale Basin 77-81
  regional-scale, controlled by crustal properties 69
  tectonic, stability of 155
landscape change, evidence demonstrating slow rate of **260-261**
landscape development, models emphasising time-dependent landforms 71
landscape differentiation, pre-Neogene, in Europe 105
landscape evolution
  part played by deep weathering 96-97
  reached by Pleistocene ice-sheets 96
  and solution subsidence 51-53
  in Southern Mongolia 201-218
  Sudetes
    a new look at 102-105
    and Sudetic Foreland, differing 96
  Tertiary, new models of 10-13
landscape stability
  eastern Taiwan 169-181
  extreme, Antarctica 255-267
landscapes, exhumed and partly buried 93
landslides, and rockfalls, triggered by heavy rain 174
Latvia
  extensive tectonic deformation 75

INDEX

short hiatus, Silurian to Devonian 82
Lebanon, landscape evolution in 143-156
  preservation of landscapes 152-153
Leith Hill 25
Lenham Beds 3, 8
Lenham Beds incursion 20
Lenham surface, controversial 3
Lewes–Medway line 30
  topography to East more subdued 36
linear inversion structures/axes/zones 13, 19, 29
  compartmentalize southern England 13-14, 34
linear zones of disturbance, Baltic Oil-Shale Basin 77
lithological control
  Chalklands 13
  scarps around Walbrzych, Middle Sudetes 103
lithosphere
  role of reflective ductile lower crust 68
  typical structure around and beneath the British Isles *67*
  typical vertical structure 67-68
lixisols 163-164
Lleyn Peninsula earthquake 70
loessic cover, Sierra de San Luis palaeolandsurface 231-232
London Basin 13
  estimating original thickness of Chalk 31
  platforms on flanks of 17-19, *19*
  Upper Chalk border 2
  warping 35
  see also Red Crag
London–Brabant Platform, no sub-horizontal reflections 68
London–Brabant/Variscan Front, a fundamental boundary 16
Long Plain Fault, a low-angle thrust 249
Longyang Gorge 194
Lower Chalk 2
lower crust, deforms along ductile shears 68
Lower Greensand outcrops 35
lower-Langhian to Messinian interval, two seismic sequences 140
Luzon Arc 169, *171*, 173

Ma Wu Gorge 188
  formation of 194
Macchiagodena sequence 115
Macigno Formation 119, *120*
Mahmoud Formation *131*, 132
Maidstone lineament 29
Mangan Formation 239
Manila Trench 169, *171*
mantle intrusion 68
marine platform, Pliocene 3, *4*, 8
marine shingles, age of changed 11
marine-cut surfaces, former, criteria for recognition of 50

Marnoso Arenacea Formation 119, *120*
mass movement, Nemegt Uul 205, *213*
Mejerda Zone *128*
Menai Straits Fault system 66, 67, 70
Menaian Surface/Platform 48-49, 70
  and the Trwyn y Parc solution-subsidence complex 57
mesas and buttes, Guide Basin 189, *189*, 195
Metlaoui Formation *130*
mica cooling ages, Larderello geothermal field 121
Mio–Pliocene Peneplain 3, 5, *6*
  concept abandoned 35
  non-existent in type area 7-8
mobile belts, British Isles involvement with 66-67
Mogugu River *187*, 188
  incision into basin deposits 194, *197-198*
  scarp above terrace 190, *191*, 195
Moho 68, 70
Mongolia, Southern, landscape evolution 201-218
monoclines 13, 29
  Isle of Wight 10, 18
morphogenic systems, as closed systems 65
morphostasis
  Tertiary episodes, southern England 33
  Wealden area 40
morphotectonic regions 14, 16, 35
morphotectonic systems 68-71
morphotectonics 105
  and Cenozoic history, Sudetic Foreland 96
Mount Lebanon 145
  northern margin 154
  truncated spur on eastern flank of 145, 153
Mount Lebanon uplift, ancestral boundary 152
Mount Marine Fault 112, *115*
mountain ranges, orogenic 248-249
Muddus Plains 88
  correlation with Palaeic surface, Southern Scandes? 89
mushroom tectonics 249
  Front Range, Colorado *248*

Napo Uplift 245
Narva–Luga Lowland (Depression) 77-79, 82
Natural Pits, Hainault, Belgium 52
Nazca Plate, subduction segment 229, 232
Nemegt Uul, Southern Mongolia 201
  a broad sigmoidal-shaped restraining bend *202*, *203*
  catchment areas 205, *210*, *211*
  flower structure geometry 204, 215
  geographic and structural setting *202*, *204*
  geographical control of sedimentary facies 217

  geomorphology 204-215
  mountain front sinuosity 205, *208*, 215
  possible differential tilting of mountain range 217
  rock types 203, 217
  structural geology 203-204
Neogene land surface
  height over Central Weald 36
  position and disposition of 33
Neogene surfaces, extrapolation of away from the London Basin 3, 5
Neotectonic Map of Italy 109
neotectonism, Plio-Pliocene 58
Netley Heath, bench a flexured facet of Sub-Palaeogene Surface 8
New Downs Member
  depositional climate 56
  organic material yields Miocene microflora 47
  represents a freshwater ecosystem 56
Niemcza Hills *94*
  etchsurfaces 100-101, *101*
North America, eastern, sediment yields 219-228
North Downs
  central, soils 8
  Kent, Lenham surface 3
North Wales, concept of the morphotectonic system 70-71
North–South Axis, Tunisia 127, *128*
  crustal and lithospheric thickness 129
Northern Scandes 86, *86*, 88
Norwich Crag 16
Nummulites vascus marker horizon *131*

oceanic–continental plate collision zone 169
Ohio River 223, *224*, *225*
oil-shale deposits, commercial, Baltic Oil-Shale Basin 76
Okinawa Trough 169, *171*
oldland-oldland geomorphological relationships, British Isles 60
ophiolite complexes
  Memegt Uul 203
  Sudetic Foreland 95
ophiolite suites, Laji Shan 185
Ordovician, Baltic Oil-Shale Basin 82
orogenesis, Andean 229
orogenies, Precambrian, Ecuador 244-245
Oued Cherichira *130*
  basal Segui Formation
    braided fluvial deposit 132
    debris-flow deposit 138
    facies change 138
  drainage system
    age of 138
    an ephemeral wadi system 127, *129*
    antecedent system maintained its base level 137, 138, 140
    not displaced by Kredija Fault 138

Oued Cherichira (*cont.*)
   time of formation 132
   generation of structural uplift 137-138, 140
   position of Cherichira Sole Thrust 137
   time of formation 132, 140
Oued Grigema *130*
   cuts through Ain Grab Formation 139
   debris-flow fan deposits of Segui Formation 132, 139
   fluvial features post-tectonic 139
   similarities to Oued Cherichira 138
   smaller ephemeral wadi system 128, *129*
   younger consequent drainage system 132, 140
outliers
   Chalk 3
   Palaeocene/Eocene 8
   see also Ballygaddy Outlier; Beacon Cottage Farm Outlier; Flimston Outlier/Flimston Pipeclays; St Agnes Outlier; St Erth Outlier
overland flow, Hortonian and saturated 174

Palaeocene
   denudation then marine encroachment 18
   variation in erosion 32
Palaeocene–Eocene boundary, affects of choice of position 10
Palaeogene
   distinction between stable areas and areas of pulsed uplift 34
   late
     evolution of the Weald during 33-35
     off-shore record 34
     time of limited erosion 34
Palaeogene Denudation Model 3-5, *37*, 38
Palaeogene residuals 3, 8, 32
Palaeogene sequences, London and Hampshire basins 31
palaeolandsurface development 232
palaeolandsurfaces
   late Palaeogene and Neogene 40
   Miocene, northern Bekaa, terminates at an erosional escarpment 152
   and neotectonics, Argentina 229-238
   and recent tectonics, central Italy 109-117
Pancanta Fault system, defined by secondary faults 235
Pandivere Heights 75, 77, *77*, *78*, 81
Pantano Negro Fault 235
Pantano Negro Fault shear zone 235
partial melting 68
passive continental margins 65
Peak Hill Gravel 11
pediments, Irangi Hills 157-158

Peleus Till, Wright Valley, preservation of upper feather edge 264
Pembroke Peninsula *see* Bosherston–Castlemartin Surface
periclines
   Cretaceous, becoming Tertiary Monoclines 13
   superimposed on Weald–Artois Anticline 25
Pewsey–London Platform inversion axis 13
   important structural divide 16-17
Philippine Sea oceanic plate 169, *171*
phosphorite layer, Baltic Oil-Shale Basin 76
Pinon Formation 239
Pinon Terrane 248
Pisayambo Formation 246
Pizzoli–Barete basin 112, *115*
plains with residual hills, Scandinavia 87, 88
   Northern Scandes 88
planated landscape elements 45
planation 55-58
   double surfaces of *see* etchplanation
planation model, contradicted in Germany 105
planation surfaces
   divide pre-planation from post-planation tectonics 250-251
   Ecuador 239-253
   Irangi Hills 160
   mark the end of tectonic regimes 248-249
   Palaeogene and Miocene 105
   separated by uplift phases, Sudetes 96
planed landscape, Palaeogene/Neogene age 58
plate tectonics
   fails to account for growth of the Andes 249-250
   and landform development 65
   Taiwan 169-173
   see also Tibet
Plateau Drift 7
platforms, Chiltern backslopes, structurally controlled 8
Pliocene
   replacement of Northern Alpine foredeep 119
   southern Chalklands coastline 3, 5, 5
Po Plain–Adriatic Sea basin, volume of Holocene sediments deposited 121
poljes
   Lebanon 155
pollen, Oligocene, Beacon Cottage Farm 47
polycyclic surface, early Tertiary 10
polygonal cracking, Dry Valleys area, preservation of volcanic ash in 259, *262*
Portezuelo Blanco Fault 234
Porth Swtan, saprolite pocket 49

Porth Wen, saprolite pocket 49
Portland–Wight inversion axis 13, 14
Portland–Wight–St Válery axis 14
post-uplift spreading *see* gravity spreading
Potomac River 220, *222*, 223, *224*, *225*, *226*
precipitation
   Dry Valleys region 256, 258
   high from tropical cyclones 173-174
pulsed tectonism 7, 10, 12, 18, 28, 34
pulsed tectonism model *see* Palaeogene Denudation Model
puna surface *see* planation surfaces, Ecuador

Qaidam Basin *184*
Qilian Shan 184, *184*, 202
Qing Shui river *187*, 190
Qinghai Lake 184, *184*
Qinghai Nan Shan *184*
Quartermain Mountains 264
quartz, *émoussés luisant* textures 57
Quaternary
   deduced rates of denudation 40-41
   differential uplift, southern England 35-36
   solutional lowering 36
   stability during 353
Quaternary movements, and Tertiary tectonic episodes 16

rectilinear slopes 259, 265
Red Crag
   deposition on marine-trimmed surface 35
   eastward tilt 16
   known extent of *15*
   marine incursion of limited geomorphological significance 9
   westward extension 9
   importance of 15-16
redbeds
   Inter-Andean Depression 242
   Sierras Pampeanas 230, 232
remnant landsurfaces, Apennines 110-111
   remnant landsurface analysis, Fucino basin 112
Reskajeage Surface 48
   planation around the Palaeogene/Neogene boundary 56
Reyueshui River *187*
Rhes-y-cae, Flintshire, solution subsidence 51-52
river/stream capture 152, 153
   Guide Basin 195
rivers
   Irangi Hills
     ephemeral 160-161
     sandbed 161
     westward-flowing 161
Roccamandolfi sequence 115
Ross Embayment 256
Roum Fault
   probably active transcurrent structure 155

seismicity focused on 145
Russian Platform 75, *76*
Ryuku Arc 169, *171*
Ryuku Arc tectonic regime 177
Ryuku collision event 171
Ryuku Trench 169, *171*

St Agnes Outlier 56, 60
  buried cliff line 47, 56
  deposits
    environment of deposition 56
    fluviatile 47
    possibly coeval to St Erth
      Outlier (Reid) 47
    terrestrial and Miocene 48
  organic material from New Downs
    Member 47
  sub-Miocene unconformity 48,
    56
St Agnes–Flimston–Trwyn y Parc
  axis, control offered by 58
St Erth Outlier, Pliocene marine
  fauna in 47
St Georges Channel Basin 70
St Ives/Mount's Bay depression 47
Salisbury Plain Chalklands 16
Santiago Terrane 248
Saouaf Formation *130, 131*, 132
  dated as Tortonian 131
  tilting of 136
saprolites 71
  Northern Apennine summits 122
  and solution-subsidence 49, 55-58,
    70
  South Swedish Dome 87
  stripped from Irangi Hills summits
    157
  stripping of, Sudetic Foreland 104
sarsens 3, 6, 10
Savana planation theory 58
Scandinavia, mountains planated
  after Caledonian orogeny 251
Scandinavian domes, uplift histories
  revealed by 85-91
  sub-Mesozoic etchsurfaces 87-88,
    87
sea-level
  eustatic fall, Middle Oligocene 34
  lowering of, late Oligocene 55
seaways, interior Antarctica, early-
  mid Pliocene 258
secondary structures, growth of in
  mid-Tertiary 5
sediment yields
  eastern North America 219-228
  world's rivers, and the global
    denudation system 226
sedimentation
  late Upper Cretaceous, Tertiary
    style 13
  Lower–Middle Pleistocene,
    Northern Apennines 122
  Oligocene, extent unknown 10
sediments
  Cenozoic 86
  marine
    Plio–Pleistocene, London Basin
      9

Turonian, at altitude in the
  Sudetes 96
Miocene, lack of 31
Oligocene 31, 34, 48, 56
Palaeogene
  estimating thickness of removed
    sediments 31
  thicknesses used in cross-
    sectional reconstruction *29*,
    32-33
  terrigenous, Middle and Upper
    Eocene, lacking beneath
    Channel 34
Segui Formation *130, 131*, 139, 140
  age 131
  back-steepened 136
  braided fluvial deposit 132, 138
  debris-flow deposit 137, 138
  near Oued Grigema watershed
    132, 139
seismic activity
  high frequency of associated with
    active collision 173
  northeast Tibet 185
seismic shaking, vulnerability of
  rocks to 173
Serdj-Ressas Line *128*
shield areas, interpretations of
  denudational history 85
Sierra de San Luis 229-236
Sierras Pampeanas
  Devonian intrusions 232
  El Realito–Mesilla del Cura area
    234, *234*
  geological setting 229-232
    bounding faults and their
      relationships *230*
  Mesilla del Cura block 234
  Neogene uplift 229
  Palaeolandsurface 232-233
  palaeotopographic surface,
    reconstruction of 233, *233*
  Rio Nogoli area 234, 235, *235*
  uplift characteristics, contribution
    of palaeolandsurface analysis
    to 235-236
silcrete formation 10
Sirius Group 264
  marine diatoms in 258
Sleza Massif(-Radunia) *94*, 95
  pre-Neogene etchsurfaces **94**,
    99-100, *100*
  imperfect correlation bedrock
    structure-topography 99
  sepentinite slope 99-100
slope failures, from torrential rain
  174
slopes, rectilinear 256, 259, 265
Snowdonia, large topographic feature
  along a major crustal hinge 70
Snowdonia block *66*, 70
Snowdonia front scarp, differing
  crustal types on either side 70
soil erosion 174
  importance of climate for 166
  Irangi Hills 157, 162, *164*, 166
soils, inheritance from former
  Palaeogene cover 8

solution features, density of on the
  Chalk 16, *17*
solution pipes
  pipe fills 3
  Trwyn y Parc 49, 57, 70
solution subsidence deposits/fills
  51-53, 57, 70
Songba Gorge 194
Souar Shale *130*
South American Craton 248
South Downs, establishment of sub-
  Palaeogene overstep 13
South Downs inversion axis 14, 29
South Swedish Dome 86, *86*
  and its palaeosurfaces 89
  last rise 87
  post-uplift development of plains
    with residual hills 87
southern England, Tertiary evolution
  of
  differential uplift 20
  importance of Neogene 36-37
  inversion tectonics framework and
    argument for pulsed
    tectonism 14
  long-term evolution of, important
    developments 13-16
  low relief in Pliocene, inherited 28
  new model of Tertiary landscape
    evolution 18-20
  views of Green/Jones/Small,
    important differences 12-13
Southern Scandes 86, *86*, 88-89
stepped terrace sequences, Apennines
  110
Sticklepath-Lustleigh Fault *66*
Sticklepath-Lustleigh Fault Zone
  67-68
storm surges 173
strath terraces 110-115
strato-volcanoes, Quaternary,
  Ecuador 246
stresses, anisotropic crust, controlled
  by lines of weakness 67, 68
strike-slip basin reactivation 140
structural basins, major
  further definition of 18-19
  Jones/Small model 10
structural compartmentalization 34
  determines morphotectonic regions
    13-14
  individual inversion axes may
    move independently 14
structural control, direct, lacking in
  Baltic Oil-Shale Basin 82
structural restlessness, mid-Jurassic
  to end Tertiary 7
structure–drainage relationship, SE
  England 7
structures, inherited, control
  geometry of recent structures
  67-68
Strzegom Hills *94*
  boreholes show unevenness of
    weathering front 97
  inselberg-like landscape 97
Strzegom-Sobótka granitoid Massif
  95

Strzegom-Sobótka granitoid Massif (*cont.*)
  change in morphology correlates with internal differentiation 97, 99
  etchsurfaces 97, **98**, 99
  granite altered to kaolinitic mantle 97
Strzelin Fault 100
Sub-Andean Zone and Oriente, Ecuador 245
sub-Cambrian peneplain
  Scandinavia 85
  South Swedish Dome 87, *87*
sub-Cenomanian surface 60
Sub-Eocene Surface 3, *4*, *6*, 11
  tectonic deformation 3
  see also Sub-Palaeogene Surface
sub-Inferior Oolite surface 60
sub-Liassic surface, Mendips 60
sub-Mesozoic unconformities 58, 60
Sub-Miocene facets 10
sub-Miocene unconformity, beneath St Agnes outlier 48, 56
Sub-Oligocene facets 10
Sub-Palaeogene Surface/unconformity 3, 17
  deduced position in relation to Chalk 29, 33
  multi-faceted, polygenetic and diachronous 10, *11*, 18
  overstep and overlap relations 7, 8, 31-32
sub-Red Crag surface, planar nature of 16
sub-Tertiary unconformity 31-32
  fossilized 10
subduction
  and the Andes 249-250
  at Manila Trench 169
subsidence, continual 68
Sudetes 93, *94*
  almost total absence of Neogene sediments 96
  basement geology similar to the Foreland 95
  inherited component of landscape 104
  a new look at landscape evolution 102-105
  sedimentary rocks in 95
  uplift
    late Cenozoic 96
    Neogene-Quaternary, and the mountain front 96
Sudetic Foreland 93, *94*, 95-96
Sudetic Marginal Fault *94*, 96
Summit Plain *see* Summit Surface
Summit Surface 5
  debate on origin of 28
  evolution under sub-tropical/warm temperate environment 12
  incomplete dissection of 12
  mantled with Clay-with-Flints 3
  survived from early Palaeogene 11
superficial deposits 3
  contrasting, separation of 11

tablelands
  developed in Upper Greensand 3
  Devon and Dorset 10-11
Taiwan
  climatic environment 173-175
  geological setting 169-173
  recent palaeoenvironment 177
  tectonic setting 169, *171*
  see also Taroko Gorge, Taiwan
talus deposits, Nemegt Uul 205, *213*
Tanzania, landforms, erosion and deposition 157-168
  earthquakes, uplift and natural erosion 161-163
  geology and climate 157-158
  recent sedimentation and denudation rates 163-166
  studies on erosion and sedimentation in degraded catchments 166
  tectonics and drainage systems 159-161
Taroko Gorge, Taiwan
  computer modelling of evolution 177-180
  formation of 176-177
  probably on boundary between two tectonic zones 177
  regional setting 175-176
  rock types 176
  sediment yield 176
  stable form along length of 177
  suspended sediment data 176
Tawonga Fault, a low angle thrust 249
Taylor Glacier *257*
tectonics
  compressional/compressive 170-171
  contribution of ancient landsurface analysis in evaluation of 109-117
  and drainage systems, Irangi Hills 159-161
  Plio-Pleistocene, mountain building 251
  see also inversion tectonics
tectonism
  Cenozoic, northeast Tibet 183
  Cretaceous 31
  in long-term land/landscape development 65
  mid-Tertiary 3, 12, 19, 33
  Miocene 14, 31
    deformation model 27-28, *37*, 38
    downgrading of 36-37
    Guide Basin 185, 192
  Neogene 232
    reduced significance of 36
  Palaeogene 18, 31
  see also deformation
Tellian Atlas 127, *128*
Tengouch Lateral Fault *130*, 131, 132, *134*
  all movement pre-Segui Formation 132
  probably formed synchronously with Cherichira Thrust 132, *134*
  reactivation of 135

  uplifted by Cherichira Anticline 135
Tengouch Thrust *130*, 132
terrace sequences
  Aterno River, uppermost reach 113, *115*
  Boiano basin 115
  Fucino basin 112, *114*
terraces
  developed above the Mogogou river 192, *195*
  development in Guide Basin may be related to climate change 195
terranes, exotic, Ecuador, case for 246, 248
Tertiary
  active stress sources 36
  southern England, evolutionary geomorphology *6*
  two periods of exhumation 14-15
  Wooldridge and Linton, emphasis on stability 3
Tertiary landscape, new models of 10-13
thrust belts, structural uplift, topographic uplift and erosion in 128-129
thrust faults 185
  Nemegt Uul 204, 208
thrust ramp angles, northern and central Tunisia 136
thrust stacking, causing regional flexural subsidence 128-129
thrust zone, Nemegt Uul 203
Tibet 143
  northeast, geomorphology and uplift 183-200
    geological structures *184*
    geomorphology south of the Yellow River 188-192
    large-scale gravitational collapse unlikely 194
    tectonic evolution, current opinions 184-185
  see also Guide Basin
Tomoko Hills, probably never deforested 158
Tornquist Zone 86
Tower Wood Gravel 11
Transantarctic Mountains 256
  structural components in the Dry Valleys area 256
transform-transform-trench triple junction 145
transgressions
  late Pliocene 8
  no evidence for in Wessex type area 9
  possibility of two 8
  St Erth 60
  Pliocene, extent controversial 19
  Red Crag 9, 15-16, *15*, 19, 35
    no morphological evidence outside the London Basin 9
tree root exposures, Africa, measurements of show erosion rates 164

Weald–Artois Anticline (*cont.*)
　uplift and denudation of western (Wealden) portion 25-43
Weald–Boulonnais–Artois horst 29-30
　possible effect on Weald uplift 32
weathering
　deep
　　of exhumed sub-Cambrian peneplain 87
　　and pre-Neogene landscape evolution 96-97
　grus weathering mantles 102
　weathering residues 3
　see also saprolites
Welsh Basin, northern edge marked by Menai Straits Fault system 70
Wesenberg Escarpment, Narva–Luga Depression 78-79
Wessex, end-Tertiary surface 36
western Britain and Ireland
　boreholes penetrating Oligocene outliers 53, 55
　broad forested plain at Palaeogene/Neogene boundary 60
　Chattian sediments in freshwater basins 58
　Chattian/Miocene sedimentary outliers and saprolites **54**, 55-58
　coastal planation surfaces 45-51

granite areas, elevated since the Eocene 105
oldland fringes represent one of oldest little-changed landscapes 60
planar sub-Mesozoic unconformities overlain by marine sequences 58, 60
Western Cordillera, Ecuador 239-240, *241*, *242*
bevelled by planation surface 245
White Chalk
　thickness at outcrop 2
　variability in three dimensions 13
Wilkes-Pensacola Basin, possible seaways across 258
Wilson cycles 250
WINCH 4 seismic profile, South Irish Sea 70
winds, low humidity from the Polar Plateau 258
Wooldridge and Linton model 1-2, 3-5, *37*, 38, *39*, 49
Woolwich and Reading Beds, overstepping 32
Wright Valley 263-264
Wuchia Landslide 174

Xining (Lake) Basin 184
　normal faulting 194

Yammouneh Fault 145-148, 150

follows valley of Wadi Chadra 147, *149*
geological setting of 143-145
northern
　drainage evolution, landscape correlations across 151-152
　inactive since the Miocene 143
　valleys traced down to Miocene palaeosurface, northern Bekaa 151-152
Yammouneh polje 147
　inherited feature 153
Yellow River
　geomorphology to south of 188-192
　in the Guide Basin 185, 198
　and the raising of Tibet 183
Yezhang Grassland, Guide Basin 188, *189*
　deposits beneath 194
　fault scarps delimit extent of 193
　northern edge, scarp with triangular facets 190, *192*, 193
Ying Yao Valley *187*, 196
　recent terraces onlap badlands *191*, 192

Zhihai Shan 185, *187*, 192
Zulova Highland/Massif *94*, 95
　inselberg landscape 97, **98**
　kaolin occurrences 97

INDEX

Treffynnon Outlier 57
tropical cyclones (typhoons),
    affecting Taiwan 173
Trwyn y Parc solution-subsidence
    complex 49, 57
    association with Menaian Surface
        57
    petrography of pipe infills 57
    saprolites
        and organic palaeosols 70
        source differs from host rock 49
Tunisia *see* Djebel Cherichira and
    Oued Grigema
Turi Formation 240
Typhoon Herb *173*
Typhoon Ophelia 174
typhoons *see* tropical cyclones
Tyrrhenian margin, volcanic activity
    111

unconformities
    angular, Sierras Pampeanas 230
    sub-Mesozoic 58, 60
    sub-Miocene 48, 56
    Sub-Palaeogene Surface/
        unconformity 3, 7, 8, 17, *29*,
        31-2, 33
    sub-Tertiary 10, 31-2
    see also Grigema Unconformity
uplift
    assessment of amount and pattern
        105
    and deforestation, causing
        accelerated erosion *164*
    differential 35
        and crustal anisotropy 65-74
        Ecuadorian Andes, vertical 249,
            251
        and gully erosion 162-163
    and inherited landscapes, Sudetic
        Foreland 93-105
    and landscape stability, Taroko,
        eastern Taiwan 169-181
    formation of Taroko Gorge 176-177
    late Cenozoic, Sudetes 96
    late Neogene–Quaternary 16
    Mount Lebanon uplift 152
    Neogene 15, 39, 105
        Sierras Pampeanas 229, 232
    Neogene–Quaternary, Sudetes 96
    Northern Apennines, timing of 122
        acceleration in Middle
            Pleistocene 123
        geomorphic impact 122-123
        mode of 122
        uplifted summit areas of low
            relief 122
    permanent, balanced by erosion 68
    Pleistocene, differences as to scale
        and significance of 12
    Plio-Pleistocene, Sierras
        Pampeanas 230
    Plio-Quaternary
        Ecuador 246
        northeast Tibet 183
    Quaternary, Wealden 16
    structural
        Djebel Cherichira 132-137

Nemegt Uul 203-204, 217-218
    in thrust belts 128-129
    tectonic, Taiwan 171, 173
    Tertiary, Scandinavia 85, 89
    topographic
        Djebel Cherichira 137-138
        in thrust belts 128-129
    transpressional, Gobi Altai 201
    see also upwarping; Weald upwarp
Upper Chalk
    deposition ceased in Maastrichtian
        18
    estimation of original thickness 30
    outcrop in Chalkland regions 2, *2*
Upper Greensand 3
upwarping 13, 19, 29
    London Basin Chalk 31

valley system evolution 151-152, *152*,
    153, *154*
Vanda, Lake 263, 264
Väner Basin 86
Viivikonna Anticlinal Zone
    structure/topography of the
        sedimentary bedrock *81*
    a zone of disturbance 79
volcanism
    Dry Valleys region
        Cenozoic 256
        volcanic ash deposits 259, *262*,
            263, *263*
    Eastern and Western Cordillera,
        Ecuador 246
    Inter-Andean Depression 240, 246
    Cuenca area 246
    Mio–Pliocene, southern Argentina
        232

Wadi Chadra *148*
    duration of eruption of Homs
        Basalt into 151
    and Jabel Akroum ridge,
        relationships between
        landscape features *154*
    limestone fractured/brecciated east
        of Yammouneh Fault 147-148,
        150
    probably uplifted 155
    section across Yammouneh Fault
        150
Wadi Charbine *148*
    no post-Miocene incision 152
Wadi ech Chaqif *148*
    headwaters captured 152
Wadi ed Deir *148*
    two-stage incision 152
Wadi Fissane *148*
    no post-Miocene incision 152
Wadi Nsoura *148*
    two-stage incision 152
Wadi Serkhane *148*, 151
    headwaters captured 152
    modern valley incised into broader
        valley 151, *152*, *153*, *154*
Walbrzych, Middle Sudetes **98**
    present day morphology 103
Wali Gong Shan 185
Walls Boundary Fault 66, 68

Wardour–Portsmouth inversion axis
    13, 14, 16, *26*, 29
warping
    late prolongation, along Anglesey–
        Arfon and Snowdonia blocks
        hinge 70-71
    London Basin 35
    Palaeogene 36
        and Neogene 14-15
    Quaternary 12
    regional, late Cretaceous 32
Weald
    background to evolutionary
        geomorphology 25-30
    catastrophist interpretations
        27-28
    fluvial erosional development 28
    inversion tectonics explanation
        28-29
    recognition of anticlinal
        structure 28
    central
        denudation patterns 38-39, *39*
        greater uplift 36
        height of Neogene land surface
            over 36
        subaerial denudation possible,
            most of Palaeogene 32
    complex horst structure 16, 36-37
    decoupled from less disturbed
        provinces 29
    established description of 28
    folds, tensional, Alpine
        compressional model
        abandoned 28
    geological reconstruction 30-37
    inversion axes, dominant and
        secondary 29
    models of evolution 37-41
    possible structural partitioning of
        26
    pulsed evolution of 41
    uplift and denudation possible,
        Eocene and Oligocene 34, 35
    variable uplift and later
        downwarping to the East 36
Weald Island/shoal, London Clay
    Sea, rejected 31
Weald upwarp 7, 13
    alternative development scenario
        33
    dating of 31
    evidence for early development
        31-32
    Neogene origin? 15
    shown to be asymmetrical in
        reconstruction 30
    shows inversion 13
    Weald-Artois upwarp decoupled
        from more stable areas 41
    well developed in early Palaeogene
        32
    western end experienced greater
        denudation, reasons for 35
Weald–Artois Anticline 10, 18
    French interpretation of events
        29-30
    removal of Chalk cover 2

Geological Society Special Publications

Series Editors

A. J. HARTLEY

R. E. HOLDSWORTH

A. C. MORTON

M. S. STOKER

It is recommended that reference to all or part of this book should be made in one of the following ways:

SMITH, B. J., WHALLEY, W. B. & WARKE, P. A. (eds) 1999. *Uplift, Erosion and Stability: Perspectives on Long-term Landscape Development*. Geological Society, London, Special Publications, **162**.

PUURA, V., VAHER, R. & TUULING, I. 1999. Pre-Devonian landscape of the Baltic Oil-Shale Basin, NW of the Russian platform. *In:* SMITH, B. J., WHALLEY, W. B. & WARKE, P. A. (eds) *Uplift, Erosion and Stability: Perspectives on Long-term Landscape Development*. Geological Society, London, Special Publications, **162**, 75–83.

Uplift, Erosion and Stability:
Perspectives on Long-term Landscape Development